METHODS IN MOLECULAR BIOLOGY

Series Editor
John M. Walker
School of Life and Medical Sciences
University of Hertfordshire
Hatfield, Hertfordshire, UK

For further volumes:
http://www.springer.com/series/7651

For over 35 years, biological scientists have come to rely on the research protocols and methodologies in the critically acclaimed *Methods in Molecular Biology* series. The series was the first to introduce the step-by-step protocols approach that has become the standard in all biomedical protocol publishing. Each protocol is provided in readily-reproducible step-by-step fashion, opening with an introductory overview, a list of the materials and reagents needed to complete the experiment, and followed by a detailed procedure that is supported with a helpful notes section offering tips and tricks of the trade as well as troubleshooting advice. These hallmark features were introduced by series editor Dr. John Walker and constitute the key ingredient in each and every volume of the *Methods in Molecular Biology* series. Tested and trusted, comprehensive and reliable, all protocols from the series are indexed in PubMed.

Ion Mobility-Mass Spectrometry

Methods and Protocols

Edited by

Giuseppe Paglia

Department of Medicine and Surgery, University of Milano-Bicocca, Vedano al Lambro, Monza e Brianza, Italy

Giuseppe Astarita

Georgetown University, Washington, DC, USA

Humana Press

Editors
Giuseppe Paglia
Department of Medicine and Surgery
University of Milano-Bicocca
Vedano al Lambro
Monza e Brianza, Italy

Giuseppe Astarita
Georgetown University
Washington, DC, USA

ISSN 1064-3745 ISSN 1940-6029 (electronic)
Methods in Molecular Biology
ISBN 978-1-0716-0029-0 ISBN 978-1-0716-0030-6 (eBook)
https://doi.org/10.1007/978-1-0716-0030-6

This Humana imprint is published by the registered company Springer Science+Business Media, LLC, part of Springer Nature.
The registered company address is: 233 Spring Street, New York, NY 10013, U.S.A.

Preface

Mass spectrometry (MS) measures the mass-to-charge ratio of ions, and it is routinely used in a variety of applications including the study of human health and disease, food and nutrition, and environmental assessments. Nevertheless, a complex chemical or biological sample often requires hyphenation of MS with other analytical technologies for the identification and characterization of its analytes. In this context, ion mobility (IM) spectrometry is an emerging technology that has come in support of some of the most difficult MS applications.

IM is a gas-phase electrophoretic technique that allows the separation of ions in the gas phase according to their charge, shape, and size. IM can be combined with traditional MS approaches, including liquid chromatography-MS, direct MS analyses, and MS imaging. Advances in IM-MS instrumentation, together with novel informatics solutions, provide unprecedented analytical advantages enabling novel qualitative and quantitative applications for the analysis of complex biological samples.

IM-MS adds three major benefits to traditional analytical approaches. First, IM-MS improves the peak capacity and signal-to-noise ratio, allowing the separation of interfering isobaric species. Second, IM-MS provides a new set of hybrid fragmentation experiments, which combine collision-induced dissociation with IM separation, improving the specificity of MS/MS-based approaches. Third, IM-MS technologies may allow the determination of the collision cross section (CCS), a unique physicochemical measure related to the conformational structure of ions in a particular gas, which can be used as an additional coordinate to maximize the confidence of analyte identification.

This book is targeted at scientists that are interested in using IM-MS instrumental and informatics approaches to improve traditional analysis of molecules. It provides fundamentals and protocols for exploiting the potential of state-of-the-art IM-MS technology for the most common analytical applications. The book has been organized into four parts, each dealing with a particular set of IM-MS applications: (1) metabolomics and lipidomics; (2) proteomics and glycomics; (3) imaging and ambient ionization IM-MS; and (4) bioinformatic solutions for analyzing IM-MS data and deriving CCS values.

Vedano al Lambro, Monza e Brianza, Italy — *Giuseppe Paglia*
Washington, DC, USA — *Giuseppe Astarita*

Contents

Contributors

GUIDO ARAUJO • *Center for Computing in Engineering and Sciences, Institute of Computing, University of Campinas, Campinas, São Paulo, Brazil*
ANDREA ARMIROTTI • *Analytical Chemistry and In-Vivo Pharmacology, Fondazione Istituto Italiano di Tecnologia, Genoa, Italy*
ERIN S. BAKER • *Department of Chemistry, North Carolina State University, Raleigh, NC, USA*
TIM J. CAUSON • *Institute of Analytical Chemistry, University of Natural Resources and Life Sciences (BOKU), Vienna, Austria*
XI CHEN • *Interdisciplinary Research Center on Biology and Chemistry, Shanghai Institute of Organic Chemistry, Chinese Academy of Sciences, Shanghai, China; University of Chinese Academy of Sciences, Beijing, China*
MICHELLE CLASQUIN • *Cambridge Laboratories, Pfizer Global Research and Development, Pfizer Inc., Cambridge, MA, USA*
EMMANUELLE CLAUDE • *Waters Corporation, Wilmslow, UK*
HELEN J. COOPER • *School of Biosciences, University of Birmingham, Birmingham, UK*
STEPHEN L. COY • *Department of Chemistry and Chemical Biology, Northeastern University, Boston, MA, USA*
CHIARA DALL'ASTA • *Department of Food and Drug, University of Parma, Parma, Italy*
KATRINA DOENGES • *Department of Pharmaceutical Sciences, School of Pharmacy and Pharmaceutical Sciences, University of Colorado, Aurora, CO, USA*
CLAIRE E. EYERS • *Department of Biochemistry, Centre for Proteome Research, Institute of Integrative Biology, University of Liverpool, Liverpool, UK*
JAMES GLICK • *Novartis Institutes for BioMedical Research, East Hanover, NJ, USA*
DAVID F. GRANT • *Department of Pharmaceutical Sciences, University of Connecticut, Storrs, CT, USA*
RIAN L. GRIFFITHS • *School of Pharmacy, University of Nottingham, Nottingham, UK*
ZEESHAN HAMID • *D3Validation, Fondazione Istituto Italiano di Tecnologia, Genoa, Italy; Scuola Superiore Sant'Anna, Pisa, Italy*
STEPHAN HANN • *Institute of Analytical Chemistry, University of Natural Resources and Life Sciences (BOKU), Vienna, Austria*
DAVID J. HARVEY • *Nuffield Department of Medicine, Target Discovery Institute, University of Oxford, Oxford, UK*
GABRIEL HEERDT • *Center for Computing in Engineering and Sciences, Institute of Chemistry, University of Campinas, Campinas, São Paulo, Brazil; Departamento de Química, Instituto de Ciências Exatas, Universidade Federal de Minas Gerais, Belo Horizonte, MG, Brazil*
DENNIS W. HILL • *Department of Pharmaceutical Sciences, University of Connecticut, Storrs, CT, USA*
KELLY M. HINES • *Department of Medicinal Chemistry, University of Washington, Seattle, WA, USA*
AMOL KAFLE • *Department of Chemistry and Chemical Biology, Northeastern University, Boston, MA, USA*

Tiina J. Kauppila • *Department of Chemistry, Finnish Institute for Verification of the Chemical Weapons Convention (VERIFIN), University of Helsinki, Helsinki, Finland*
Klaudia I. Kocurek • *Department of Chemistry, Texas A&M University, College Station, TX, USA*
Ruwan T. Kurulugama • *Agilent Technologies, Santa Clara, CA, USA*
James I. Langridge • *Waters Corporation, Wilmslow, UK*
Amy Li • *Department of Medicinal Chemistry, University of Washington, Seattle, WA, USA*
Jody C. May • *Department of Chemistry, Center for Innovative Technology, Institute of Chemical Biology, Vanderbilt University, Nashville, TN, USA; Vanderbilt-Ingram Cancer Center, Institute for Integrative Biosystems Research and Education, Vanderbilt University, Nashville, TN, USA*
Michael McCullagh • *Waters Corporation, Wilmslow, UK*
John A. McLean • *Department of Chemistry, Center for Innovative Technology, Institute of Chemical Biology, Vanderbilt University, Nashville, TN, USA; Vanderbilt-Ingram Cancer Center, Institute for Integrative Biosystems Research and Education, Vanderbilt University, Nashville, TN, USA*
Cole Michel • *Department of Pharmaceutical Sciences, School of Pharmacy and Pharmaceutical Sciences, University of Colorado, Aurora, CO, USA*
Caleb B. Morris • *Department of Chemistry, Center for Innovative Technology, Institute of Chemical Biology, Vanderbilt University, Nashville, TN, USA; Vanderbilt-Ingram Cancer Center, Institute for Integrative Biosystems Research and Education, Vanderbilt University, Nashville, TN, USA*
Melanie T. Odenkirk • *Department of Chemistry, North Carolina State University, Raleigh, NC, USA*
James C. Poland • *Department of Chemistry, Center for Innovative Technology, Institute of Chemical Biology, Vanderbilt University, Nashville, TN, USA; Vanderbilt-Ingram Cancer Center, Institute for Integrative Biosystems Research and Education, Vanderbilt University, Nashville, TN, USA*
Kevin Quinn • *Department of Pharmaceutical Sciences, School of Pharmacy and Pharmaceutical Sciences, University of Colorado, Aurora, CO, USA*
Nichole Reisdorph • *Department of Pharmaceutical Sciences, School of Pharmacy and Pharmaceutical Sciences, University of Colorado, Aurora, CO, USA*
Rick Reisdorph • *Department of Pharmaceutical Sciences, School of Pharmacy and Pharmaceutical Sciences, University of Colorado, Aurora, CO, USA*
Laura Righetti • *Department of Food and Drug, University of Parma, Parma, Italy*
Milinda A. Samaraweera • *Department of Pharmaceutical Sciences, University of Connecticut, Storrs, CT, USA*
Munir S. Skaf • *Center for Computing in Engineering and Sciences, Institute of Chemistry, University of Campinas, Campinas, São Paulo, Brazil*
Paulo C. T. Souza • *Center for Computing in Engineering and Sciences, Institute of Chemistry, University of Campinas, Campinas, São Paulo, Brazil; Faculty of Mathematics and Natural Sciences, University of Groningen, Groningen, The Netherlands*
Sylwia A. Stopka • *The George Washington University, Washington, DC, USA*
Weston B. Struwe • *Chemistry Research Laboratory, Department of Chemistry, University of Oxford, Oxford, UK; Department of Biochemistry, Oxford Glycobiology Institute, University of Oxford, Oxford, UK*
Lauren J. Tomlinson • *Department of Biochemistry, Centre for Proteome Research, Institute of Integrative Biology, University of Liverpool, Liverpool, UK*

NICHOLAS B. VERA • *Cambridge Laboratories, Pfizer Global Research and Development, Pfizer Inc., Cambridge, MA, USA; Department of Chemistry and Chemical Biology, Northeastern University, Boston, MA, USA*
AKOS VERTES • *The George Washington University, Washington, DC, USA*
JOHANNES P. C. VISSERS • *Waters Corporation, Wilmslow, UK*
PAUL VOUROS • *Department of Chemistry and Chemical Biology, Northeastern University, Boston, MA, USA; Barnett Institute of Chemical and Biological Analysis, Northeastern University, Boston, MA, USA*
LIBIN XU • *Department of Medicinal Chemistry, University of Washington, Seattle, WA, USA*
LEANDRO ZANOTTO • *Center for Computing in Engineering and Sciences, Institute of Chemistry, University of Campinas, Campinas, São Paulo, Brazil*
ZHIWEI ZHOU • *Interdisciplinary Research Center on Biology and Chemistry, Shanghai Institute of Organic Chemistry, Chinese Academy of Sciences, Shanghai, China; University of Chinese Academy of Sciences, Beijing, China*
ZHENG-JIANG ZHU • *Interdisciplinary Research Center on Biology and Chemistry, Shanghai Institute of Organic Chemistry, Chinese Academy of Sciences, Shanghai, China*

Chapter 1

Fundamentals of Ion Mobility-Mass Spectrometry for the Analysis of Biomolecules

Caleb B. Morris, James C. Poland, Jody C. May, and John A. McLean

Abstract

Ion mobility-mass spectrometry (IM-MS) combines complementary size- and mass-selective separations into a single analytical platform. This chapter provides context for both the instrumental arrangements and key application areas that are commonly encountered in bioanalytical settings. New advances in these high-throughput strategies are described with description of complementary informatics tools to effectively utilize these data-intensive measurements. Rapid separations such as these are especially important in systems, synthetic, and chemical biology in which many small molecules are transient and correspond to various biological classes for integrated omics measurements. This chapter highlights the fundamentals of IM-MS and its applications toward biomolecular separations and discusses methods currently being used in the fields of proteomics, lipidomics, and metabolomics.

Key words Ion mobility, Ion mobility-mass spectrometry, Biomolecules, Omics

1 Origins of Biomolecular Analysis by Ion Mobility-Mass Spectrometry

Ion mobility-mass spectrometry (IM-MS) is an integrated chemical separation technique which combines complementary size- and mass-selective separations into a single analytical platform. IM-MS is capable of separating individual chemical compounds on the millisecond time scale and as such is considered a high-throughput technique, and IM-MS is currently used in numerous chemical analysis applications including those of biological origin. A timeline of historical developments in IM and IM-MS is provided in Fig. 1. Ion mobility traces its roots back to experiments by Rutherford and Thomson in the late 1890s [1] and was further improved by Tyndall in the 1920s using pure drift gases [2, 3], before being paired with mass spectrometry in the 1960s [4, 5]. Electrospray ionization (ESI) and laser ionization were first used with IM in 1968 [6, 7] and 1982 [8], respectively, with Bowers demonstrating matrix-assisted laser desorption/ionization (MALDI) with IM-MS

Giuseppe Paglia and Giuseppe Astarita (eds.), *Ion Mobility-Mass Spectrometry: Methods and Protocols*, Methods in Molecular Biology, vol. 2084, https://doi.org/10.1007/978-1-0716-0030-6_1,

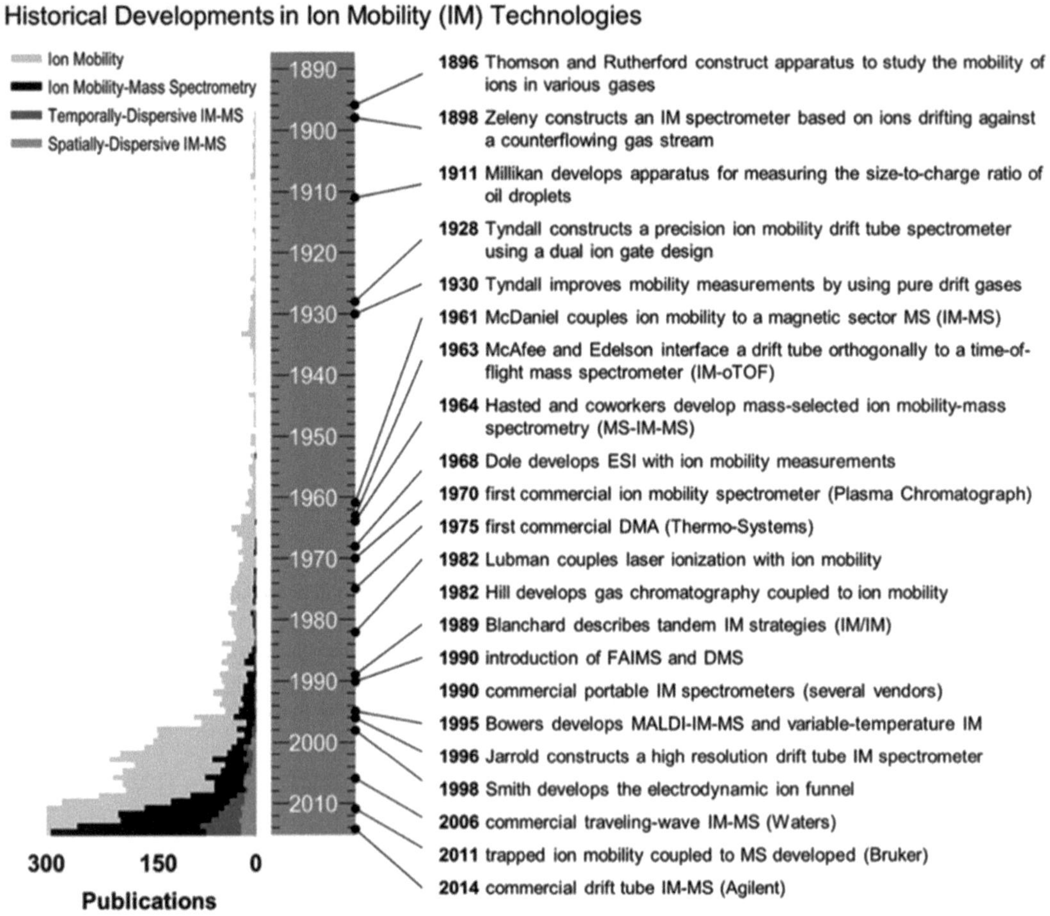

Fig. 1 A timeline and histogram summarizing the number of publications published between 1890 and 2014 as pertaining to ion mobility and ion mobility-mass spectrometry. Specific historical milestones in the development of ion mobility and IM-MS instrumentation are summarized at the right of the timeline (Figure from May et al. [5])

in the mid-1990s [9]. In the late 1990s onward, both MALDI and ESI ionization techniques were utilized with custom- and home-built IM-MS instruments by several laboratories, establishing IM-MS as highly sensitive technique for biomolecular analysis [10–14]. These soft ionization techniques were readily paired with IM-MS and enabled the analysis of large, intact molecules [15], and the growth in the application of IM-MS to biomolecules is illustrated in the timeline in Fig. 2 [16]. In 2006, a commercially available ion mobility-mass spectrometer well suited for biomolecular analysis was developed by Waters Corporation, named the Synapt HDMS ("High Definition" MS), which combined the soft ionization of ESI and liquid chromatography-ESI with the rapid separation of traveling wave-based IM technology and time-of-flight MS. The traveling wave IM technique was improved

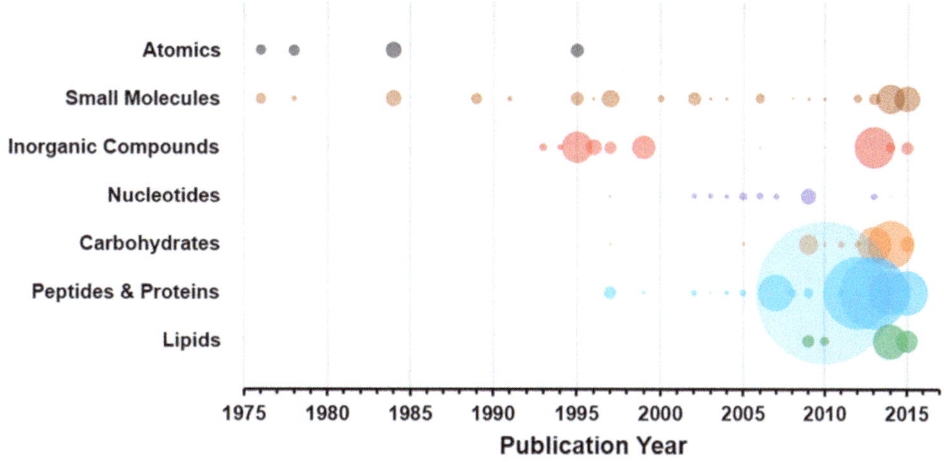

Fig. 2 A bubble plot summarizing the number of CCS values reported over time for seven chemical compound classes. The size of each bubble represents the relative number of CCS values in each respective year (Figure from May et al. [16])

significantly in 2011 and the platform subsequently supported other optional analytical capabilities including MALDI and MS imaging, as well as integration with a variety of other ion sources and accessories [17–19]. This first commercial offering enabled a larger scientific community access to IM-MS which otherwise was only available to researchers with the capability to build their own instruments. Subsequently, other vendors entered the market with their own commercial IM-MS offerings, including Agilent Technologies in 2014 (drift tube IM) [20] and Bruker Corporation in 2016 (trapped IM) [21]. With the availability of versatile commercial instrumentation, biomolecular analysis using IM-MS flourished and continues to be a rapidly growing field providing fundamental insight into some of our most important biological questions.

2 Biomolecular Separation and Analytical Utility of IM-MS

IM-MS has found great utility for biomolecular separation due to its broad sample compatibility, resolution, and sensitivity. It has been used to separate and analyze a wide range of masses, from small molecules to large protein complexes [22–29]. The analytical importance of IM is summarized in Table 1 [16]. IM provides a broad range of analytically valuable enhancements, including increasing peak capacity when coupled with other separation stages (chromatography, MS, etc.), enhancing low abundance signals of importance by reducing interference from chemical noise, and allowing for structural characterization of analytes through the measurement of their gas-phase cross-sectional area. Furthermore, it has been demonstrated that biomolecular classes maintain

Table 1
Three analytical uses of ion mobility

Analytical use of ion mobility	Description	Additional requirements	Example application areas
1. Chemical separation	partition signal from chemical noise and increase peak capacity of the analysis	none	detection of illicit compounds (e.g., drugs and explosives) and screening of exogenous metabolites (e.g., pesticides and industrial chemicals)
2. Analyte identification and characterization	use CCS measurement to characterize unknown by correlation	reference values from databases and libraries incorporating normalized drift times, reduced mobilities, and/or CCS	emerging omic and small molecule discovery initiatives
3. Structural analysis	Utilize the experimental CCS to infer structural information	Computational methods to link theoretical structure (s) to the experimental CCS	Insights into protein complex arrangements and structure

From May et al. [16]

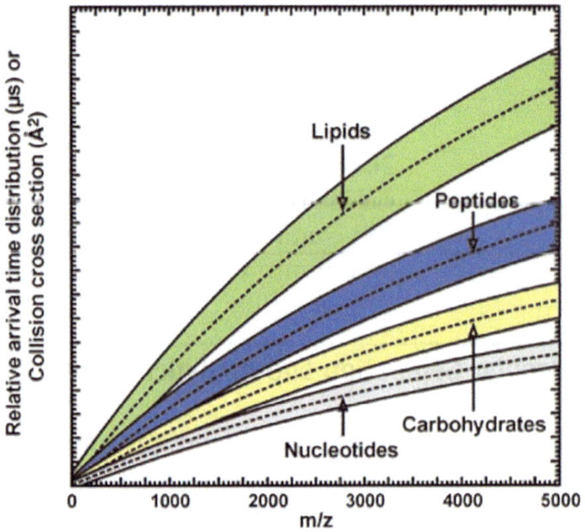

Fig. 3 A conceptual depiction of where various biomolecular classes are observed in IM–MS conformational space (Figure from Fenn et al. [30])

different molecular packing efficiencies in the gas phase, which results in class-specific correlations in 2-dimensional IM-MS spectra (Fig. 3) [20, 30, 31]. Notably, the canonical biochemical classes (lipids, peptides, carbohydrates, and nucleotides) align themselves into unique regions of IM-MS analytical space, which reflect their conformational preferences. Within each of these broad biochemical class correlations, subclass trends may also exist. As a prominent example, lipids have a characteristic behavior in IM-MS indicative of their primary class and structure [32]. Specifically, lipid packing efficiency is affected by both headgroup identity and modifications within their acyl tail(s), forming different correlations of gas-phase collision cross section (CCS) versus mass. Current research is considering the effects of degree of unsaturation and carbon chain length on lipid packing efficiency and correlation behavior in IM-MS [33–35]. Novel approaches such as ozonolysis paired with IM-MS are now also being used in lipidomics to allow for determination of double-bond location in lipids [36, 37]. For broad-scale identification of unknowns, empirically derived databases of mass and CCS are being assembled for a variety of biomolecular compounds [16, 24, 29, 38–44]. In addition to specific chemical class information, IM can address the challenge of resolving and identifying isobaric compounds that are unresolvable by conventional MS. In many cases, isobaric species including isomers and conformers can be distinguished in the IM dimension due to the different conformations they adopt in the gas phase, whether peptides, carbohydrates, or lipids [45–50]. Collectively, the capability to rapidly resolve biochemicals based on both structure and mass has made IM-MS an invaluable analytical tool for biomolecular analysis.

3 Instrumentation Considerations

3.1 Chromatography

The inclusion of a chromatographic separation prior to IM-MS provides added peak capacity while enhancing the analytical sensitivity of both the IM and MS stages by reducing ion suppression resulting from the simultaneous infusion of multiple analytes. Liquid chromatography (LC) is commonly implemented with IM-MS due to its ability to use a wide variety of flow rates, column choices, and solvents that pair well with ESI ion generation [51, 52]. Since the IM separation occurs postionization, all of the LC conditions which work in LC-MS are also compatible with IM-MS, including both normal- and reverse-phase LC columns and numerous solvent systems [53]. Autosampler systems are also commonly utilized with LC to facilitate automated, high-throughput LC-IM-MS workflows [54]. Other chromatographic separation techniques such gas chromatography (GC) and supercritical fluid chromatography (SFC) have also been demonstrated with IM-MS [55–57], although these are less common due to the more limited analytical space that these techniques encompass (e.g., volatile and nonpolar analytes, respectively). Figure 4 illustrates the compatibility of timescales for chromatography with IM-MS [5]. While chromatography operates on a timescale of minutes, the downstream analytical

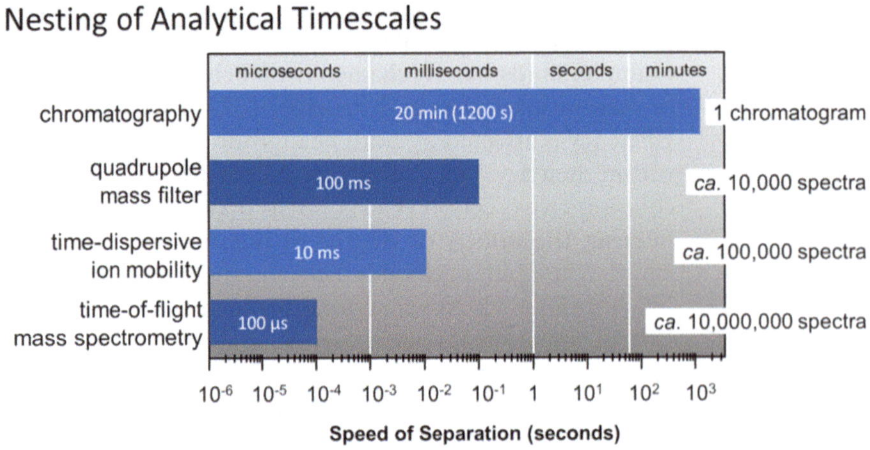

Fig. 4 The relationship of various analytical timescales based on the respective speed of each separation. The total number of possible spectra obtained through nesting the subsequent analytical separation dimensions is summarized to the right (Figure from May et al. [5])

strategies such as IM and MS operate on the order of milliseconds to microseconds, allowing further separation of components eluting from the chromatographic dimension without affecting the overall sample throughput.

3.2 Ion Sources

Two of the most common methods for ionization of biomolecules are MALDI and ESI. These so called "soft" ionization techniques impart minimal energy to molecules during the ionization process, which helps to maintain the analyte's structural integrity during transfer to the gas phase, which is crucial for the analysis of fragile biomolecules [9, 15]. MALDI is performed on solid samples using a laser to desorb and generate ions and tends to generate low charge states in a narrow distribution, which simplifies the mass spectra [58, 59]. The use of a laser also facilitates operating MALDI in an imaging mode, allowing spatially resolved mass information to be obtained [60]. MALDI-based MS imaging (MSI) has been utilized to determine analyte location in tissue samples [61], and MSI has also been coupled with IM-MS to provide a high-dimensional separation strategy that can simultaneously separate analytes based on spatial location, size, and mass [62, 63]. MALDI, however, is not conducive to direct analysis of liquid samples such as the effluent stream originating from LC, which requires offline sample fraction collection for MALDI analysis. Liquid sample analysis is, however, readily facilitated by ESI, which generates a continuous flow of ions from liquid-phase samples. ESI has enabled high-throughput LC-MS experiments and has been used to combine LC with IM-MS. In contrast to the low charge states observed in MALDI, ESI typically generates multiply charged ions. Since mass spectrometers measure ions as a mass-to-charge ratio, higher charge states effectively increase the practical mass range accessible to a mass spectrometer, facilitating analysis of proteins and other high mass analytes alongside small molecules in a complex biological sample [64, 65]. Multiply charged ions have also improved detection limits for Fourier transform MS (FTMS) techniques [66, 67], improved the ion fragment yields for traditional collisional activation experiments (collision-induced dissociation, CID) [68], and have enabled electron-induced fragmentation techniques such as electron capture dissociation (ECD) and electron transfer dissociation (ETD) to be implemented [69, 70]. Specifically for IM analysis, the various charge states generated from ESI are readily resolved into distinct correlations of cross section and mass, which facilitates the separation and identification of multiply charged species [10].

3.3 Time-Dispersive Ion Mobility Techniques

Two ion mobility techniques that are commonly used are drift tube ion mobility spectrometry (DTIMS) and traveling wave ion mobility spectrometry (TWIMS) (Fig. 5) [5]. TWIMS was one of the first IM techniques to be used in a commercially available IM-MS platform. In TWIMS, ion separations result from a series of dynamically pulsed voltages (an electrodynamic field) which creates a traveling wave potential that transfers ions through the drift region in a mobility-selective mode. TWIMS separations are similar to DTIMS separations and can access similar resolving powers [71]. Although DTIMS combined with MS was implemented commercially several years after TWIMS, it is an older technique and considered the standard and classically most accurate method for CCS determination. Unlike TWIMS, DTIMS separates ions using a constant voltage gradient (a uniform electric field) which allows the measured ion drift times to be related directly to CCS via the kinetic theory of gases, commonly referred to as the Mason–Schamp relationship [72, 73]. On the other hand, TWIMS-derived CCS values are determined via an empirical calibration relationship between TWIMS measured drift times and known CCS values obtained from DTIMS [27, 32, 74]. Due to the fact that TWIMS does not require high voltages for separations, it is readily scalable to longer path lengths, which fundamentally improves instrument resolution. In contrast, high operational voltages must be utilized in order to increase the path length in DTIMS. Next-generation TWIMS instruments are taking advantage of this scalability, producing high resolving power platforms based on serpentine and cyclic designs [45, 75].

3.4 Mass Spectrometry Considerations

Mass spectrometry is routinely conducted after IM separations to provide mass measurement on ion mobility separated ions. In this IM-MS configuration, most of the chemical separation occurs in

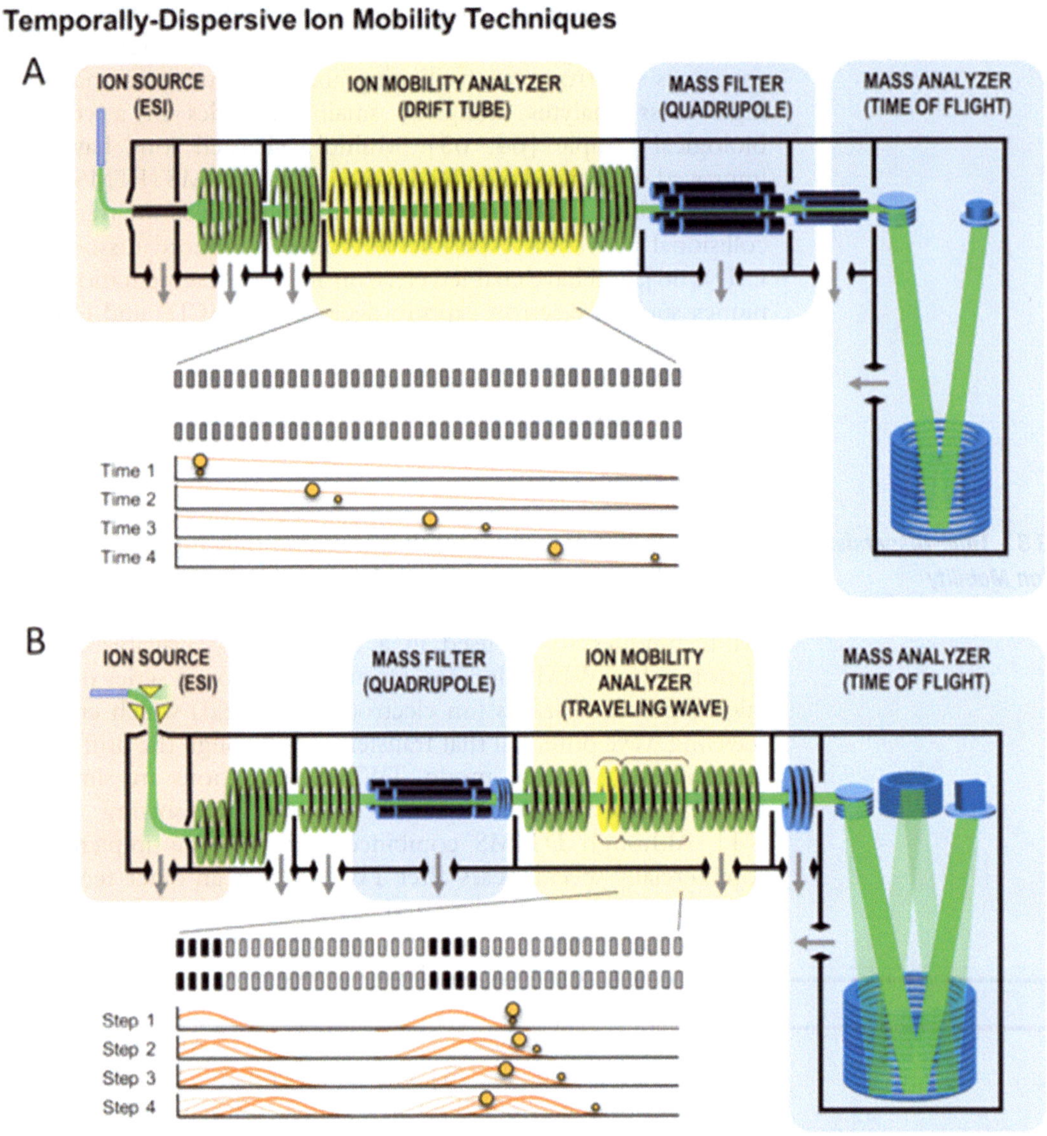

Fig. 5 Two conceptual schematic diagrams of commercially-available time-dispersive IM-MS instrumentation. (**a**) An electrostatic drift tube ion mobility spectrometer (DTIMS) coupled to a quadrupole-time-of-flight (QTOF) mass spectrometer, and (**b**) an electrodynamic traveling wave ion mobility spectrometer (TWIMS) coupled with a quadrupole mass filter and a QTOF. Conceptual experimental sequences are shown in the insets of each panel and illustrate the mechanism giving rise to the temporal separation of smaller and larger collision cross section ions experimentally observed in each technique (Figure from May et al. [5])

the mass dimension due to the high resolving power (>10,000) accessible by modern mass spectrometers; however, the added IM dimension provides improved peak capacity and the capability for resolving isomeric compounds based on structural differences [76]. IM is commonly coupled with time-of-flight (TOF) mass spectrometry due to its ability to rapidly analyze a wide mass range simultaneously and with high resolution. The timescale of the TOF is on the order of microseconds, pairing well with an upstream IM which is on the order of milliseconds per separation. Quadrupole MS is also commonly used with IM-MS as an added mass-filtering stage for tandem MS/MS experiments, which can further aid in analyte identification and characterization.

Fragmentation methods have been utilized with great success in tandem mass spectrometry and in IM-MS, where it can be initiated between the IM and MS stages (IM-MS/MS) or prior to IM-MS either with a front-end mass filter, or without, as is the case with in-source ion activation [77]. Collision-induced dissociation (CID) is commonly used to fragment ions as it is readily implemented with existing ion optics. CID is implemented through ion collisions with an inert background gas (such as nitrogen or argon) under elevated voltage conditions. CID is considered an ergodic process where energy is restributed within the molecule and the weakest bonds are preferentially cleaved, which leads to predictable and reproducible fragmentation of ions. The analysis of CID fragmentation spectra has been used with IM-MS to identify isobaric species which exhibit different bond energies. The CID method is commonly used in proteomics to determine the amino acid sequence of peptides. While highly predictable, CID does not preserve weak bonds typical of noncovalent complexes and post-translational modifications, and is less effective at fragmenting large molecules, such as intact proteins. To address these deficiencies, electron capture dissociation (ECD) and electron transfer dissociation (ETD) have been used, both of which fragment the ion in a nonergodic process and thus creating different but informative fragments compared to CID. ECD requires simultaneous trapping of ions and free electrons and has only been implemented in ion cyclotron resonance MS. ETD and CID have been used simultaneously with IM-MS for glycoproteomics [78–80]. This allows a variety of fragments to be determined for a given precursor ion, in particular for preserving informative labile bonds when obtaining fragmentation information. Combining the ion drift time, precursor mass, and fragment mass information can facilitate confident identification of the precursor molecular.

IM-MS holds an important and ever increasing role in biomolecular analyses. The ability of IM-MS to integrate with a wide variety of chromatographic techniques, ion generation methods, fragmentation methods, and mass analyzers results in a versatile and power technique for biomolecular analysis, particularly in the omic sciences (i.e., proteomics, lipidomics, metabolomics, etc.) where deriving biological answers requires high sensitivity and high-confidence identifications. Additionally, IM improves peak capacity compared to standalone MS and is compatible with other analytical stages such as LC separations and tandem ion fragmentation. In this context, IM can differentiate isobaric species and provides an additional descriptor (ion drift times or gas phase CCS) for the identification of unknowns. The following sections describe current trends in specific applications of IM-MS to biomolecular analyses.

4 Current Trends in IM-MS for Biomolecular Analyses

The confident identification of small molecules continues to be one of the most difficult challenges in omic studies. While advances in proteomics have streamlined IM-MS-based identification efforts for peptides and proteins [38, 81, 82], metabolomic and lipidomic identification capabilities have generally lagged behind [83, 84]. One reason for this is that while the structure of peptides consists of translationally assembled amino acid building blocks which can often be elucidated through ion fragmentation strategies, metabolites and lipids are less predictable and contain a wide array of substructural units. Additionally, the prevalence of isobaric species in lipidomics complicates feature identification, and metabolomics studies often encounter features of redundant mass that lack unique fragmentation patterns, which further confounds attempts at identification by MS alone. Therefore, a combination of analytical techniques is required for high-confidence lipidomics and metabolomics. As shown in Fig. 6, gas chromatography, liquid chromatography, or another front-end separations can be readily combined with IM and MS analysis to provide high-dimensional data sets which can be partitioned into specific omics workflows [53, 84]. The IM analysis provides charge state and chemical class-specific separations based on differences in intrinsic gas phase packing, as well as quantitative size information via CCS measurement which can be used as a reproducible measurement for identification purposes [84].

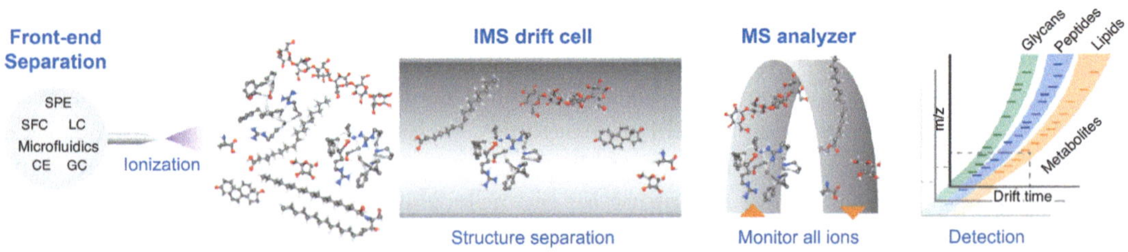

Fig. 6 A conceptual IM-MS analytical workflow incorporating various front-end separation techniques (*SPE*, solid phase extraction; *SFC*, super critical fluidic chromatography; *LC*, liquid chromatography; *CE*, capillary electrophoresis *GC*, gas chromatography). The resulting IM-MS analysis yields drift time vs. m/z spectra which can be interpreted for relative size-mass information intrinsic to different biochemical classes (Figure adapted from Zhang et al. [84])

5 Integrating Ion Mobility for Omics Analyses

IM-MS provides a fast separation of chemically unique biological groups with the added benefit of measuring collision cross section concurrently with mass-to-charge ratio [31, 82, 85]. This capability is important for omics studies utilizing complex biological samples which routinely require extensive sample purification strategies to isolate molecules of interest from undesirable compounds that would otherwise make MS analysis difficult (e.g., salts, detergents, and other types of chemical noise) [81, 86]. Sample preparation strategies have the potential to chemically alter molecules of interest, for example, by oxidation, reduction, conversion to a secondary metabolite, or loss of a post-translational modification in peptides. Integrating IM with MS can help offset some of the burden of chemical separation and alleviate the need for extensive sample handling. In certain cases where fragmentation occurs postmobility, either intrinsically or intentionally, IM-MS also allows the alignment of precursor and product ions which can further aid in identification [83, 87]. The following sections provide examples of how ion mobility has been integrated in proteomic, lipidomic, and metabolomic analyses. Although each analysis is discussed separately, it should be noted that ion mobility allows for simultaneous analysis of these individual omic classes via chemical class separation, providing a truly integrated multiomic experiment.

5.1 Proteomics

Proteomics, the large-scale study of proteins, has been a driving force in systems biology and has increased our understanding of diseases and human health. Proteins serve as the machines for all cellular processes; thus proteomic studies are one of the most crucial tasks in systems biology. MS-based proteomics is a major component to the advancements in the field [88, 89]. Proteins encompass a large dynamic range of concentrations, necessitating separation techniques to enhance lower abundance species prior to mass analysis. Common separation methods for proteomics include gel electrophoresis, liquid chromatography, and, in a growing number of instances, ion mobility [90–92].

Ion mobility has been utilized extensively in structural proteomics [82, 93]. Proteins of similar or exact mass, such as protein conformers, can be separated by IM due to differences in their gas phase structure [94]. Figure 7 illustrates IM separations for ions of protein complexes of different sizes and configurations, but similar mass-to-charge ratio [93]. The majority of multiprotein complexes have been analyzed on TWIMS instruments, and it has been demonstrated that incorporation of TWIMS with MS analysis can increase proteome coverage by up to 60% [95].

In addition to separating proteins, IM is used to probe structural information [93]. Temperature-controlled ESI sources and

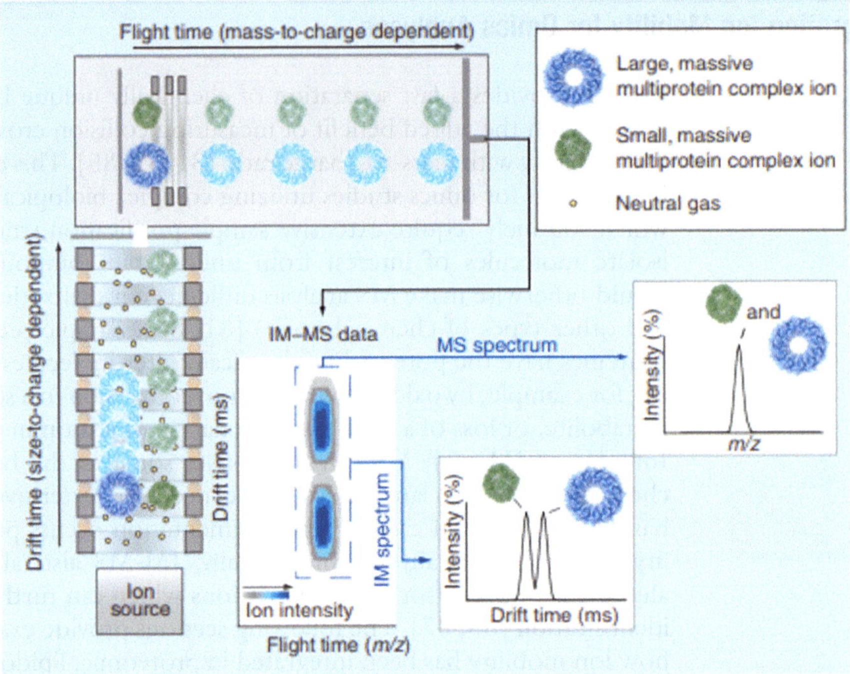

Fig. 7 Basic principles of IM-MS data acquisition and the resulting spectra. Ions generated in the ion source (lower left) are introduced into an IM spectrometer, which is effectively an ion guide consisting of a background of neutral gas molecules and a weak electric field. Following IM analysis, the ions are introduced into a mass analyzer operated under vacuum. The resulting IM and MS data are 3-dimensional, containing ion intensity, size, and mass information, and can be projected either as a contour plot (middle, bottom), or 2D selections in either drift time or *m*/*z* space (lower right) (Figure from Zhang et al. [93])

heated ion transfer capillaries have been used prior to IM-MS to rapidly heat proteins and monitor their controlled denaturation [11, 96, 97]. In addition, thermally induced protein conformational transformations as well as protein–ligand interactions are able to be observed. In these ways and others, IM-MS progresses from a separation strategy to a structural measurement technique to broaden our understanding of how protein clusters are formed and stabilized [96, 97].

5.2 Lipidomics

Lipids comprise a large portion of the small molecules extracted from organisms. They have three major functions in biological systems: energy storage, cellular signaling, and structural functions. Lipids can be divided into eight major categories (fatty acids, glycerolipids, glycerophospholipids, sphingolipids, sterols, prenols, saccharolipids, and polyketides) with many different structural motifs ranging from fused cyclic molecules to long-chain fatty acids. Lipids cover a large range of m/z ratios, and while mass spectrometry has been a powerful tool in lipidomics, the numerous isomeric lipid species which can exist make lipid structural

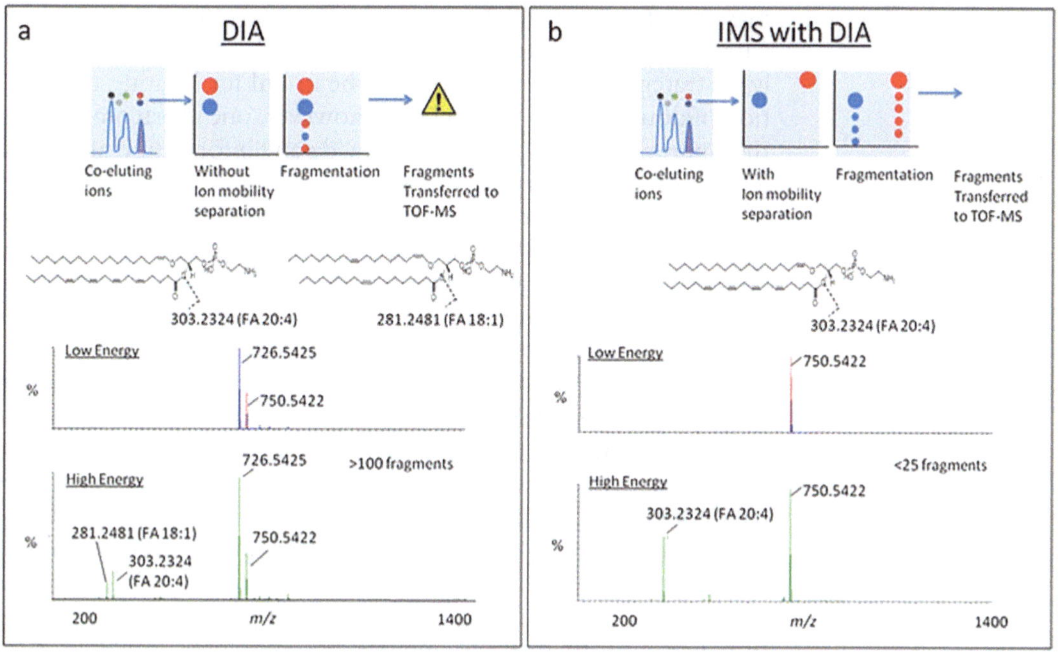

Fig. 8 Two examples of data-independent acquisition (DIA) using IM-MS instrumentation. (**a**) Conventional DIA where ions are comprehensively fragmented following chromatography, (MS^E) and (**b**) MS^E coupled with IM ($HDMS^E$). DIA combined with IM separation enables mobility-aligned fragmentation spectra, which can increase the specificity and confidence in identifying complex lipids (Figure from Paglia et al. [103])

characterization by MS challenging [98–100]. Contributing to lipid structural complexity are the numerous potential double-bond positions, geometric (*cis/trans*), constitutional (linear and branched), and stereochemical orientations that a lipid can adopt which are all isobaric in mass. Identifying complex lipid structures has been a struggle in the field of lipidomics, one that ion mobility is well-suited to address [99–102]. Lipid identifications in MS-based lipidomics utilize the exact mass measurement, as well as characteristic fragmentation patterns obtained from tandem MS/MS experiments to assign confidence to structural identifications. As many lipids are chemically and structurally similar, lipids can be challenging to separate using LC alone, making it difficult to correlate fragment ions with their precursor ion forms. As demonstrated by Paglia et al., ion mobility can be used to align fragmentation spectra with precursor parent ions to increase the confidence in lipid identification (Fig. 8). In this study, ion mobility was also demonstrated to be useful in separating coeluting lipid structures with the same *m/z* ratios [103].

Ozone-induced dissociation has been recently demonstrated with IM-MS to elucidate the location of double bonds in the acyl tail region of lipids. Two separate strategies have been described: solution-phase ozonolysis of lipids prior to being introduced to the mass spectrometer [104], and gas-phase ozonolysis of lipid ions within the MS, the latter technique termed OzID. These ozonolysis strategies have been shown to be useful for locating the position of double bonds within lipids; however, one shortcoming with this approach is that it does not provide any information about the geometry (*cis/trans*) of the double bond prior to ozonolysis [104, 105]. This emphasizes the strength of IM-MS analysis, where ion mobility allows for the differentiation of some geometric lipid isomers, such as *cis* versus *trans*, even when present in a complex biological mixture. Groessl et al. demonstrated that CCS differences of 1% or more are sufficient to baseline separate lipids in DTIMS [101]. Although the possibility to use IM for identification purposes was also discussed, it was noted that high precision and accuracy are needed to create and populate reliable reference data libraries. Recently, Leaptrot et al. reported on an empirical CCS database of various sphingolipids and phospholipids incorporating 456 high-precision DTIMS measurements (global average RSD of 0.18%), which provides a quantitative means of identifying unknown lipids, including isomers, by IM-derived CCS information [33]. As these CCS libraries become more available, it is expected that lipidomics using IM-MS will continue to grow.

5.3 Metabolomics

Metabolomics is the measurement of the thousands of small molecules in a biological system. Unlike genomics and proteomics, metabolomics encompasses a large amount of chemical diversity as it consists of molecules from many different biological classes, such as carbohydrates, amino acids, hormones, and lipids [106]. There are generally two approaches to MS-based metabolomics: (1) targeted analysis, in which a panel of metabolites are selected for chemical analysis, and (2) untargeted analysis, in which all small molecules present in the sample are analyzed simultaneously [106]. Both approaches have their advantages. Targeted studies allows for semiquantitative analysis of small molecules on a curated list. Isotope standards can be analyzed concurrently with unknowns and calibrated against empirically measured response curves (calibration curves) in order to determine the concentration of specific metabolites within a sample. Since only a small number of m/z values are prioritized, targeted studies are commonly conducted on MS instrumentation with selective ion monitoring capabilities, such as triple quadrupole and ion trap instruments where a select m/z can be tuned and monitored in a continuous duration, greatly increasing the instrument sensitivity and response reproducibility, which improves quantitative results. While targeted approaches provide quantitative information regarding metabolites of interest, these studies do not provide detailed information for other small molecules present in the sample. On the other hand, untargeted approaches focus on separating and comprehensively measuring all of the small molecules present in the sample but lacks robust means of quantifying these signals. Also, untargeted studies generally utilize analytical methods and settings that attempt to measure a large breadth of molecules, and thus can be less sensitive to a specific class or pool of analytes [106–108]. While untargeted studies can detect a large number of molecules present in a sample, identifying the oftentimes thousands of metabolites observed in a single untargeted study can be an arduous task. For a single *m/z* feature, there can be hundreds of matches to known compounds in various metabolomic databases, making an absolute identification difficult [107–109]. Thus in order to improve confidence in metabolite identification, multiple dimensions of analytical information are generally obtained in untargeted experiments, which can include measurements from front-end chromatographic separations (analyte polarity and retention times), as well as post-ionization techniques such as ion mobility (likely chemical class and collision cross sections) and tandem MS (ion stability and fragmentation information). Ion mobility in particular can provide additional peak capacity, isomeric differentiation, and an additional molecular descriptor (CCS) which can help to alleviate some of the difficulties associated with confident metabolite identification [107–111].

As noted in the previous section, IM can improve fragmentation identification in lipidomics, and this advantage applies to metabolomics as well. With such chemical diversity found in the metabolome, ion isolation is complicated by coeluting species, and thus aligning precursor ions with their fragments originating from ion activation experiments is challenging. Using IM prior to ion fragmentation allows coeluting small molecules to be further resolved following chromatography [107–109]. For example, Wickramasekara et al. classified coeluting lipid species based on the differences in their IM drift times (Fig. 9) [112]. In addition, the drift time-extracted spectra can be used to align fragment ions to the corresponding parent ion, as these signals have identical drift times when conducting the fragmentation postmobility. This capability to mobility-align ion precursors and fragments can help to collate specific ion fragment information which can then be used with the accurate mass measurement to match unknowns to database entries, thus increasing the confidence in assigning metabolite identifications.

In addition to utilizing the enhanced separation and fragment alignment capabilities from IM, CCS measurements derived from IM experiments can improve metabolite validation [109]. CCS is linked to an intrinsic molecular property of the analyte (the microscopic molecular cross section) and thus is considered more robust than other conditional measurement parameters such as the chromatographic retention time. This property makes CCS useful as an additional molecular descriptor that can be used in metabolomic studies along with accurate mass and fragmentation information.

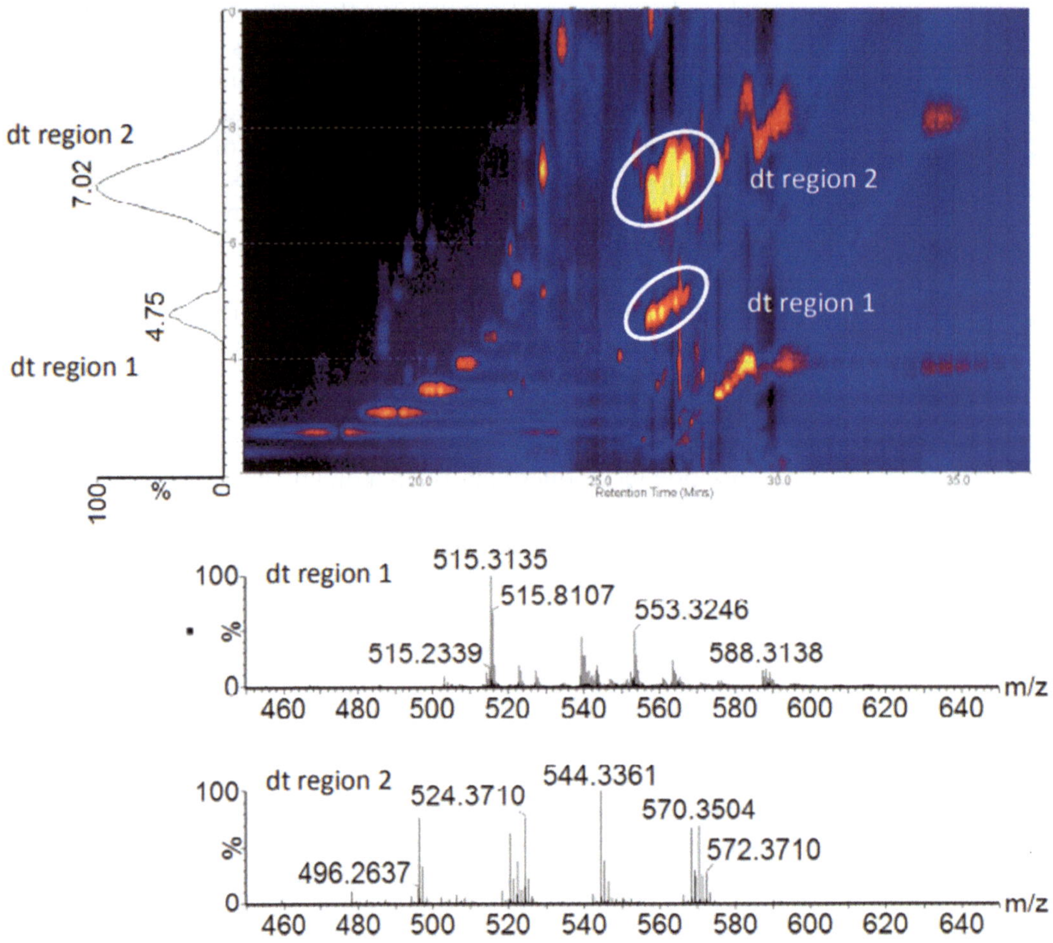

Fig. 9 An example 2D LC-IM spectrum (drift time vs. retention time) showing IM separation of different compound classes observed in rat plasma. The encircled regions represent compound classes that elute at a similar retention time (26–28 min). Extracted IM spectra (bottom) reveal that these two regions of ion signal represent different lipid classes, namely Lyso-PC and SM lipids (sphingosine phosphocholines) (Figure from Wickramasekara et al. [112])

Currently, there are several laboratories attempting to use CCS in metabolite identification workflows. For example, Paglia et al. have described a robust analytical workflow incorporating CCS for both metabolite and lipid identifications and report a ca. 2% interlaboratory reproducibility of the TWIMS-derived CCS [42, 83]. Stow et al. utilized standardized DTIMS instrumentation deployed across several international laboratories to achieve an interlaboratory CCS reproducibility of better than 0.5% [113], and recent work by Nichols et al. describe the utility of DTIMS CCS measurements as a molecular descriptor in untargeted studies of primary human metabolites [111]. As more research shifts toward incorporating CCS into metabolomic analysis, there is a need for database expansion to support bioinformatic strategies. Recent efforts for developing CCS databases to support metabolite identifications have included pesticides, pollutants, xenobiotics, and steroids [114–117]. Leveraging the standardization efforts for DTIMS, Picache et al. have recently described a "Unified CCS Compendium" which compiles over 3800 CCS measurements obtained from different studies into a single, self-consistent resource with a global average CCS precision of 0.25% RSD [43]. These and other efforts will allow rapid and reliable metabolite identification and quantification, which becomes increasingly important as the field shifts toward comprehensively characterizing individual metabolomic pathways.

6 Accelerating Advancements in IM-MS Technology

Innovation in IM-MS instrumentation continues at a rapid pace. Various improvements have been suggested and novel IM-MS technologies now being described allow for ingenious solutions for challenges encountered in biomolecular analysis. For example, a novel instrument design approach now being actively developed for IM-MS utilizes a scalable ion optical architecture consisting of electrode pads on a printed circuit board (PCB) and driven with electrodynamic (RF) fields that confine ions to a predefined ion optical path. This approach, named by R.D. Smith and colleagues as "structures for lossless ion manipulations" (SLIM), utilizes a 2-dimensional electrode geometry that is both modular and scalable such that various different experiments can be achieved on the same instrument platform [45, 118–121]. In SLIM, two PCBs with a mirrored electrode symmetry are placed above and below one another to create the ion path of travel in between the boards, and dynamic electric fields are used to both contain and guide the ions through the SLIM device [122]. SLIM allows high ion transfer transmission through elevated pressure regions, and SLIM-based ion mobility separations based on both DTIMS and TWIMS have been demonstrated [123, 124]. The ability to print electrodes on a two-dimensional surface allows for various ion manipulation modules to be fabricated, including modules to move ions at 90° angles (elbows and tees) [123, 125]. This facilitates cyclic racetrack and serpentine geometries to be fabricated for long path length, high-resolution ion mobility separations, and incorporation of "tee" junctions allows selection of a discreet ion mobility region for further tandem analysis by either IM or MS. An example of SLIM-based ion mobility instrumentation is shown in Fig. 10 [124]. Current designs have created instruments with some of the highest IM resolution currently available [71].

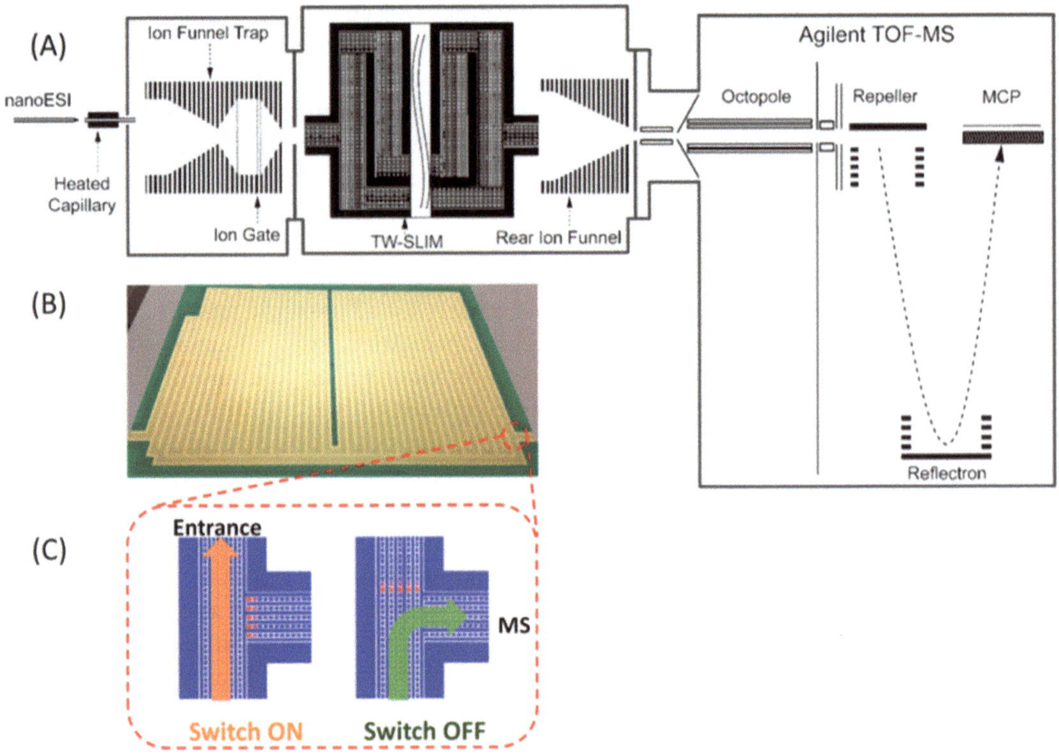

Fig. 10 (**a**) A schematic of the multipass SLIM SUPER IM-MS instrument. (**b**) Photograph of one of the two SLIM surfaces which are stacked in mirrored symmetry in the instrumentation, and (**c**) an illustration of an ion switch which is used to direct ions through various stages of the experiment (Figure from Deng et al. [124])

In addition to TWIMS and DTIMS, a relatively new ion mobility technique called trapped ion mobility spectrometry (TIMS) is currently available in commercial instrumentation [126, 127]. TIMS performs ion mobility separations by selectively releasing ions trapped in a mobility "analyzer" cell combining a gas flow and an opposing electric field [128, 129]. The ions trapped in TIMS are released slowly by lowering the electric field barrier, which allows a mobility spectrum to be obtained and subsequent MS analysis to be performed. The rate at which the electric field is lowered corresponds with IM resolution, with slower scan rates leading to higher resolution. TIMS instruments are capable of high IM resolution and are very versatile due to its variable scan rate [130]. Either high-throughput or high-resolution scan rates may be chosen as needed, or a scan rate may be variable during analysis to allow high resolution only for a certain range of mobilities, enabling targeted high-resolution experiments to be conducted. This is in contrast to current high-resolution approaches based on cyclic or racetrack IM technologies, which relies on selecting a narrow population of mobilities for high-resolution analysis, at a cost of rejecting other ions outside of this mobility window.

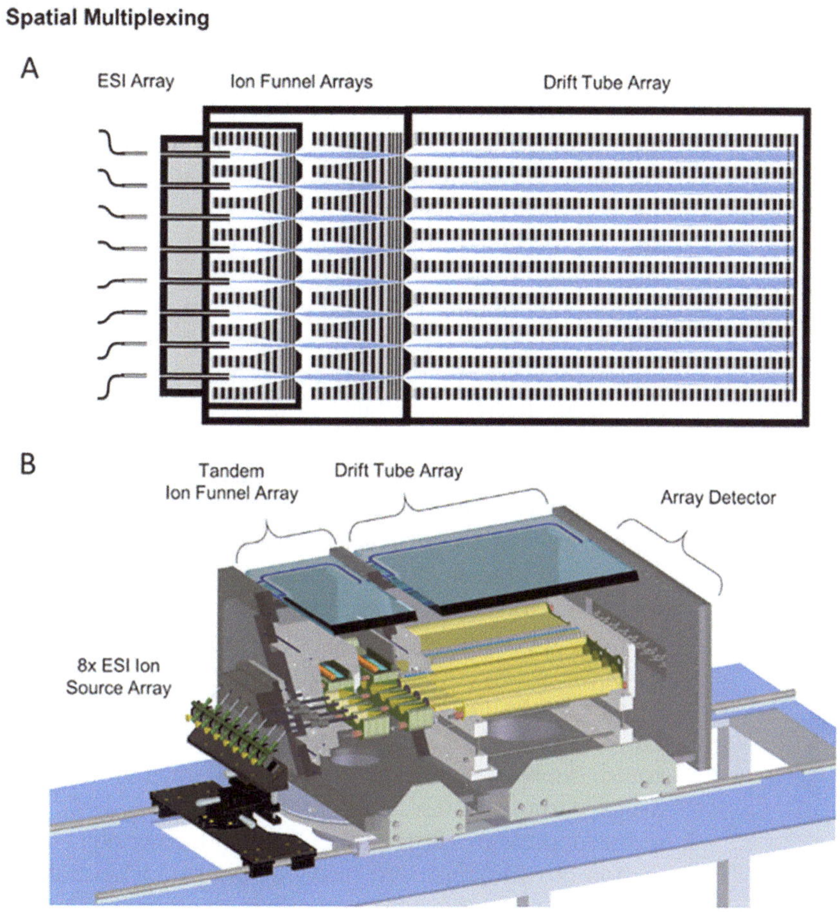

Fig. 11 An example of a spatial multiplexing strategy for DTIMS using eight individual IM analysis channels: (**a**) simulation of ion trajectories through the interfacing ion funnels and the drift tube array, and (**b**) a cutaway showing component details of the spatially multiplexed IM instrument (Figure from May et al. [5])

While most efforts have focused on improving IM resolution, some approaches have sought to improve sensitivity and throughput. An example of such an approach is a multichannel IM spectrometer shown in Fig. 11 [5]. This instrument utilizes eight discrete ion optical paths to perform eight ion mobility separations in parallel [5, 131]. Each ion channel can act independently, analyzing eight unique samples at a time, improving throughput. Alternatively, the multichannel instrument can analyze the same sample simultaneously across the eight ion optical paths, increasing the sensitivity of the instrument. As the analytical community pushes for rapid extraction of more information from complex samples, advancements in high-throughput instrument designs remain crucial.

Additional strategies for improving ion mobility separation have focused on increasing the chemical selectivity of existing IM instrumentation by using alternate drift gases [132]. The understanding and application of the effect of drift gas on IM separations and the associated CCS measurement are still in early development, but there is now mounting evidence that the use of more polarizable drift gases (e.g., CO_2, N_2O, NO_2) can increase the resolution for certain ion species [132, 133]. While the hard-sphere interactions between the ion and drift gas tend to predominate the mobility of ions in the IM experiment [134], long-range interactions also play a role in the observed IM separations and are exploited by varying the drift gas polarizability. Improved selectivity can occur between certain ionic species depending upon their susceptibility to long-range interactions [135]. The effect on separation efficiency by varying the drift gas composition is similar to the effect of varying the solvent conditions in capillary electrophoresis to affect separation selectivity. It is common to alter the drift gas in some high-field IM techniques, such as high-field asymmetric waveform ion mobility spectrometry (FAIMS) and differential mobility spectrometry (DMS) [136–141]. In low-field IM techniques, this is less common, but there have been a number of significant examples [142, 143]. Using an ambient pressure drift tube, Hill and coworkers showed that varying the drift gas polarizability could improve selectivity for various small molecules including drugs, carbohydrates, and peptides [132, 133, 144, 145]. Using a reduced pressure drift tube, Yost and coworkers reported that carbon dioxide improved resolving power for several isobaric steroids [146]. Eberlin and coworkers demonstrated that replacing nitrogen with carbon dioxide in a TWIMS instrument could improve separation of a number of analytes such as carbohydrates and isomeric haloanilines [50, 147–149]. However, it is common for observations of more polarizable drift gases to report minimal improvement to overall IM peak capacity or resolution compared to nitrogen or helium [23, 135, 150–152]. For example, Fjeldsted and coworkers investigated the separation of various small molecule pesticides, isomeric carbohydrates, and fluoroalkyl phosphazenes in a wide variety of drift gases, including helium, nitrogen, argon, carbon dioxide, nitrous oxide, and sulfur hexafluoride [23]. Generally, it was observed that helium and nitrogen had the highest resolution and resolving power, with some of the more polarizable drift gases demonstrating better selectivity when comparing certain analyte pairs [23]. To fully evaluate drift gas effect, a broader range of masses and biological classes needs to be reported in a variety of drift gases across multiple platforms and laboratories. Recently, Morris et al. utilized DTIMS operated in a variety of drift gases (helium, nitrogen, argon, and carbon dioxide) to study several different classes of compounds (quaternary ammoniums, phosphazenes, polypeptides, and carbohydrates) and provided recommended experimental parameters that allow for comparable CCS values to be determined among multiple laboratories [153]. Currently, the majority of CCS measurements have been reported in either helium or nitrogen, hindering the evaluation of IM separation performance in other alternate drift gases [16]. As the methods and parameters to perform these variable drift gas experiments become standardized, normalized measurements such as CCS will be vital for allowing direct comparison of separations conducted on different platforms (e.g., drift tube or traveling wave) [16]. The potential analytical importance of drift gas composition on improving IM resolution and separation selectivity is significant, and currently this area of research is largely unexplored.

7 Accelerating Advancements in Integrated Omics Analyses

When utilized in various omics studies, IM has been used primarily to partition analytes of interest from chemical noise originating from complex samples, resolve ambiguities within coeluting features, and align precursor and fragmentation data acquired in data-independent strategies. While the majority of IM-MS applications in biomolecular analysis have focused on specific omic fields (e.g., proteomics, lipidomics, metabolomics), there is increasing interest in utilizing IM-MS for untargeted, multiomic studies which examine a large breadth of molecule types simultaneously [154]. A truly, multiomic analytical workflow will facilitate the development of system maps which connect relationships between molecule types and allow perturbed pathways to be elucidated. To illustrate this concept, Fig. 12 displays work from Paglia et al. on building pathway maps to track metabolites being shuttled between mitochondria and the cytosol [155]. An area that could significantly benefit from this type of analysis is the microbiome and chemical communications fields. The microbiome has experienced recent and significant attention aimed at understanding the integral role that commensal bacteria plays on human health [156–158]. This focus, in large part, is due to advancements in sequencing of bacterial communities allowing for whole populations to be analyzed simultaneously, facilitating the comparison of healthy versus disease states [157, 158]. One challenge that remains in microbiome research is the understanding of the mechanisms that lead to disease, which can be addressed at least in part by building biochemical inventories of small molecule metabolites observed within the samples. MS-based metabolomics can be applied to a variety of sample types, can measure thousands of metabolites, requires minute sample quantities, and thus has high potential for integration into multiomics investigations. In particular, the ability of IM-MS experiments to simultaneously separate and detect multiple biological classes and chemical motifs can allow perturbed

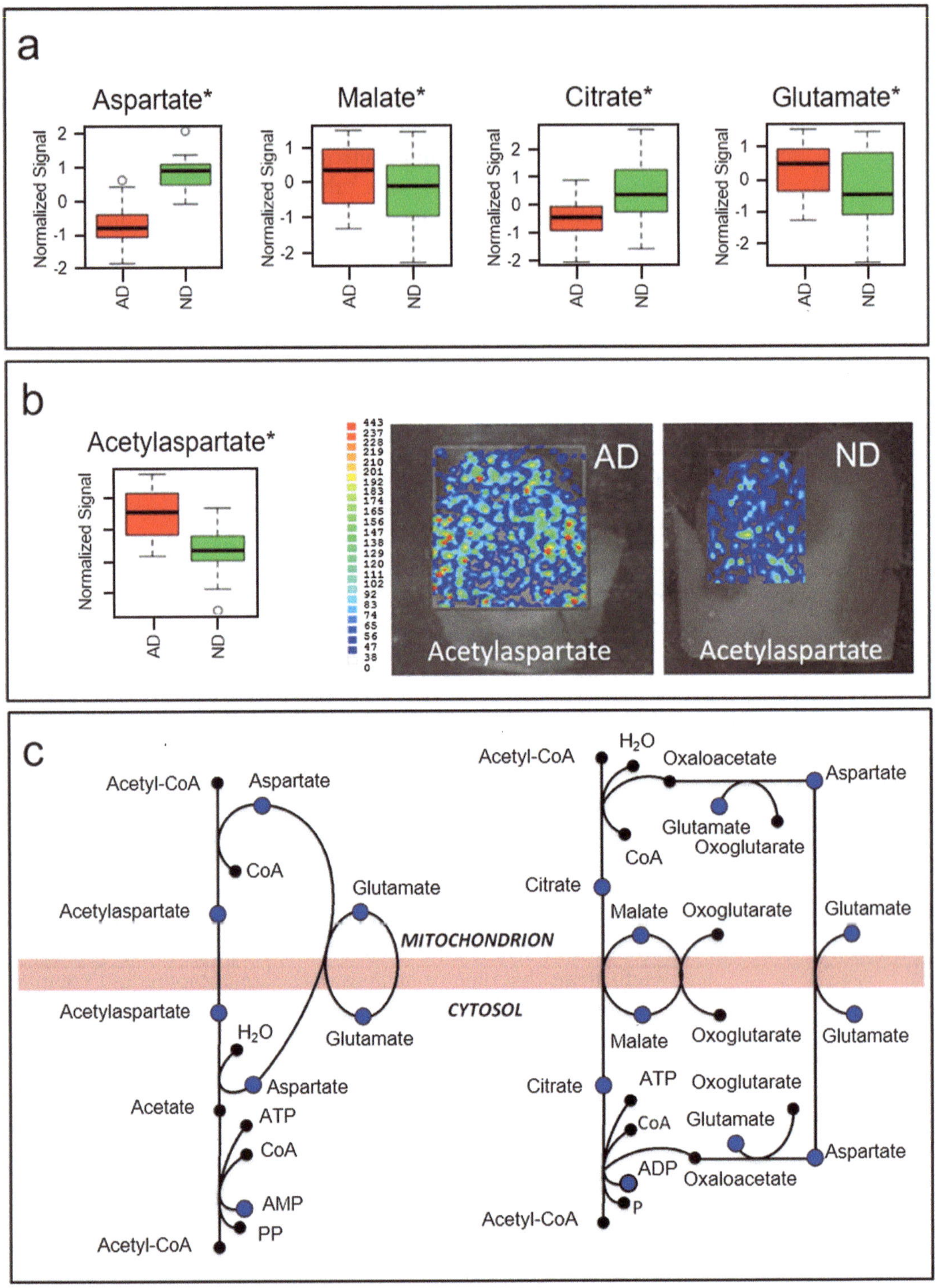

Fig. 12 (**a**) Bar charts summarizing the normalized ion signal observed in IM-MS experiments for aspartate, malate, citrate, and glutamate obtained from AD (red) and ND control subjects (green). (**b**) N-acetylaspartate (NAA) levels between AD (red) and control subjects (green) and MS images obtained from brain sections of both AD and control subjects. (**c**) Mitochondrial shuttles describing the exchange of metabolites between the cytosol and mitochondria and the metabolites quantified in the IM-MS experiments (blue dots) of Paglia et al. $*p < 0.05$ (*t* test) (Figure from Paglia et al. [155])

metabolites to be observed along with changes in the bacterial community. Uncovering changes in the metabolomic profile can provide insight into the role that particular bacteria plays within the complex interplay of biomolecules associated with the microbiome and/or host organisms.

Throughout this chapter, biomolecular analysis by various IM-MS technologies has been highlighted. Currently, IM is used primarily to enhance the separation of molecules present within complex samples. However, with higher quality IM measurements now being reported throughout the community [43], the role of IM for supporting biomolecular identifications is poised to make a significant impact in the omics sciences, particularly within the metabolomics and lipidomics communities. As technological and experimental advances are continuing at an accelerated pace, IM-MS will play an increasingly important role in providing deeper levels of information in large-scale biomolecular analysis initiatives including those in systems, synthetic, and chemical biology.

Acknowledgements

This work was supported in part using the resources of the Center for Innovative Technology at Vanderbilt University. The authors gratefully acknowledge financial support for this work provided by the National Institutes of Health (NIH NIGMS R01GM092218 and NCI R03CA222452) and the U.S. Environmental Protection Agency under Assistance Agreement No. 83573601. This work has not been formally reviewed by the EPA. The views expressed in this document are solely those of the authors and do not necessarily reflect those of the funding agencies and organizations. The U.S. Government does not endorse any products or commercial services mentioned in this publication.

References

1. Thomson JJ, Rutherford E (1896) On the passage of electricity through gases exposed to Roentgen rays. Phil Mag Ser 5 42 (258):392–407
2. Tyndall AM, Powell CF (1930) The mobility of ions in pure gases. Proc R Soc Lond Ser A 129(809):162–180
3. Tyndall AM, Starr LH, Powell CF (1928) The mobility of ions in air. Part IV. Investigations by two new methods. Proc R Soc Lond Ser A 121(787):172–184
4. Barnes WS, Martin DW, McDaniel EW (1961) Mass spectrographic identification of the ion observed in hydrogen mobility experiments. Phys Rev Lett 6(3):110–111
5. May JC, McLean JA (2015) Ion mobility-mass spectrometry: time-dispersive instrumentation. Anal Chem 87(3):1422–1436
6. Dole M, Mack LL, Hines RL, Mobley RC, Ferguson LD, Alice MB (1968) Molecular beams of macroions. J Chem Phys 49 (5):2240–2249
7. Dole M, Hines RL, Mack LL, Mobley RC, Ferguson LD, Alice MB (1968) Gas phase macroions. Macromolecules 1(1):96–97
8. Lubman DM, Kronick MN (1982) Plasma chromatography with laser-produced ions. Anal Chem 54(9):1546–1551
9. von Helden G, Wyttenbach T, Bowers MT (1995) Inclusion of a MALDI ion source in

the ion chromatography technique: conformational information on polymer and biomolecular ions. Int J Mass Spectrom Ion Process 146-147:349–364

10. Wittmer D, Chen YH, Luckenbill BK, Hill HH Jr (1994) Electrospray ionization ion mobility spectrometry. Anal Chem 66 (14):2348–2355
11. Clemmer DE, Jarrold MF (1997) Ion mobility measurements and their applications to clusters and biomolecules. J Mass Spectrom 32(6):577–592
12. Hoaglund CS, Valentine SJ, Sporleder CR, Reilly JP, Clemmer DE (1998) Three-dimensional ion mobility/TOFMS analysis of electrosprayed biomolecules. Anal Chem 70(11):2236–2242
13. Henderson SC, Valentine SJ, Counterman AE, Clemmer DE (1999) ESI/ion trap/ion mobility/time-of-flight mass spectrometry for rapid and sensitive analysis of biomolecular mixtures. Anal Chem 71(2):291–301
14. Gillig KJ, Ruotolo B, Stone EG, Russell DH, Fuhrer K, Gonin M, Schultz AJ (2000) Coupling high-pressure MALDI with ion mobility/orthogonal time-of-flight mass spectrometry. Anal Chem 72(17):3965–3971
15. Smith RD, Loo JA, Loo RRO, Busman M, Udseth HR (1991) Principles and practice of electrospray ionization–mass spectrometry for large polypeptides and proteins. Mass Spectrom Rev 10(5):359–452
16. May JC, Morris CB, McLean JA (2017) Ion mobility collision cross section compendium. Anal Chem 89(2):1032–1044
17. Giles K, Williams JP, Campuzano I (2011) Enhancements in travelling wave ion mobility resolution. Rapid Commun Mass Spectrom 25(11):1559–1566
18. Giles K, Pringle SD, Worthington KR, Little DLW, Bateman RH (2004) Applications of a traveling wave-based radio-frequency-only stacked ring ion guide. Rapid Commun Mass Spectrom 18(20):2401–2414
19. Pringle SD, Giles K, Wildgoose JL, Williams JP, Slade SE, Thalassinos K, Bateman RH, Bowers MT, Scrivens JH (2007) An investigation of the mobility separation of some peptide and protein ions using a new hybrid quadrupole/travelling wave IMS/oa-ToF instrument. Int J Mass Spectrom 261 (1):1–12
20. May JC, Goodwin CR, Lareau NM, Leaptrot KL, Morris CB, Kurulugama RT, Mordehai A, Klein C, Barry W, Darland E, Overney G, Imatani K, Stafford GC, Fjeldsted JC, McLean JA (2014) Conformational ordering of biomolecules in the gas phase: nitrogen collision cross sections measured on a prototype high resolution drift tube ion mobility-mass spectrometer. Anal Chem 86 (4):2107–2116
21. Silveira JA, Ridgeway ME, Park MA (2014) High resolution trapped ion mobility spectrometry of peptides. Anal Chem 86 (12):5624–5627
22. Zhang F, Guo S, Zhang MY, Zhang ZX, Guo YL (2015) Characterizing ion mobility and collision cross section of fatty acids using electrospray ion mobility mass spectrometry. J Mass Spectrom 50(7):906–913
23. Kurulugama RT, Darland E, Kuhlmann F, Stafford G, Fjeldsted J (2015) Evaluation of drift gas selection in complex sample analyses using a high performance drift tube ion mobility-QTOF mass spectrometer. Analyst 14(20):6834–6844
24. Paglia G, Williams JP, Menikarachchi L, Thompson JW, Tyldesley-Worster R, Halldorsson S, Rolfsson O, Moseley A, Grant D, Langridge J, Palsson BO, Astarita G (2014) Ion mobility derived collision cross sections to support metabolomics applications. Anal Chem 86(8):3985–3993
25. Allen SJ, Schwartz AM, Bush MF (2013) Effects of polarity on the structures and charge states of native-like proteins and protein complexes in the gas phase. Anal Chem 85(24):12055–12061
26. Campuzano I, Bush MF, Robinson CV, Beaumont C, Richardson K, Kim H, Kim HI (2012) Structural characterization of drug-like compounds by ion mobility mass spectrometry: comparison of theoretical and experimentally derived nitrogen collision cross sections. Anal Chem 84(2):1026–1033
27. Bush MF, Campuzano IDG, Robinson CV (2012) Ion mobility mass spectrometry of peptide ions: effects of drift gas and calibration strategies. Anal Chem 84 (16):7124–7130
28. Shah AR, Agarwal K, Baker ES, Singhal M, Mayampurath AM, Ibrahim YM, Kangas LJ, Monroe ME, Zhao R, Belov ME, Anderson GA, Smith RD (2010) Machine learning based prediction for peptide drift times in ion mobility spectrometry. Bioinformatics 26 (13):1601–1607
29. McLean JA (2009) The mass-mobility correlation redux: the conformational landscape of anhydrous biomolecules. J Am Soc Mass Spectrom 20(10):1775–1781
30. Fenn LS, McLean JA (2008) Biomolecular structural separations by ion-mobility-mass

spectrometry. Anal Bioanal Chem 391 (3):905–909

31. Fenn LS, Kliman M, Mahsut A, Zhao SR, McLean JA (2009) Characterizing ion mobility-mass spectrometry conformation space for the analysis of complex biological samples. Anal Bioanal Chem 394 (1):235–244
32. Hines KM, May JC, McLean JA, Xu LB (2016) Evaluation of collision cross section calibrants for structural analysis of lipids by traveling wave ion mobility-mass spectrometry. Anal Chem 88(14):7329–7336
33. Leaptrot KL, May JC, Dodds JN, McLean JA (2019) Ion mobility conformational lipid atlas for high confidence lipidomics. Nat Commun 10(1):985
34. Baker ES, Burnum-Johnson KE, Jacobs JM, Diamond DL, Brown RN, Ibrahim YM, Orton DJ, Piehowski PD, Purdy DE, Moore RJ, Danielson WF, Monroe ME, Crowell KL, Slysz GW, Gritsenko MA, Sandoval JD, LaMarche BL, Matzke MM, Webb-Robertson B-JM, Simons BC, McMahon BJ, Bhattacharya R, Perkins JD, Carithers RL, Strom S, Self SG, Katze MG, Anderson GA, Smith RD (2014) Advancing the high throughput identification of liver fibrosis protein signatures using multiplexed ion mobility spectrometry. Mol Cell Proteomics 13 (4):1119–1127
35. Liu X, Plasencia M, Ragg S, Valentine SJ, Clemmer DE (2004) Development of high throughput dispersive LC-ion mobility-TOFMS techniques for analysing the human plasma proteome. Brief Funct Genomics Proteomics 3(2):177–186
36. Poad BLJ, Zheng XY, Mitchell TW, Smith RD, Baker ES, Blanksby SJ (2018) Online ozonolysis combined with ion mobility-mass spectrometry provides a new platform for lipid isomer analyses. Anal Chem 90 (2):1292–1300
37. Harris RA, May JC, Stinson CA, Xia Y, McLean JA (2018) Determining double bond position in lipids using online ozonolysis coupled to liquid chromatography and ion mobility-mass spectrometry. Anal Chem 90 (3):1915–1924
38. Valentine SJ, Counterman AE, Clemmer DE (1999) A database of 660 peptide ion cross sections: use of intrinsic size parameters for bona fide predictions of cross sections. J Am Soc Mass Spectrom 10(11):1188–1211
39. Tao L, McLean JR, McLean JA, Russell DH (2007) A collision cross-section database of singly-charged peptide ions. J Am Soc Mass Spectrom 18(9):1727–1728
40. Dilger JM, Valentine SJ, Glover MS, Ewing MA, Clemmer DE (2012) A database of alkali metal-containing peptide cross sections: influence of metals on size parameters for specific amino acids. Int J Mass Spectrom 330–332:35–45
41. Dilger JM, Valentine SJ, Glover MS, Clemmer DE (2013) A database of alkaline-earth-coordinated peptide cross sections: insight into general aspects of structure. J Am Soc Mass Spectrom 24(5):768–779
42. Paglia G, Angel P, Williams JP, Richardson K, Olivos HJ, Thompson JW, Menikarachchi L, Lai S, Walsh C, Moseley A, Plumb RS, Grant DF, Palsson BO, Langridge J, Geromanos S, Astarite G (2015) Ion mobility-derived collision cross section as an additional measure for lipid fingerprinting and identification. Anal Chem 87(2):1137–1144
43. Picache JA, Rose BS, Balinski A, Leaptrot KL, Sherrod SD, May JC, McLean JA (2019) Collision cross section compendium to annotate and predict multi-omic compound identities. Chem Sci 10(4):983–993
44. Ellis HW, Thackston MG, McDaniel EW, Mason EA (1984) Transport properties of gaseous ions over a wide energy range. Part III. At Data Nucl Data Tables 31(1):113–151
45. Deng LL, Ibrahim YM, Baker ES, Aly NA, Hamid AM, Zhang X, Zheng XY, Garimella SVB, Webb IK, Prost SA, Sandoval JA, Norheim RV, Anderson GA, Tolmachev AV, Smith RD (2016) Ion mobility separations of isomers based upon long path length structures for lossless ion manipulations combined with mass spectrometry. Chem Select 1 (10):2396–2399
46. Dodds JN, May JC, McLean JA (2017) Investigation of the complete suite of the leucine and isoleucine isomers: toward prediction of ion mobility separation capabilities. Anal Chem 89(1):952–959
47. Dwivedi P, Bendiak B, Clowers BH, Hill HH Jr (2007) Rapid resolution of carbohydrate isomers by electrospray ionization ambient pressure ion mobility spectrometry-time-of-flight mass spectrometry (ESI-APIMS-TOFMS). J Am Soc Mass Spectrom 18 (7):1163–1175
48. Fenn LS, McLean JA (2011) Structural resolution of carbohydrate positional and structural isomers based on gas-phase ion mobility-mass spectrometry. Phys Chem Chem Phys 13(6):2196–2205
49. Groessl M, Graf S, Knochenmuss R (2015) High resolution ion mobility-mass spectrometry for separation and identification of isomeric lipids. Analyst 14(20):6904–6911

50. Lalli PM, Corilo YE, Fasciotti M, Riccio MF, de Sa GF, Daroda RJ, Souza GHMF, McCullagh M, Bartberger MD, Eberlin MN, Campuzano IDG (2013) Baseline resolution of isomers by traveling wave ion mobility mass spectrometry: investigating the effects of polarizable drift gases and ionic charge distribution. J Mass Spectrom 48(9):989–997
51. Valentine SJ, Kulchania M, Barnes CAS, Clemmer DE (2001) Multidimensional separations of complex peptide mixtures: a combined high-performance liquid chromatography/ion mobility/time-of-flight mass spectrometry approach. Int J Mass Spectrom 212(1–3):97–109
52. Matz LM, Dion HM, Hill HH Jr (2002) Evaluation of capillary liquid chromatography–electrospray ionization ion mobility spectrometry with mass spectrometry detection. J Chromatogr A 946(1–2):59–68
53. Lareau NM, May JC, McLean JA (2015) Non-derivatized glycan analysis by reverse phase liquid chromatography and ion mobility-mass spectrometry. Analyst 140 (10):3335–3338
54. Nichols CM, May JC, Sherrod SD, McLean JA (2018) Automated flow injection method for the high precision determination of drift tube ion mobility collision cross sections. Analyst 143(7):1556–1559
55. Gerhardt N, Schwolow S, Rohn S, Perez-Cacho PR, Galan-Soldevilla H, Arce L, Weller P (2019) Quality assessment of olive oils based on temperature-ramped HS-GC-IMS and sensory evaluation: comparison of different processing approaches by LDA, kNN, and SVM. Food Chem 278:720–728
56. Donato P, Giuffrida D, Oteri M, Inferrera V, Dugo P, Mondello L (2018) Supercritical fluid chromatography x ultra-high pressure liquid chromatography for red chilli pepper fingerprinting by photodiode array, quadrupole-time-of-flight and ion mobility mass spectrometry (SFC x RP-UHPLC-PDA-Q-ToF MS-IMS). Food Anal Methods 11 (12):3331–3341
57. Hill HH, Stlouis RH, Morrissey MA, Shumate CB, Siems WF, McMinn DG (1992) A detection method for unified chromatography-ion mobility monitoring. J High Resol Chromatogr 15(7):417–422
58. Hillenkamp F, Karas M, Beavis RC, Chait BT (1991) Matrix-assisted laser desorption/ionization mass spectrometry of biopolymers. Anal Chem 63(24):1193A–1203A
59. Karas M, Bahr U, Gießmann U (1991) Matrix-assisted laser desorption ionization mass spectrometry. Mass Spectrom Rev 10 (5):335–357
60. Stoeckli M, Chaurand P, Hallahan DE, Caprioli RM (2001) Imaging mass spectrometry: a new technology for the analysis of protein expression in mammalian tissues. Nat Med 7(4):493
61. Cornett DS, Reyzer ML, Chaurand P, Caprioli RM (2007) MALDI imaging mass spectrometry: molecular snapshots of biochemical systems. Nat Methods 4(10):828
62. McLean JA, Ridenour WB, Caprioli RM (2007) Profiling and imaging of tissues by imaging ion mobility-mass spectrometry. J Mass Spectrom 42(8):1099–1105
63. Jackson SN, Ugarov M, Egan T, Post JD, Langlais D, Albert Schultz J, Woods AS (2007) MALDI-ion mobility-TOFMS imaging of lipids in rat brain tissue. J Mass Spectrom 42(8):1093–1098
64. Fenn JB, Mann M, Meng CK, Wong SF, Whitehouse CM (1989) Electrospray ionization for mass spectrometry of large biomolecules. Science 246(4926):64–71
65. Fenn JB, Mann M, Meng CK, Wong SF, Whitehouse CM (1990) Electrospray ionization–principles and practice. Mass Spectrom Rev 9(1):37–70
66. Belov ME, Gorshkov MV, Udseth HR, Anderson GA, Smith RD (2000) Zeptomole-sensitivity electrospray ionization—fourier transform ion cyclotron resonance mass spectrometry of proteins. Anal Chem 72(10):2271–2279
67. Hardman M, Makarov AA (2003) Interfacing the orbitrap mass analyzer to an electrospray ion source. Anal Chem 75(7):1699–1705
68. Tang XJ, Thibault P, Boyd RK (1993) Fragmentation reactions of multiply-protonated peptides and implications for sequencing by tandem mass spectrometry with low-energy collision-induced dissociation. Anal Chem 65(20):2824–2834
69. Zubarev RA, Kelleher NL, McLafferty FW (1998) Electron capture dissociation of multiply charged protein cations. A nonergodic process. J Am Chem Soc 120(13):3265–3266
70. Syka JE, Coon JJ, Schroeder MJ, Shabanowitz J, Hunt DF (2004) Peptide and protein sequence analysis by electron transfer dissociation mass spectrometry. Proc Natl Acad Sci 101(26):9528–9533
71. Dodds JN, May JC, McLean JA (2017) Correlating resolving power, resolution, and collision cross section: unifying cross-platform assessment of separation efficiency in ion

mobility spectrometry. Anal Chem 89 (22):12176–12184

72. Mason EA, McDaniel EW (1988) Transport properties of ions in gases. John Wiley & Sons, New York, NY, p 560
73. Siems WF, Viehland LA, Hill HH Jr (2012) Improved momentum-transfer theory for ion mobility. 1. Derivation of the fundamental equation. Anal Chem 84(22):9782–9791
74. Bush MF, Hall Z, Giles K, Hoyes J, Robinson CV, Ruotolo BT (2010) Collision cross sections of proteins and their complexes: a calibration framework and database for gas-phase structural biology. Anal Chem 82 (22):9557–9565
75. Deng LL, Ibrahim YM, Hamid AM, Garimella SVB, Webb IK, Zheng XY, Prost SA, Sandoval JA, Norheim RV, Anderson GA, Tolmachev AV, Baker ES, Smith RD (2016) Ultra-high resolution ion mobility separations utilizing traveling waves in a 13 m serpentine path length structures for lossless ion manipulations module. Anal Chem 88 (18):8957–8964
76. May JC, McLean JA (2016) Advanced multidimensional separations in mass spectrometry: navigating the big data deluge. Ann Rev Anal Chem 9:387–409
77. May JC, Goodwin CR, McLean JA (2011) Gas-phase ion mobility-mass spectrometry (IM-MS) and tandem IM-MS/MS strategies for metabolism studies and metabolomics. Encyclopedia Drug Metab Interact. https://doi.org/10.1002/9780470921920.edm099
78. Kolli V, Schumacher KN, Dodds ED (2017) Ion mobility-resolved collision-induced dissociation and electron transfer dissociation of N-glycopeptides: gathering orthogonal connectivity information from a single mass-selected precursor ion population. Analyst 142(24):4691–4702
79. Lermyte F, Verschueren T, Brown JM, Williams JP, Valkenborg D, Sobott F (2015) Characterization of top-down ETD in a travelling-wave ion guide. Methods 89:22–29
80. Williams JP, Pringle S, Richardson K, Gethings L, Vissers JPC, De Cecco M, Houel S, Chakraborty AB, Yu YQ, Chen WB, Brown JM (2013) Characterisation of glycoproteins using a quadrupole time-of-flight mass spectrometer configured for electron transfer dissociation. Rapid Commun Mass Spectrom 27(21):2383–2390
81. Ruotolo BL, Benesch JLP, Sandercock AM, Hyung S, Robinson C (2008) Ion mobility-mass spectrometry analysis of large protein complexes. Nat Protoc 3:1139–1152
82. McLean JA, Ruotolo BT, Gillig KJ, Russel DH (2005) Ion mobility-mass spectrometry: a new paradigm for proteomics. Int J Mass Spectrom 240(3):301–315
83. Paglia G, Astarita G (2017) Metabolomics and lipidomics using traveling-wave ion mobility mass spectrometry. Nat Protoc 12:797–813
84. Zhang X, Quinn K, Cruickshank-Quinn C, Reisdorph R, Reisdorph N (2018) The application of ion mobility mass spectrometry to metabolomics. Curr Opin Chem Biol 42:60–66
85. May JC, Goodwin CR, McLean JA (2015) Ion mobility mass spectrometry strategies for untargeted systems, synthetic, and chemical biology. Curr Opin Biotechnol 31:117–121
86. Dunn WB, Broadhurst D, Begley P, Zelena E, Francis-McIntyre S, Anderson N, Brown M, Knowles JD, Halsall A, Haselden JN, Nicholls AW, Wilson ID, Kell DB, Goodacre R, The Human Serum Metabolome Consortium (2011) Procedures for large-scale metabolic profiling of seruma and plasma using gas chromatography and liquid chromatraphy coupled to mass spectrometry. Nat Protoc 6:1060–1083
87. Hoaglund-Hyzer CS, Li J, Clemmer DE (2000) Mobility labeling for parallel CID of ion mixtures. Anal Chem 72(13):2737–2740
88. Taguchi F, Solomon B, Gregorc V, Roder H, Gray R, Kashara K, Nisho M, Brahmer J, Spreafico A, Ludovini V, Massion PP, Dziadziuszho R, Schiller J, Grigorieva J, Tsypin M, Hunsucker SW, Caprioli R, Duncan MW, Hirsch FR, Bunn PA, Carbone DP (2007) Mass spectrometry to classify non-small-lung cell lung cancer patients for clinical outcome after treatment with epidermal growth factor receptor tyrosine kinase inhibitors: a multicohort cross-institutional study. J Natl Cancer Instit 99(11):838–846
89. Mori H, Takio K, Ogawara M, Selkoe DJ (1992) Mass spectrometry of purified amyloid beta protein in Alzheimer's disease. J Biochem Chem 267:17082–17086
90. Djidja MC, Francese S, Loadman PM, Sutton CW, Scriven P, Claude E, Snel MF, Franck J, Salzet M, Clench MR (2009) Detergent addition to tryptic digests and ion mobility separation prior to MS/MS improves peptide yield and protein identification for *in situ* proteomics investigation of frozen and formalin-fixed paraffin-embedded adenocarcinoma tissue sections. Proteomics 9(10):2750–2763
91. Moon MH, Myung S, Plasencia M, Hilderbrand AE, Clemmer DE (2003) Nanoflow

LC/Ion mobility/CID/TOF for proteomics: analysis of a human urinary proteome. J Proteome Res 2(6):589–597

92. Thalassinos K, Grabenauer M, Slade SE, Hilton GR, Bowers MT, Scrivens JH (2009) Characterization of phosphorylated peptides using traveling wave-based and drift cell ion mobility mass spectrometry. Anal Chem 81 (1):248–254
93. Zhong Y, Hyung S, Ruotolo BT (2012) Ion mobility-mass spectrometry for structual proteomics. Expert Rev Proteomics 9(1):47–58
94. May JC, McLean JA (2015) A uniform field ion mobility study of melittin and implications of low-field mobility for resolving fine cross-sectional detail in peptide and protein experiments. Proteomics 15(16):2862–2871
95. Shliaha PV, Bond NJ, Gatto L, Liley KS (2013) Effects of traveling wave ion mobility separation of data independent acquistion in proteomics studies. J Proteome Res 12 (6):2323–2339
96. Wang G, Abzalimov RR, Kaltashov IA (2011) Direct monitoring of heat-stressed biopolymers with temperature-controlled electrospray ionization mass spectrometry. Anal Chem 83(8):2870–2876
97. El-Baba TJ, Woodall DW, Raab SA, Fuller DR, Langanowsky A, Russell DH, Clemmer DE (2017) Melting Proteins: evidence for multiple stable structures upon thermal denaturation of native ubiquitin from ion mobility spectrometry-mass spectrometry measurements. J Am Chem Soc 139(18):6306–6309
98. Han X, Yang K, Gross RW (2011) Multidimensional mass spectrometry-based shotgun lipidomics and novel strategies for lipidomic analysis. Mass Spectrom Rev 31 (1):134–178
99. Kliman M, May JC, McLean JA (2011) Lipid analysis and lipidomics by structurally selective ion mobility-mass spectrometry. Biochim Biophys Acta 1811(11):935–945
100. Paglia G, Kliman M, Claude E, Geromanos S, Astarita G (2015) Application of ion-mobility mass spectrometry forl lipids. Anal Bioanal Chem 407(17):4995–5007
101. Grossel M, Graf S, Knochenmuss R (2015) High resolution ion mobility-mass spectrometry for separation and identification of isomeric lipids. Analyst 140(20):6904–6911
102. Di Giovanni JP, Barkley RM, Jones DNM, Hankins JA, Murphy RC (2018) Tandem mass spectrometry and ion mobility reveals structural insight into eicosanoid product ion formation. J Am Soc Mass Spectrom 29 (6):1231–1241
103. Paglia G, Shrestha B, Astarita G (2017) Ion-mobility mass spectrometry for lipidomics applications. In: Wood P (ed) Lipidomics. Humana, New York, NY, pp 61–79
104. Thomas MC, Mitchell TW, Harman DG, Deeley JM, Murphy RC, Blanksby SJ (2007) Elucidation of double bond position in unsaturated lipids by ozone electrospray ionization mass spectrometry. Anal Chem 79 (13):5013–5022
105. Sun C, Zhao Y, Curtis JM (2014) Elucidation of phosphatidylcholine isomers using two dimensional liquid chromatography coupled in-line with ozonolysis mass spectrometry. J Chromatogr A 1351:37–45
106. Dettmer K, Aronov PA, Hammock BD (2007) Mass spectrometry-based metabolomics. Mass Spectrom Rev 26(1):51–78
107. Zhang X, Quinn K, Cruickshank-Quinn C, Reisdorph R, Reisdorph N (2018) Applications of ion mobility mass spectrometry to metabolomics. Curr Opin Chem Biol 42:60–66
108. Sinclair E, Hollywood KA, Yan C, Blankley R, Rainer B, Barran P (2018) Mobilising ion mobility mass spectrometry for metabolomics. Analyst 19:4783–4788
109. Schrimpe-Rutledge AC, Codreanu SG, Sherrod SD, McLean JA (2016) Untargeted metabolomics strategies—challenges and emerging directions. J Am Soc Mass Spectrom 27(12):1897–1905
110. May JC, Gant-Branum RL, McLean JA (2016) Targeting the untargeted in molecular phenomics with structurally selective ion mobility-mass spectrometry. Curr Opin Biotechnol 39:192–197
111. Nichols CM, Dodds JN, Rose BS, Picache JA, Morris CB, Codreanu SG, May JC, Sherrod SD, McLean JA (2018) Untargeted molecular discovery in primary metabolism: collision cross section as a molecular descriptor in ion mobility-mass spectrometry. Anal Chem 90 (24):14484–14492
112. Wickramasekara SI, Zandkarimi F, Morré J, Kirkwood J, Legette L, Jiang Y, Gombart AF, Stevens JF, Maier CS (2013) Electrospray quadrupole travelling wave ion mobility time-of-flight mass spectrometry for the detection of plasma metabolome changes caused by xanthohumol in obese zucker (fa/fa) rats. Meta 3:701–717
113. Stow MS, Causon TJ, Zheng X, Kurulugama RT, Mairinger T, May JC, Rennie EE, Smith RD, McLean JA, Hann S, Fjeldsted JC (2017) An interlaboratory evaluation of drift

tube ion mobility-mass spectrometry collision cross section measurements. Anal Chem 89 (17):9048–9055

114. Regueiro J, Negreira N, Berntssen MH (2016) Ion-mobility-derived collision cross section as an additional identification point for multiresidue screening of pesticides in fish feed. Anal Chem 88(22):11169–11177
115. Stephan S, Hippler J, Köhler T, Deeb AA, Schmidt TC, Schmitz OJ (2016) Contaminant screening of wastewater with HPLC-IM-qTOF-MS and LC+ LC-IM-qTOF-MS using a CCS database. Anal Bioanal Chem 408(24):6545–6555
116. Zheng X, Aly NA, Zhou Y, Dupuis KT, Bilbao A, Paurus VL, Orton DJ, Wilson R, Payne SH, Smith RD, Baker ES (2017) A structural examination and collision cross section database for over 500 metabolites and xenobiotics using drift tube ion mobility spectrometry. Chem Sci 8(11):7724–7736
117. Hernández-Mesa M, Le Bizec B, Monteau F, García-Campaña AM, Dervilly-Pinel G (2018) Collision cross section (CCS) database: an additional measure to characterize steroids. Anal Chem 90(7):4616–4625
118. Chen TC, Ibrahim YM, Webb IK, Garimella SVB, Zhang X, Hamid AM, Deng LL, Karnesky WE, Prost SA, Sandoval JA, Norheim RV, Anderson GA, Tolmachev AV, Baker ES, Smith RD (2016) Mobility-selected ion trapping and enrichment using structures for lossless ion manipulations. Anal Chem 88 (3):1728–1733
119. Ibrahim YM, Hamid AM, Deng LL, Garimella SVB, Webb IK, Baker ES, Smith RD (2017) New frontiers for mass spectrometry based upon structures for lossless ion manipulations. Analyst 142(7):1010–1021
120. Zhang XY, Garimella SVB, Prost SA, Webb IK, Chen TC, Tang KQ, Tolmachev AV, Norheim RV, Baker ES, Anderson GA, Ibrahim YM, Smith RD (2015) Ion trapping, storage, and ejection in structures for lossless ion manipulations. Anal Chem 87 (12):6010–6016
121. Chouinard CD, Nagy G, Webb IK, Shi TJ, Baker ES, Prost SA, Liu T, Ibrahim YM, Smith RD (2018) Improved sensitivity and separations for phosphopeptides using online liquid chromatography coupled with structures for lossless ion manipulations ion mobility-mass spectrometry. Anal Chem 90 (18):10889–10896
122. Garimella SVB, Ibrahim YM, Webb IK, Tolmachev AV, Zhang XY, Prost SA, Anderson GA, Smith RD (2014) Simulation of electric potentials and ion motion in planar electrode structures for lossless ion manipulations (SLIM). J Am Soc Mass Spectrom 25 (11):1890–1896
123. Webb IK, Garimella SVB, Tolmachev AV, Chen TC, Zhang XY, Cox JT, Norheim RV, Prost SA, LaMarche B, Anderson GA, Ibrahim YM, Smith RD (2014) Mobility-resolved ion selection in uniform drift field ion mobility spectrometry/mass spectrometry: dynamic switching in structures for lossless ion manipulations. Anal Chem 86 (19):9632–9637
124. Deng LL, Webb IK, Garimella SVB, Hamid AM, Zheng XY, Norheim RV, Prost SA, Anderson GA, Sandoval JA, Baker ES, Ibrahim YM, Smith RD (2017) Serpentine ultralong path with extended routing (SUPER) high resolution traveling wave ion mobility-MS using structures for lossless ion manipulations. Anal Chem 89(8):4628–4634
125. Garimella SVB, Ibrahim YM, Webb IK, Ipsen AB, Chen TC, Tolmachev AV, Baker ES, Anderson GA, Smith RD (2015) Ion manipulations in structures for lossless ion manipulations (SLIM): computational evaluation of a 90 degrees turn and a switch. Analyst 14 (20):6845–6852
126. Fernandez-Lima FA, Kaplan DA, Suetering J, Park MA (2011) Gas-phase separation using a trapped ion mobility spectrometer. Int J Ion Mobil Spectrom 14(2–3):93–98
127. Adams KJ, Montero D, Aga D, Fernandez-Lima F (2016) Isomer separation of polybrominated diphenyl ether metabolites using nanoESI-TIMS-MS. Int J Ion Mobil Spectrom 19(2–3):69–76
128. Michelmann K, Silveira JA, Ridgeway ME, Park MA (2015) Fundamentals of trapped ion mobility spectrometry. J Am Soc Mass Spectrom 26(1):14–24
129. Hernandez DR, DeBord JD, Ridgeway ME, Kaplan DA, Park MA, Fernandez-Lima F (2014) Ion dynamics in a trapped ion mobility spectrometer. Analyst 139(8):1913–1921
130. Silveira JA, Danielson W, Ridgeway ME, Park MA (2016) Altering the mobility-time continuum: nonlinear scan functions for targeted high resolution trapped ion mobility-mass spectrometry. Int J Ion Mobil Spectrom 19 (2–3):87–94
131. May JC, Leaptrot KL, Sundarapandian S, McLean JA. In theoretical evaluation and performance characterization of an 8-channel spatially multiplexed ion mobility-mass spectrometer, 60th annual ASMS conference on mass spectrometry and allied topics, Vancouver, BC, May 2012

132. Asbury GR, Hill HH (2000) Using different drift gases to change separation factors (alpha) in ion mobility spectrometry. Anal Chem 72 (3):580–584
133. Matz LM, Hill HH Jr, Beegle LW, Kanik I (2002) Investigation of drift gas selectivity in high resolution ion mobility spectrometry with mass spectrometry detection. J Am Soc Mass Spectrom 13(4):300–307
134. Berant Z, Karpas Z (1989) Mass-mobility correlation of ions in view of new mobility data. J Am Chem Soc 111(11):3819–3824
135. Howdle MD, Eckers C, Laures AMF, Creaser CS (2010) The effect of drift gas on the separation of active pharmaceutical ingredients and impurities by ion mobility–mass spectrometry. Int J Mass Spectrom 298 (1–3):72–77
136. Purves RW, Ozog AR, Ambrose SJ, Prasad S, Belford M, Dunyach JJ (2014) Using gas modifiers to significantly improve sensitivity and selectivity in a cylindrical FAIMS device. J Am Soc Mass Spectrom 25(7):1274–1284
137. Waraksa E, Perycz U, Namiesnik J, Sillanpaa M, Dymerski T, Wojtowicz M, Puton J (2016) Dopants and gas modifiers in ion mobility spectrometry. TRAC Trends Anal Chem 82:237–249
138. Schneider BB, Covey TR, Nazarov EG (2013) Investigation of the chemical orthogonality effect of transport gas modifiers on DMS separations. Abstr Pap Am Chem Soc 246:1
139. Kafle A, Coy SL, Wong BM, Fornace AJ, Glick JJ, Vouros P (2014) Understanding gas phase modifier interactions in rapid analysis by differential mobility-tandem mass spectrometry. J Am Soc Mass Spectrom 25 (7):1098–1113
140. Levin DS, Vouros P, Miller RA, Nazarov EG, Morris JC (2006) Characterization of gas-phase molecular interactions on differential mobility ion behavior utilizing an electrospray ionization-differential mobility-mass spectrometer system. Anal Chem 78 (1):96–106
141. Porta T, Varesio E, Hopfgartner G (2013) Gas-phase separation of drugs and metabolites using modifier-assisted differential ion mobility spectrometry hyphenated to liquid extraction surface analysis and mass spectrometry. Anal Chem 85(24):11771–11779
142. Fernandez-Maestre R, Wu C, Hill HH (2012) Buffer gas modifiers effect resolution in ion mobility spectrometry through selective ion-molecule clustering reactions. Rapid Commun Mass Spectrom 26(19):2211–2223
143. Garabedian A, Leng FF, Ridgeway ME, Park MA, Fernandez-Lima F (2018) Tailoring peptide conformational space with organic gas modifiers in TIMS-MS. Int J Ion Mobil Spectrom 21(1–2):43–48
144. Beegle LW, Kanik I, Matz L, Hill HH (2001) Electrospray ionization nigh-resolution ion mobility spectrometry for the detection of organic compounds, 1. Amino acids. Anal Chem 73(13):3028–3034
145. Beegle LW, Kanik I, Matz L, Hill HH (2002) Effects of drift-gas polarizability on glycine peptides in ion mobility spectrometry. Int J Mass Spectrom 216(3):257–268
146. Chouinard CD, Beekman CR, Kemperman RHJ, King HM, Yost RA (2017) Ion mobility-mass spectrometry separation of steroid structural isomers and epimers. Int J Ion Mobil Spectrom 20(1–2):31–39
147. Fasciotti M, Lalli PM, Klitzke CF, Corilo YE, Pudenzi MA, Pereira RCL, Bastos W, Daroda RJ, Eberlin MN (2013) Petroleomics by traveling wave ion mobility–mass spectrometry using CO_2 as a drift gas. Energy Fuel 27 (12):7277–7286
148. Fasciotti M, Sanvido GB, Santos VG, Lalli PM, McCullagh M, de Sá GF, Daroda RJ, Peter MG, Eberlin MN (2012) Separation of isomeric disaccharides by traveling wave ion mobility mass spectrometry using CO2 as drift gas. J Mass Spectrom 47 (12):1643–1647
149. Bataglion GA, Souza G, Heerdt G, Morgon NH, Dutra JDL, Freire RO, Eberlin MN, Tata A (2015) Separation of glycosidic catiomers by TWIM-MS using CO2 as a drift gas. J Mass Spectrom 50(2):336–343
150. Ruotolo BT, McLean JA, Gillig KJ, Russell DH (2004) Peak capacity of ion mobility mass spectrometry: the utility of varying drift gas polarizability for the separation of tryptic peptides. J Mass Spectrom 39(4):361–367
151. Jurneczko E, Kalapothakis J, Campuzano IDG, Morris M, Barran PE (2012) Effects of drift gas on collision cross sections of a protein standard in linear drift tube and traveling wave ion mobility mass spectrometry. Anal Chem 84(20):8524–8531
152. Davidson KL, Bush MF (2017) Effects of drift gas selection on the ambient-temperature, ion mobility mass spectrometry analysis of amino acids. Anal Chem 89 (3):2017–2023
153. Morris CB, May JC, Leaptrot KL, McLean JA (2019) Evaluating separation selectivity and collision cross section measurement reproducibility in helium, nitrogen, argon, and

carbon dioxide drift gases for drift tube ion mobility-mass spectrometry. J Am Soc Mass Spectrom 30(6):1059–1068

154. Norris JL, Farrow MA, Gutierrez DB, Palmer LD, Muszynski N, Sherrod SD, Pino JC, Allen JL, Spraggins JM, ALR L, Jordan A, Burns W, Poland JC, Romer C, Manier ML, Nei Y, Prentice BM, Rose KL, Hill S, Van de Plas R, Tsui T, Braman NM, Keller MR, Rutherford A, Lobdell N, Lopez CF, Lacy DB, JA ML, Wikswo JP, Skaar EP, Caprioli RM (2017) Integrated, high-throughput, multiomics platform enables data-driven construction of cellular responses and reveals global drug mechanisms of action. J Proteome Res 16(3):1364–1375
155. Paglia G, Stocchero M, Cacciato S, Lai S, Angel P, Alam MT, Keller M, Ralser M, Astarita G (2016) Unbiased metabolomic investigation of Alzheimer's disease brain points to dysregulation of mitochondrial aspartate metabolism. J Proteome Res 15:608–618
156. Morgan XC, Segata N, Huttenhower C (2013) Biodiversity and functional genomics in the human microbiome. Trends Genet 29 (1):51–58
157. Weir TL, Manter DK, Brittany BA, Heuberger AL, Ryan EP (2013) Stool microbiome and metabolome differences between colorectal cancer patients and healthy adults. PLoS One 8(8):e70803
158. Raman M, Ahmed I, Gillevet PM, Probert CS, Ratcliffe NM, Smith S, Greenwood R, Sikaroodi M, Lam V, Crotty P, Bailey J, Meyes RP, Rious KP (2013) Fecal microbiome and volatile organic compound metabolome in obese humans with nonalcoholic fatty liver disease. Clin Gastroenterol Hepatol 11(7):868–875

Part I

Metabolomics and Lipidomics

Chapter 2

Utilizing Drift Tube Ion Mobility Spectrometry for the Evaluation of Metabolites and Xenobiotics

Melanie T. Odenkirk and Erin S. Baker

Abstract

Metabolites and xenobiotics are small molecules with a molecular weight that often falls below 600 Da. Over the last few decades, multiple small molecule databases have been curated listing structures, masses, and fragmentation spectra possible in metabolomic and exposomic measurements. To date only a small portion of the spectra in these databases are experimentally derived due to the high expense of obtaining, synthesizing, and analyzing standards. A vast majority of spectra have thus been created using theoretical programs to fit the available experimental data. The errors associated with theoretical data have however caused problems with current small molecule identifications, and accurate quantitation as searching the databases using just one or two analysis dimensions (i.e., chromatography retention times and mass spectrometry (MS) *m/z* values) results in numerous annotations for each experimental feature. Additional analysis dimensions are therefore needed to better annotate and identify small molecules. Drift tube ion mobility spectrometry coupled with MS (DTIMS-MS) is a promising technique to address this challenge as it is able to perform rapid structural evaluations of small molecules in complex matrices by assessing the collision cross section values for each in addition to their *m/z* values. The use of IMS in conjunction with other separation techniques such as gas or liquid chromatography and MS has therefore enabled more accurate identifications for the small molecules present in complex biological and environmental samples. Here, we present a review of relevant parameter considerations for DTIMS application with emphasis on xenobiotics and metabolomics isomer separations.

Key words Ion mobility spectrometry (IMS), Drift tube ion mobility spectrometry (DTIMS), Metabolomics, Exposomics, Collision cross section (CCS)

1 Introduction

Ion mobility spectrometry (IMS) is a gas phase technique widely used for rapid structural separations. IMS by itself is a useful analytical technique, but its coupling with mass spectrometry (IMS-MS) and other front-end separations has been extremely beneficial for separating molecules that cannot be rapidly distinguished with other techniques. IMS separates ions based on their interactions with an electric field and a drag force caused by

Giuseppe Paglia and Giuseppe Astarita (eds.), *Ion Mobility-Mass Spectrometry: Methods and Protocols*, Methods in Molecular Biology, vol. 2084, https://doi.org/10.1007/978-1-0716-0030-6_2,

collisions with buffer gas molecules [1, 2]. Different variations of the electric field and stationary state of the buffer gas have given rise to multiple IMS-based platforms such as drift tube IMS (DTIMS) [3–6], traveling wave IMS (TWIMS) [7], trapped IMS (TIMS) [8], overtone IMS (OIMS) [9, 10], differential IMS (DIMS) [11], field asymmetric IMS (FAIMS) [12–14], and transversal modulation IMS (TM-IMS) [15]. While each has been used in small molecule applications, this chapter will focus on the DTIMS methods and considerations used to determine the CCS values for more than 500 metabolites and xenobiotics in the manuscript *A structural examination and collision cross section database for over 500 metabolites and xenobiotics using drift tube ion mobility spectrometry* [16] and in other small molecule studies currently taking place with DTIMS.

2 DTIMS Fundamentals and Its Coupling with Other Techniques

In DTIMS, ions are pulsed into a drift tube filled with buffer gas at a specific interval which is often in the range of 50–200 μs [17]. The ions then travel through the drift tube under the influence of a weak electric field (E), while colliding with the stationary buffer gas. These interactions cause compact ions to drift faster than more extended ions, which have a higher probability of colliding with the buffer gas molecules. The ion's motion therefore results in a unique drift velocity (v_d) dependent upon its mobility (K) through the drift tube (Fig. 1 and Eq. 1) [18]. v_d can also be written in terms of drift tube length (L) per drift time (t_d) [19].

$$v_d = K \cdot E = \frac{L}{t_d}. \tag{1}$$

When a constant E is applied to all analytes, an ion's mobility (K) is dependent on the strength of the decelerating force (number of gas-analyte collisions) and influenced by the gas number density (N), temperature (T), and pressure (p) of the drift cell [19]. Variations in the gas number density result in significant drift time shifts with changes in the abundance of gas molecules, significantly altering the probability of ion-neutral collisions in the drift tube. To account for changes in the friction force, gas number density is normalized to standard gas number density (n_0) using standard temperature (T_0 = 273.16 K) and pressure (p_0 = 760 Torr) for better comparison between experiments (Eq. 2) [19].

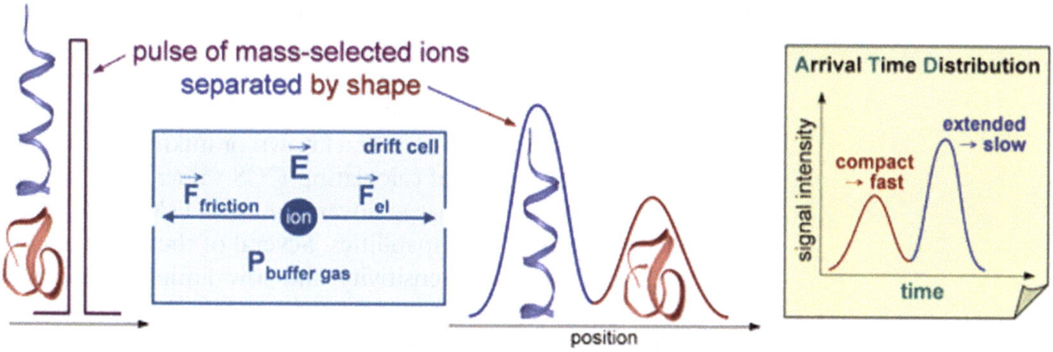

Fig. 1 A schematic of an ion mobility spectrometry separation showing how an extended (blue) and compact (red) molecule traverses the drift tube. Upon pulsing into the drift tube, an electric force (F_{el}) projects ions forward. A force in the opposite direction ($F_{friction}$) results from ion-neutral collisions of the analyte with buffer gas molecules. More extended conformations are more likely to undergo a greater number of ion nuetral interactions to result in a larger $F_{friction}$ and longer drift time (t_A). (Figure was obtained from Bowers CCS theoretical website [])

$$K_0 = K\cdot\left(\frac{n}{n_0}\right) = K\cdot\left(\frac{T_0}{T}\right)\cdot\left(\frac{p}{p_0}\right) \text{ and } v_d = \left(\frac{V\,K_0}{L\,p}\right)\cdot\left(\frac{760T}{273.16}\right) \quad (2)$$

DTIMS experiments are performed under the low field limit where the ratio of E/N is small. In this region, it can be assumed that v_d is small relative to the thermal velocity of collision gas molecules in order for the collisions to be effective in reducing mobility with respect to analyte size and shape [19]. Therefore, an ion's mobility can be expressed in terms of the Mason–Schamp equation, where μ is the reduced mass, k_B is Boltzmann's constant, and ez is ion charge (Eq. 3).

$$K_0 = \left(\frac{3ez}{16N}\right)\cdot\left(\frac{2\pi}{\mu k_B T}\right)^{1/2}\cdot\left(\frac{1}{\Omega}\right) \quad (3)$$

In this equation, Ω represents the momentum transfer collision integral which is often termed the collision cross section (CCS). CCS represents the momentum transfer in ion-neutral collisions over the relative thermal velocities [18] and is different in definition than Ω. However, measuring Ω for a nonspherical system is difficult, and CCS values offer a good approximation [18]. Equation 3 also shows that the charge of an ion has an effect on its mobility since the electric field pulls ions with more charges faster through the drift tube. Therefore, DTIMS effectively separates analytes by their shape and mass relative to their charge (*e.g.*, Ω/z) [18].

CCS values are instrument-independent for DTIMS as the length, pressure, temperature, and electric field can all be accounted for in Eq. 3. For these reasons, CCS values are universal, intrinsic, highly descriptive physical properties specific for analytes [20]. The experimental CCS values can also be utilized for comparison with molecular modeling calculations, ultimately leading to potential structure identification for a known or unknown molecule [21–23]. While the utility of calculating CCS values directly from an ion's arrival time is one large advantage of DTIMS analyses, it has many other promising capabilities. Several of the advantages of DTIMS include (1) high sensitivity and low limits of detection (LoD) in the ppb range; (2) separations for sequence and structural isomers that are indistinguishable with MS measurements [21, 22, 24–26]; (3) the ability to be easily coupled with additional chromatographic separations and time-of-flight (TOF) MS instruments for nested data acquisition of both shape and *m/z* information in a single DTIMS-MS experiment (Fig. 2) [4, 25, 27, 28]; and (4) when combined with MS a reduction in mass spectral peak congestion by spreading ions out by their shapes and charge states (*see* **Note 1**). DTIMS-MS has been coupled with numerous different front-end separation techniques such as gas chromatography (GC), supercritical fluid chromatography (SFC), liquid chromatography (LC), solid phase extractions (SPE), capillary electrophoresis (CE), field asymmetric ion mobility spectrometry (FAIMS), and microfluidic devices (Table 1) [29]. The data from these multidimensional analyses have proven much richer than each technique coupled with MS alone, in addition to providing greatly improved confidence in molecular identifications due to the multi-characteristic matching [30, 31].

3 Multidimensional DTIMS-MS Measurements

The capabilities of performing multidimensional analyses with IMS is of great interest for many molecule types, but for small molecules in particular. In metabolomic analyses, small molecule metabolites, intermediates, and products of metabolism are studied in environmental and biological systems [32]. However, the presence of xenobiotic molecules in metabolomic evaluations is also possible as they can be introduced through various sources such as environment pollutants, consumption of foods and drugs, or contact with personal-care products. The measurement of xenobiotic molecules is often termed exposomics but is synonymous with measuring exogenous metabolites. However, studying xenobiotics can require special sample preparations and analyses depending on their molecular composition (e.g., polyaromatic hydrocarbons will not be observed in polar extractions or with electrospray ionization). Thus, the complexity of different biological and environmental matrices coupled with the occurrence of thousands of various small molecules makes the detection of as many compounds as

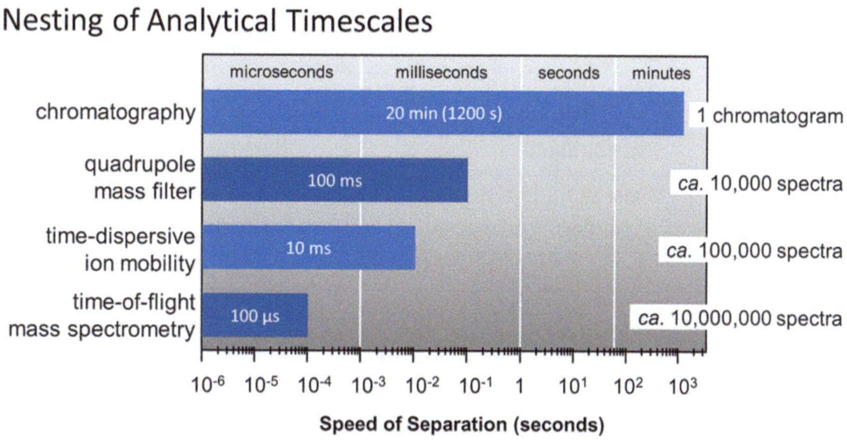

Fig. 2 A schematic of common analytical timescales. IMS measurements on the millisecond timescale are easily nested between chromatography (minute timscale) and high-throughput mass spectrometry (i.e., TOF) analyses (microsecond timescale). These LC-IMS-MS measurements increase the confidence in analyte identifications by providing more unique features for identification, but also greatly increase the amount of data obtained from each experiment. (Reprinted with permission from May, J. C. & McLean, J. A. Ion mobility-mass spectrometry: time-dispersive instrumentation. *Anal Chem* **87**, 1422–1436, doi:https://doi.org/10.1021/ac504720m (2015) [26]. Copyright 2015 American Chemical Society)

possible extremely difficult in untargeted analyses. In particular, the small *m/z* window for the estimated million metabolites possible in humans makes their accurate identification extremely difficult if multiple techniques are not combined, and additional separations are not performed [33]. Analyses with LC-IMS-MS therefore show great promise for small molecule untargeted analyses due to its speed and multidimensional assessments. IMS-MS and LC-IMS-MS have been utilized in numerous small molecule studies to improve both selectivity and coverage of the metabolome by providing additional structural information compared to routine MS or LC-MS-based methods [34]. Since different molecular classes fall on distinct trend lines based on their CCS and *m/z* values, IMS has been extremely valuable for separating out many different types of small molecules (Fig. 3) [35] (*see* **Note 2**). To date there have been various CCS databases released for different substance classes such as peptides [36–38], N-glycans [39], drug-like compounds [38], metabolites [16, 20], lipids [40–42], environmental toxins [43], and various biomolecules [44], ultimately aiding in their classification through distinct structural and *m/z* groupings (Fig. 4) [16]. These classifications are also being used in machine learning and molecular dynamics calculations to provide ways for obtaining CCS values for small molecules that currently have no standards available [45, 46]. Even with the many distinct molecular classifications observed, some isomers still result in very similar or identical CCS values. In these cases, experimental considerations

Table 1
Advantages and disadvantages to front-end separations used in conjunction with DTIMS.

Technique	Analysis time	Advantages	Disadvantages
GC	Minutes to hours	Allows high peak capacity small-molecule analyses; chromatography is highly reproducible	Is limited to volatile compounds; high temperature can induce decomposition
SFC	Minutes	Provides fast separations; is orthogonal to reverse-phase LC	Works best for nonpolar molecules
LC	Minutes to hours	Is applicable to many molecule types; allows high peak capacity small-molecule analyses	Has limited throughput because long separation times provide the best measurements
SPE	Seconds to minutes	Provides ultrafast separations; is ideal for sample desalting and molecular class extraction	Has low or no peak capacity, so it is similar to direct infusion
CE	Minutes	Has low sample consumption	Lacks orthogonality with DTIMS
FAIMS	Milliseconds to seconds	Allows high-throughput analyses; reduces noise and interferences	Experiences large ion losses due to its scanning characteristics; separation occurs after ionization, which does not help ionization suppression problems
Microfluidics	Seconds to hours	Is automatable; allows high-throughput analyses; requires minimal sample	Loading large amounts of samples is difficult, resulting in lower sensitivity for abundant samples

Reprinted with permission from Zheng, X. et al. Coupling Front-End Separations, Ion Mobility Spectrometry, and Mass Spectrometry For Enhanced Multidimensional Biological and Environmental Analyses. *Annu Rev Anal Chem (Palo Alto Calif)* **10**, 71-92, doi:https://doi.org/10.1146/annurev-anchem-061516-045212 (2017) [29]. Copyright 2017 Annual Review of Analytical Chemistry

CE capillary electrophoresis, *DTIMS* drift tube IMS, *FAIMS* field asymmetric IMS, *GC* gas chromatography, *IMS* ion mobility spectrometry, *LC* liquid chromatography, *SFC* supercritical fluid chromatography, *SPE* solid-phase extraction

must be examined to see if the molecules can be separated using specific parameters such as adducts or drift gas modifiers.

4 DTIMS Single and Stepped-Field Analyses

DTIMS analyses can be performed in two different ways depending on the type of measurement needed. In DTIMS analyses, ions travel a specific distance in the drift tube before arriving at the ion optics or hitting the detector [27]. Because most DTIMS-MS instruments lack a grid or way to end the drifting length, some of the mobility the ions have carries over to the MS ion optics [47]. Without an accurate drift time, accurate CCS measurements cannot be made [47], so it is essential to evaluate an ion's arrival

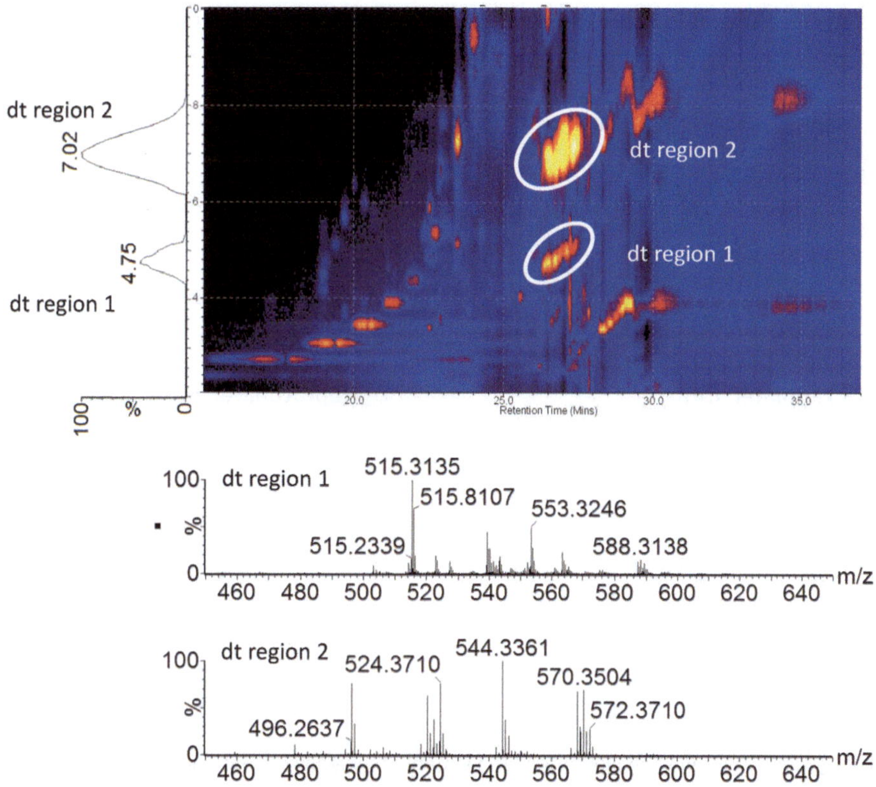

Fig. 3 The separation of different ion classes using DTIMS. In the top panel, the drift time versus retention time for a lipidomoic analysis of rat plasma is demonstrated. In the bottom panel, the extracted drift time regions show distinct clusters of the lysoPC (4.75 ms) and sphingomyelins (7.02 ms) lipid classes. (Reprinted with permission from Wickramasekara, S. I. et al. Electrospray Quadrupole Travelling Wave Ion Mobility Time-of-Flight Mass Spectrometry for the Detection of Plasma Metabolome Changes Caused by Xanthohumol in Obese Zucker (fa/fa) Rats. *Metabolites* **3**, 701–717, doi:https:/doi.org/10.3390/metabo3030701 (2013) [35]. Copyright 2013 MDPI)

time (t_A) as a function of both the time inside (t_D) and time outside (t_0) the drift region (Eq. 4).

$$t_A - t_0 = t_D = \left(\frac{L^2}{K_0}\right) \cdot \left(\frac{273.16}{760\ T}\right) \cdot \left(\frac{p}{V}\right) \tag{4}$$

To evaluate t_A and t_0, stepped-field analyses are often performed (Fig. 5) [48]. Here, experiments are evaluated at different electric field strengths by changing the voltage applied to the drift tube (usually five different voltages for one CCS experiment), but keeping all other parameters constant (p, T, and L) [48]. Voltages are often selected over a narrow V window to ensure high reproducibility and avoid significant variation in magnitude and precision of the CCS values [48]. Plotting the observed arrival time versus P/V results in a linear relationship with the y-axis intercept representative of t_0 and $K_0 = \left(\frac{L^2}{\text{slope}}\right) \cdot \left(\frac{273.16}{760T}\right)$ from Eq. 4

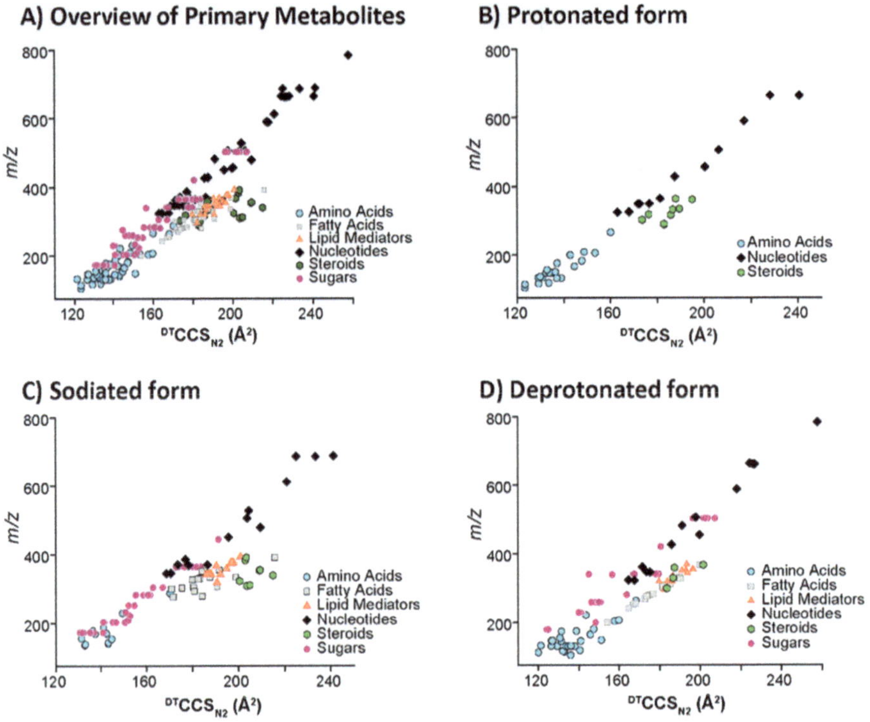

Fig. 4 The CCS and *m/z* trend lines illustrated for a variety of molecules in different ionization forms. Grouping was observed for the different molecules in all ionization forms based on how cyclic (smaller) or linear (larger) their structures were. (Reprinted with permission from Zheng, X. et al., A structural examination and collision cross section database for over 500 metabolites and xenobiotics using drift tube ion mobility spectrometry. *Chem Sci* **8**, 7724–7736, doi:https://doi.org/10.1039/c7sc03464d (2017) [16]. Copyright 2017 Chemical Science)

[47]. Knowledge of these parameters allows CCS values to be empirically determined and requires no calibrations, making it ideal for comparison among different groups and DTIMS instrument designs [47]. The limitation of stepped field is the additional time required for collecting multiple voltage data to determine the CCS values [47]. Unfortunately, this timescale is too slow to be coupled with LC analyses when the stepped-field analysis is on a second timescale [47]. To overcome this, the t_A values from single-field DTIMS experiments can be converted to CCS values using a calibration equation [49]. In single-field measurements, calibrant ions of known CCS values are analyzed at specific applied voltages (electric fields) to determine the regression relating slope, beta, and the intercept (t_{fix}) of the standardized CCS values (Eq. 5) [47].

$$t_A = \left[\left(\frac{\beta}{z}\right)\cdot\left(\frac{m_i}{m_B + m_i}\right)\cdot(\mathrm{CCS})\right] + t_{fix} \quad (5)$$

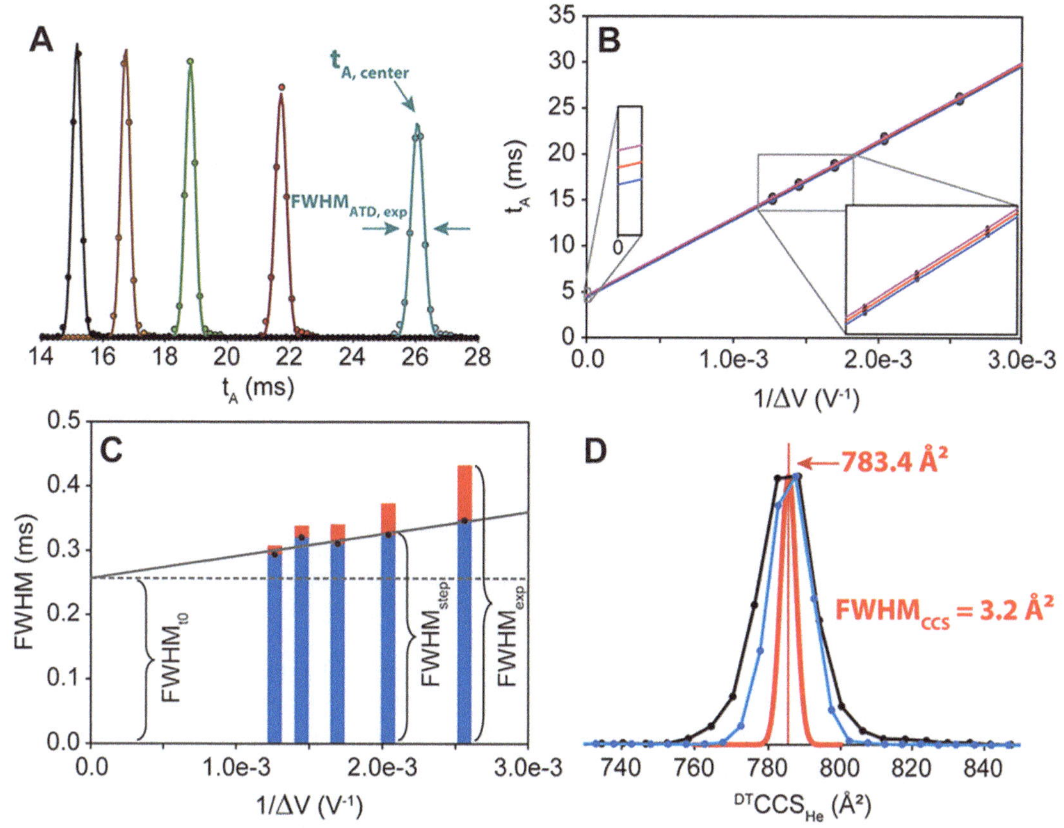

Fig. 5 The stepped-field experimental procedure workflow. In stepped-field measurements, arrival time distributions of analytes are measured over a series of different initial drift velocities at constant gas number density, temperature, and pressure normalized by area (**a**). t_a values are plotted against $1/V$ to form a straight line to determine the peak center ±½ fwhm for the stepped-field experiment (**b**). Experimental peak width (full bar) with contributing variation of $1/V$ at half maximum (blue portion of bar) (**c**). CCS distributions are then reconstructed using Eq. 5 (**d**). (Reprinted with permission from Marchand, A., Livet, S., Rosu, F. & Gabelica, V. Drift Tube Ion Mobility: How to Reconstruct Collision Cross Section Distributions from Arrival Time Distributions? *Anal Chem* **89**, 12674–12681, doi:https://doi.org/10.1021/acs.analchem.7b01736 (2017) [48]. Copyright 2017 American Chemical Society)

Subsequent CCS and t_A values in single-field measurements can then be calculated for unknown compounds under the same electric field strength [47]. The caveats for single-field IMS arise from its dependence on calibration and conditions being reproducible [47]. However, an interlaboratory study comparing DTIMS values for stepped and single field yielded average biases of 0.34% and 0.54%, suggesting only small variation between methods [47] (*see* **Note 3**).

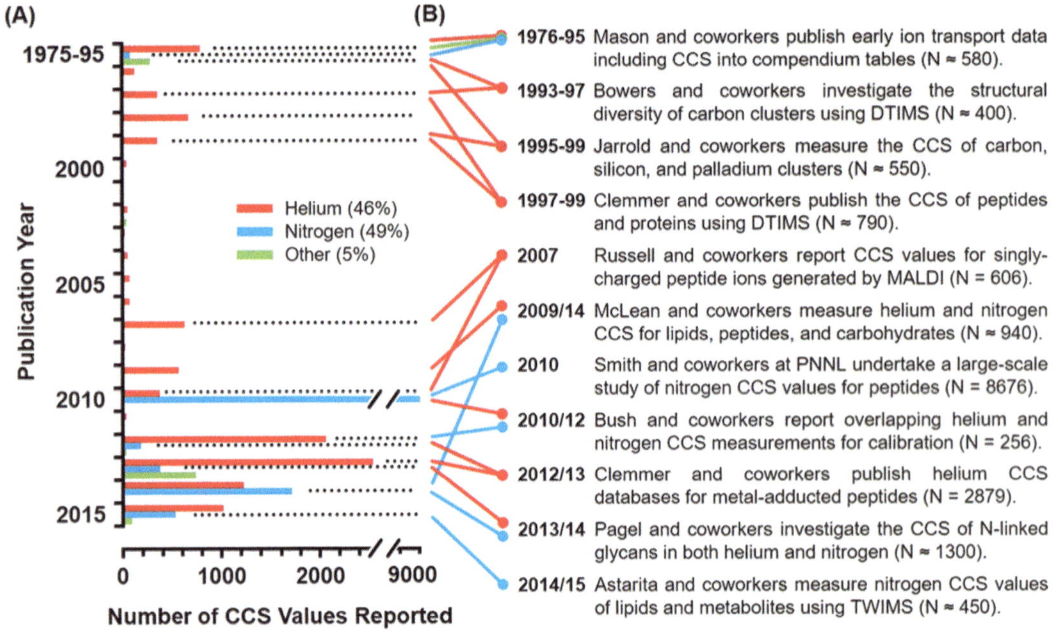

Fig. 6 The different types of drift gasses used over the last 40 years (**a**) and the study they are associated with (**b**). (Reproduced with permission from May, J. C., Morris, C. B. & McLean, J. A. Ion Mobility Collision Cross Section Compendium. *Anal Chem* **89**, 1032–1044, doi:https:/doi.org/10.1021/acs.analchem.6b04905 (2017) [52]. Copyright 2017 American Chemical Society)

5 DTIMS Drift Gas Selection

Separation in DTIMS experiments is largely accomplished by the frictional force of analyte collisions with the drift gas [30]. While ions with a constant applied voltage have different kinetic energies dependent upon their mass, their resolution in complex matrices would be inadequate for analyte separation [50]. In DTIMS, an ion's mobility (and CCS) is also dependent upon the buffer gas used [47]. As drift gas size and/or polarizability increases, the stronger frictional forces decelerating the ions results in longer retention times within the drift gas and smaller mobilities [51]. Ideally, the ion-neutral interactions vary based on polarizability and mass and can be used to manipulate the IMS selectivity similarly to changing the stationary phase in LC and GC separations [17]. Historically, the most popular drift gas choices have been helium (He, 46%) and nitrogen (N_2, 49%) (Fig. 6) [52]. While other drift gases such as CO_2 and argon have been successfully implemented in IMS experiments, He and N_2 are most often used because of their inert and relatively simply structures [52]. Many scientists often ask the question as to whether He or N_2 is better. He is the smallest noble gas on the periodic table, making it the least likely to have energetically significant collisions with analytes because of the low

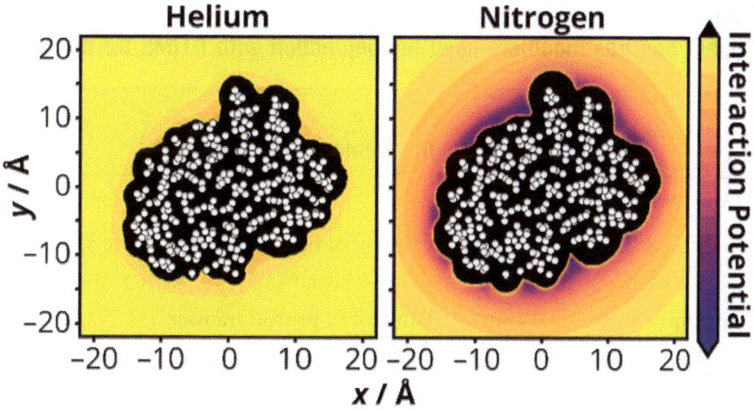

Fig. 7 Interaction potential measurements for He (left) and N_2 (right). Nitrogen has a higher interaction potential to result in more energetic ion-neutral collisions. Helium's smaller interaction potential results in less energetic interactions that can afford better isomer separation but be negligible for separation of larger molecules. (Reprinted with permission form Canzani, D., Laszlo, K. J. & Bush, M. F. Ion Mobility of Proteins in Nitrogen Gas: Effects of Charge State, Charge Distribution, and Structure. *J Phys Chem A* **122**, 5625–5634, doi:https:/doi.org/10.1021/acs.jpca.8b04474 (2018) [54]. Copyright 2018 American Chemical Society)

ion-neutral polarization and its potential reduction of mass-mobility discrimination. This relationship allows a more direct sampling of ion size as opposed to mass, which is beneficial for isomer resolution [53]. For large molecules, however, He's small ion-neutral collisions, while high in counts, could be negligible and result in poor separation. The simplified interactions of ion-gas molecules in a helium drift cell also make it simpler to theorize correlation of He CCS values and yield more accurate theoretical predictions [54]. Ion transmission has also been shown to improve with He in DTIMS and TWIMS analysis [55]. N_2 has slowly been replacing He in many DTIMS instruments as it is cheaper, renewable, and theoretical parameterization has recently been implemented. N_2 yields a higher ion-neutral interaction potential compared to He, yielding more energetic collisions between N_2 and analytes (Fig. 7) [54]. Alternatively, drift gas composition can be selectively altered to manipulate the IMS separation capabilities and enhance the resolving power of ions. The addition of drift gas modifiers (which aid in ion transport) and dopants (which block the ionization of interferences) that have been used to optimize analyte separations for specific applications are given in Table 2 [56] for both DTIMS and DMS [56, 57]. In many cases, these modifiers and dopants have successfully separated isomers that were indistinguishing without their addition (*see* **Note 4**).

Table 2
Dopant and gas modifiers used in conjunction with DTIMS for enhanced analyte separations

	IMS mode	Mechanism of interaction	Detected analytes
Dopants for DT-IMS			
Ammonia	Positive	Control of proton transfer, clusterization	Amines, CWA, narcotics, explosives
Acetone	Positive	Control of proton transfer	OPC, CWA, pesticides
Higher ketones, e.g., 5-nonanone, 4-heptanone	Positive	Clusterization, control of proton transfer	Monomethyl hydrazine, hydrazine, alkanolamines
Amides	Positive	Control of proton transfer	Narcotics
Nicotinamide	Negative	Production of stable adducts	CWA, peroxide explosives
Amines, e.g., *n*-nonylamine	Positive	Control of proton transfer, clusterization	Biogenic amines
OPC, e.g., TEP, DMMP	Positive	Control of proton transfer, clusterization	Biogenic amines, ammonia
Halogens	Negative	Production of stable adducts	Explosives, propofol
Methyl salicylate	Negative	Selective creation of clusters	Acid gases
NOx	Positive	Charge exchange, adduct formation, hydride abstraction	Analytes with low PA, e.g., alkanes and aromatic hydrocarbons
Gas modifiers for DT-IMS			
Chiral modifiers: S-(+)-2-butanol R-(−)-2-butanol	Positive	Changing of collision cross-section for ions moving in drift section, analyte-modifier cluster formation	Stereoisomers
Ketones, e.g., 5-nonanone	Positive	Analyte-modifier cluster formation	Separation of hydrazines and ammonia
Ammonia	Positive	Analyte-modifier cluster formation	Separation of amine derivatives, and 2,4-lutidine
Nitrobenzene	Positive	Analyte-modifier cluster formation	Separation of amine derivatives
Ethyl lactate	Positive	Analyte-modifier cluster formation	Separation of atenolol, arginine, histidine, lysine, caffeine, and glucosamine
Methanol	Positive	Analyte-modifier cluster formation	Separation of TTEA, asparagine, and valine

Reprinted with permission from Waraksa, E. et al. Dopants and gas modifiers in ion mobility spectrometry. *TrAC Trends in Analytical Chemistry* **82**, 237–249, doi:https://doi.org/10.1016/j.trac.2016.06.009 (2016) [56].

6 Ionization Techniques Compatible with DTIMS

Not only is DTIMS compatible with various buffer gases and front-end separation, but also a variety of ionization sources. For biological samples, the most common ionization techniques are often ESI [58] and MALDI [59] which have both been demonstrated in a variety of applications. More application-specific ionization sources have also been used, such as atmospheric chemical ionization (APCI) and atmospheric pressure ionization (APPI) for isomeric separation of a variety of hydrocarbons [43, 60]. ^{63}Ni β ionization, for example, has been coupled to IMS for quantification of the depth of anesthesia for patients during surgical procedures [61]. Comparisons of CCS values across sources have shown general agreement, with MALDI and nESI differing by <2% [43, 60]; implying IMS is amenable to a variety of sources and sample delivery rates [62]. The limitless compatibility of IMS with different sources is perhaps best personified by mass spectrometry imaging (MSi) experiments where chromatography is not an option as data are generated on a faster timescale than chromatographic separations or material is limited [63]. Without separation prior to analysis, current MSi methodology lacks the capability to fully separate analytes, often leading to ion or detector suppression [64, 65]. IMS is unique in that it can be integrated, so that imaging data acquisition rates are not adversely affected while also enabling in-depth profiling of isomeric analytes. Various ablation sources coupled with IMS for imaging applications have shown an increase in detected molecules in biological and environmental samples for lipids, proteins, and metabolites (Fig. 8) [64, 65]. Drawbacks to IMS-MSi applications however currently arise from the combination of these techniques being in its infancy, resulting in complex data analysis and a lack of software programs able to handle the data [66].

7 Concluding Remarks

While the incorporation of IMS in metabolomic and exposomic measurements is still in its infancy, it has the ability to bring these workflows to another dimension quite literally. The analyte-specific CCS values calculated from IMS measurements enable isomer separation, compound identification, and the reduction of interfering species that has previously been extremely difficult or impossible using other analysis techniques. Isomer interference is a significant problem in the quantification of specific analytes in small molecule studies [67]. Often LC and fragmentation patterns cannot resolve this issue, leading to an incomplete picture of how isomers change under different perturbations. By grouping several isomers

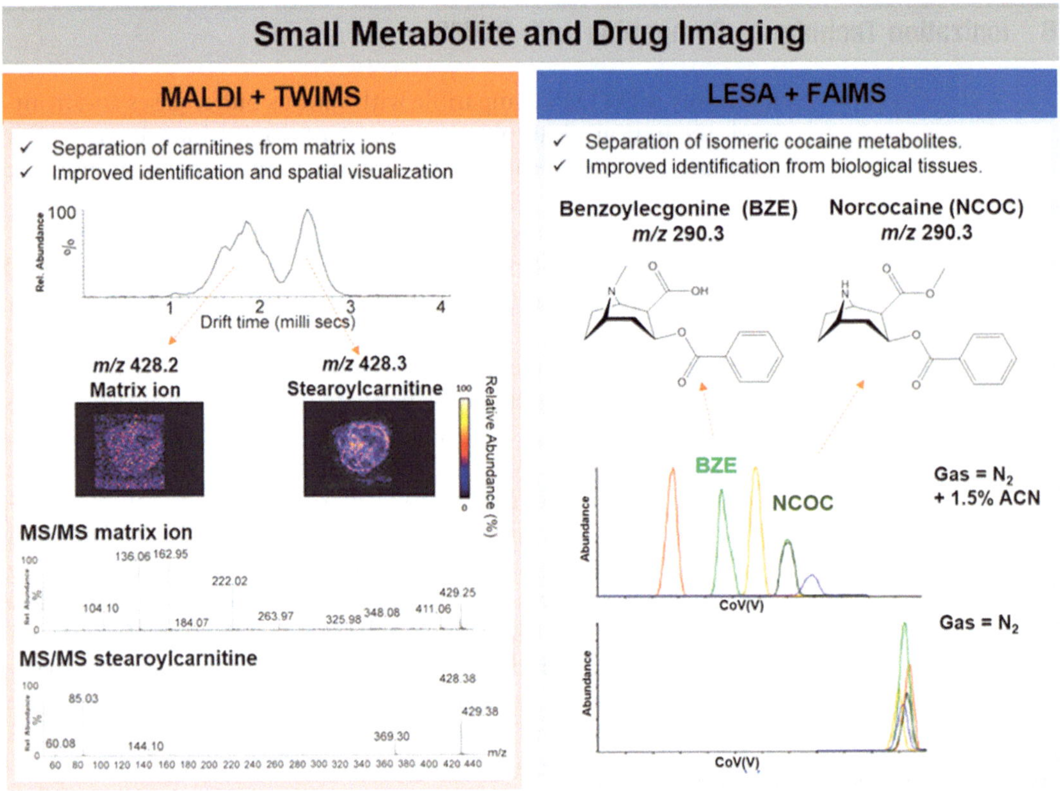

Fig. 8 Examples of IMS applications for small molecule identification in a variety of biological samples with mass spectrometry imaging. Here, carnitines, nucleotides, and drug metabolites were effectively separated with TWIMS and FAIMS to improve spatial distribution of the molecules analyzed. (Figure reprinted with permission from Sans, M., Feider, C. L. & Eberlin, L. S. Advances in mass spectrometry imaging coupled to ion mobility spectrometry for enhanced imaging of biological tissues. *Curr Opin Chem Biol* **42**, 138–146, doi: https:/doi.org/10.1016/j.cbpa.2017.12.005 (2018) [65]. Copyright 2018 Science Direct)

together in the results, the inaccurate quantification causes problems in fully understanding the changes in the samples analyzed [67]. In metabolomic and exposomic measurements, since most molecules are <600 Da, their distinction in complex biological and environmental mixtures can be extremely difficult as many isomers and nominal mass isobars exist. The high selectivity possible with IMS can differentiate isomers with ≥0.5% difference in CCS for most systems [50], and this spatial separation pulls many of the isobars to different drift time areas [31, 68–70]. When isomers cannot be readily separated with DTIMS, the trapping region prior to the drift cell has been utilized to perform chemical interaction processes that will both form and break the isomers [69]. Ozonolysis, for example, has been combined with DTIMS to elucidate isomers in lipid classes (e.g., phosphatidylethanolamines) that were previously indistinguishable in DTIMS separations [71, 72] (*see* **Note 5**).

IMS separations are also showing great potential in untargeted analyses. As most small molecule evaluations are only able to identify ~10% of the features observed, new strategies are needed to identify the statistically significant unknown molecules. Currently, small molecule databases for matching LC, GC, MS, and MS/MS features contain a majority of theoretical data and only a small portion of experiment data. For example, the human metabolome database (HMDB) houses mass information for over 114,100 metabolites and xenobiotics, but only ~30,000 have experimental MS/MS spectra available [73]. This lack of molecular descriptors has caused problems with small molecule identifications and accurate quantitation. For example, if cholesterol (*m/z* 386.3548) is searched in the HMDB, it yields 13 hits, and if unit mass resolution is used (1 Da), 93 hits result [73]. Thus, identifying metabolites in complex samples requires information other than just mass alone (e.g., fragmentation, retention time standards, CCS) to distinguish between similar compounds. The empirical calculation of CCS values from DTIMS measurements offers the possibility to identify substances by the combination of their exact mass and their specific size. By measuring CCS values for available standards and compiling databases, rapid targeted analyses and identification of small molecules are conceivable. CCS database values are also being incorporated into computational methods such as theoretical modeling of structures [74] and machine-based approaches [75, 76]. These approaches are enabling molecular insight when no standards are available such as the prediction of possible identities or molecular classification by mapping to specific molecular trend lines. Theoretical approaches therefore have the potential to enable the annotation of many more small molecules, even though only ~30,000 in databases to date have experimentally measured MS/MS spectra. Thus, the use of IMS measurements in metabolomic and exposomic studies for the distinction of isomers, separation of interferences, acquisition of experimental CCS values, and comparison with theoretical CCS values is expected to greatly escalate in the coming years.

8 Notes

1. The evaluation of an ion's collisional cross section (CCS) is an especially beneficial aspect of IMS since it provides information on the analyte's conformational size relative to its charge. When coupled with MS analyses, the IMS separation increases the confidence of each molecular identification, reduces MS peak congestion, improves measurement sensitivity and allows structural isomer distinction. Furthermore, the rapid millisecond timescale of IMS allows its coupling with numerous

separation techniques for more comprehensive metabolomic and exposomic analyses (e.g. LC-IMS-MS, GC-IMS-MS, etc.)

2. Different classes of small molecules separate into distinct trend lines when CCS versus m/z values are plotted. This property allows for the class evaluation molecules in untargeted studies and the theoretical calculation of CCS values for molecules in library databases without IMS information.
3. CCS can be evaluated using either single or stepped-field methods. While stepped-field is a more direct evaluation, only single-field is compatible with the timescale required for nested LC acquisitions, making it the most common way to perform global CCS analyses. Both stepped field and single field methods have shown reproducibility across several laboratories, making each amenable to metabolomic and exposomic applications.
4. DTIMS separations are dictated by buffer gas composition, drift tube length, pressure, temperature, and electric field. DTIMS separation has further been improved with the addition of dopant or gas modifiers, but these can be difficult to fully remove from the drift tube after analysis. We therefore recommend altering the measurement temperature or pressure prior to changing buffer gas composition.
5. IMS has been integrated into numerous application areas such as high-throughput screening and mass spectrometry imagining to reap the benefits of an added dimensionality when chromatography is not possible. The trapping region prior to the drift tube has also been used to perform chemical reactions to further elucidate isomers when DTIMS alone could not differentiate the species.

References

1. Mason E, McDaniel E (1988) Transport properites of ions in gases. Wiley, Hoboken, NJ
2. Guevremont R, Siu KW, Wang J, Ding L (1997) Combined ion mobility/time-of-flight mass spectrometry study of electrospray-generated ions. Anal Chem 69:3959–3965. https://doi.org/10.1021/ac970359e
3. Cohen MJ, Karasek FW (1970) Plasma chromatography™—a new dimension for gas chromatography and mass spectrometry. J Chromatogr Sci 8:330–337. https://doi.org/10.1093/chromsci/8.6.330
4. Hoaglund CS, Valentine SJ, Sporleder CR, Reilly JP, Clemmer DE (1998) Three-dimensional ion mobility/TOFMS analysis of electrosprayed biomolecules. Anal Chem 70:2236–2242. https://doi.org/10.1021/ac980059c
5. Wyttenbach T, Kemper PR, Bowers MT (2001) Design of a new electrospray ion mobility mass spectrometer. Int J Mass Spectrom 212:13–23. https://doi.org/10.1016/S1387-3806(01)00517-66
6. Tang K et al (2005) High-sensitivity ion mobility spectrometry/mass spectrometry using electrodynamic ion funnel interfaces. Anal Chem 77:3330–3339. https://doi.org/10.1021/ac048315a
7. Pringle SD et al (2007) An investigation of the mobility separation of some peptide and protein ions using a new hybrid quadrupole/travelling wave IMS/oa-ToF instrument. Int J

Mass Spectrom 261:1–12. https://doi.org/10.1016/j.ijms.2006.07.021

8. Michelmann K, Silveira JA, Ridgeway ME, Park MA (2015) Fundamentals of trapped ion mobility spectrometry. J Am Soc Mass Spectrom 26:14–24. https://doi.org/10.1007/s13361-014-0999-4
9. Ewing MA, Conant CRP, Zucker SM, Griffith KJ, Clemmer DE (2015) Selected overtone mobility spectrometry. Anal Chem 87:5132–5138. https://doi.org/10.1021/ac504555u
10. Zucker SM, Ewing MA, Clemmer DE (2013) Gridless overtone mobility spectrometry. Anal Chem 85:10174–10179. https://doi.org/10.1021/ac401568r
11. Rus J et al (2010) IMS–MS studies based on coupling a differential mobility analyzer (DMA) to commercial API–MS systems. Int J Mass Spectrom 298:30–40. https://doi.org/10.1016/j.ijms.2010.05.008
12. Brown LJ, Creaser CS (2013) Field asymmetric waveform ion mobility spectrometry analysis of proteins and peptides: a review. Curr Anal Chem 9:192–198
13. Guevremont R (2004) High-field asymmetric waveform ion mobility spectrometry: a new tool for mass spectrometry. J Chromatogr A 1058:3–19. https://doi.org/10.1016/j.chroma.2004.08.119
14. Kolakowski BM, Mester Z (2007) Review of applications of high-field asymmetric waveform ion mobility spectrometry (FAIMS) and differential mobility spectrometry (DMS). Analyst 132:842–864. https://doi.org/10.1039/b706039d
15. Vidal-de-Miguel G, Macía M, Cuevas J (2012) Transversal modulation ion mobility spectrometry (TM-IMS), a new mobility filter overcoming turbulence related limitations. Anal Chem 84:7831–7837. https://doi.org/10.1021/ac301127u
16. Zheng X et al (2017) A structural examination and collision cross section database for over 500 metabolites and xenobiotics using drift tube ion mobility spectrometry. Chem Sci 8:7724–7736. https://doi.org/10.1039/c7sc03464d
17. Cumeras R, Figueras E, Davis CE, Baumbach JI, Gracia I (2015) Review on ion mobility spectrometry. Part 1: current instrumentation. Analyst 140:1376–1390. https://doi.org/10.1039/c4an01100g
18. Bowers MT (2018) The bowers group-university of california santa barbara: theory/analysis ion mobility theorty introduction. Available at: https://labs.chem.ucsb.edu/bowers/michael/theory_analysis/ion-mobility/index.shtml. (Accessed: 9th November 2018)
19. Gabelica V, Marklund E (2018) Fundamentals of ion mobility spectrometry. Curr Opin Chem Biol 42:51–59. https://doi.org/10.1016/j.cbpa.2017.10.022
20. Paglia G et al (2015) Ion mobility-derived collision cross section as an additional measure for lipid fingerprinting and identification. Anal Chem 87:1137–1144. https://doi.org/10.1021/ac503715v
21. Shelimov KB, Hunter JM, Jarrold MF (1994) Small carbon rings – dissociation, isomerization, and a simple-model based on strain. Int J Mass Spectrom Ion Proc 138:17–31. https://doi.org/10.1016/0168-1176(94)80007-3
22. Hunter JM, Jarrold MF (1995) Drift-tube studies of large carbon clusters – new isomers and the mechanism of giant fullerene formation. J Am Chem Soc 117:10317–10324. https://doi.org/10.1021/ja00146a016
23. Henderson SC, Valentine SJ, Counterman AE, Clemmer DE (1999) ESI/ion trap/ion mobility/time-of-flight mass spectrometry for rapid and sensitive analysis of biomolecular mixtures. Anal Chem 71:291–301. https://doi.org/10.1021/ac9809175
24. Von Helden G, Hsu MT, Kemper PR, Bowers MT (1991) Structures of carbon cluster ions from 3 to 60 atoms – linears to rings to fullerenes. J Chem Phys 95:3835–3837. https://doi.org/10.1063/1.460783
25. Von Helden G, Hsu MT, Gotts N, Bowers MT (1993) Carbon cluster cations with up to 84 atoms – structures, formation mechanism, and reactivity. J Phys Chem 97:8182–8192. https://doi.org/10.1021/j100133a011
26. May JC, McLean JA (2015) Ion mobility-mass spectrometry: time-dispersive instrumentation. Anal Chem 87:1422–1436. https://doi.org/10.1021/ac504720m
27. Fernandez-Maestre R, Wu C, Hill HH Jr (2012) Buffer gas modifiers effect resolution in ion mobility spectrometry through selective ion-molecule clustering reactions. Rapid Commun Mass Spectrom 26:2211–2223. https://doi.org/10.1002/rcm.6335
28. Shvartsburg AA (2010) Ion mobility spectrometry (IMS) and mass spectrometry (MS), 2nd edn. ScienceDirect, Amsterdam, pp 1140–1148
29. Zheng X et al (2017) Coupling front-end separations, ion mobility spectrometry, and mass spectrometry for enhanced multidimensional biological and environmental analyses.

Annu Rev Anal Chem (Palo Alto Calif) 10:71–92. https://doi.org/10.1146/annurev-anchem-061516-045212
30. Baker ES et al (2015) Enhancing bottom-up and top-down proteomic measurements with ion mobility separations. Proteomics 15:2766–2776. https://doi.org/10.1002/pmic.201500048
31. Rainville PD et al (2017) Ion mobility spectrometry combined with ultra performance liquid chromatography/mass spectrometry for metabolic phenotyping of urine: Effects of column length, gradient duration and ion mobility spectrometry on metabolite detection. Anal Chim Acta 982:1–8. https://doi.org/10.1016/j.aca.2017.06.020
32. Fiehn O (2002) Metabolomics – the link between genotypes and phenotypes. Plant Mol Biol 48:155–171
33. Schrimpe-Rutledge AC, Codreanu SG, Sherrod SD, McLean JA (2016) Untargeted metabolomics strategies-challenges and emerging directions. J Am Soc Mass Spectrom 27:1897–1905. https://doi.org/10.1007/s13361-016-1469-y
34. Ortmayr K, Causon TJ, Hann S, Koellensperger G (2016) Increasing selectivity and coverage in LC-MS based metabolome analysis. TrAC Trends Anal Chem 82:358–366. https://doi.org/10.1016/j.trac.2016.06.011
35. Wickramasekara SI et al (2013) Electrospray quadrupole travelling wave ion mobility time-of-flight mass spectrometry for the detection of plasma metabolome changes caused by xanthohumol in obese zucker (fa/fa) rats. Meta 3:701–717. https://doi.org/10.3390/metabo3030701
36. Bush MF et al (2010) Collision cross sections of proteins and their complexes: a calibration framework and database for gas-phase structural biology. Anal Chem 82:9557–9565. https://doi.org/10.1021/ac1022953
37. Valentine SJ, Counterman AE, Clemmer DE (1999) A database of 660 peptide ion cross sections: use of intrinsic size parameters for bona fide predictions of cross sections. J Am Soc Mass Spectrom 10:1188–1211. https://doi.org/10.1016/S1044-0305(99)00079-3
38. Stephan S et al (2016) Contaminant screening of wastewater with HPLC-IM-qTOF-MS and LC+LC-IM-qTOF-MS using a CCS database. Anal Bioanal Chem 408:6545–6555. https://doi.org/10.1007/s00216-016-9820-5
39. Hofmann J et al (2014) Estimating collision cross sections of negatively charged N-glycans using traveling wave ion mobility-mass spectrometry. Anal Chem 86:10789–10795. https://doi.org/10.1021/ac5028353
40. Zhou Z et al (2019) LipidIMMS analyzer: integrating multi-dimensional information to support lipid identification in ion mobility – mass spectrometry based lipidomics. Bioinformatics 35:698–700. https://doi.org/10.1093/bioinformatics/bty661
41. Kyle JE et al (2018) Evaluating lipid mediator structural complexity using ion mobility spectrometry combined with mass spectrometry. Bioanalysis 10:279–289. https://doi.org/10.4155/bio-2017-0245
42. Paglia G, Kliman M, Claude E, Geromanos S, Astarita G (2015) Applications of ion-mobility mass spectrometry for lipid analysis. Anal Bioanal Chem 407:4995–5007. https://doi.org/10.1007/s00216-015-8664-8
43. Zheng XY et al (2018) Utilizing ion mobility spectrometry and mass spectrometry for the analysis of polycyclic aromatic hydrocarbons, polychlorinated biphenyls, polybrominated diphenyl ethers and their metabolites. Anal Chim Acta 1037:265–273. https://doi.org/10.1016/j.aca.2018.02.054
44. May JC et al (2014) Conformational ordering of biomolecules in the gas phase: nitrogen collision cross sections measured on a prototype high resolution drift tube ion mobility-mass spectrometer. Anal Chem 86:2107–2116. https://doi.org/10.1021/ac4038448
45. Bijlsma L et al (2017) Prediction of collision cross-section values for small molecules: application to pesticide residue analysis. Anal Chem 89:6583–6589. https://doi.org/10.1021/acs.analchem.7b00741
46. Plante PL et al (2019) Predicting ion mobility collision cross-sections using a deep neural network: DeepCCS. Anal Chem 91:5191–5199. https://doi.org/10.1021/acs.analchem.8b05821
47. Stow SM et al (2017) An interlaboratory evaluation of drift tube ion mobility-mass spectrometry collision cross section measurements. Anal Chem 89:9048–9055. https://doi.org/10.1021/acs.analchem.7b01729
48. Marchand A, Livet S, Rosu F, Gabelica V (2017) Drift tube ion mobility: how to reconstruct collision cross section distributions from arrival time distributions? Anal Chem 89:12674–12681. https://doi.org/10.1021/acs.analchem.7b01736
49. Alex Mordehai RK, Darland E, Stafford G, Fjeldsted J (2015) Single field direct drift time to CCS calibration for a linear drift tube ion

mobility mass spectrometer, in ASMS. Agilent Technologies, Santa Clara, CA

50. Dodds JN, May JC, McLean JA (2017) Correlating resolving power, resolution, and collision cross section: unifying cross-platform assessment of separation efficiency in ion mobility spectrometry. Anal Chem 89:12176–12184. https://doi.org/10.1021/acs.analchem.7b02827
51. Matz LM, Hill HH, Beegle LW, Kanik I (2002) Investigation of drift gas selectivity in high resolution ion mobility spectrometry with mass spectrometry detection. J Am Soc Mass Spectrom 13:300–307. https://doi.org/10.1016/S1044-0305(01)00366-X
52. May JC, Morris CB, McLean JA (2017) Ion mobility collision cross section compendium. Anal Chem 89:1032–1044. https://doi.org/10.1021/acs.analchem.6b04905
53. Kiss A, Heeren RM (2011) Size, weight and position: ion mobility spectrometry and imaging MS combined. Anal Bioanal Chem 399:2623–2634. https://doi.org/10.1007/s00216-010-4644-1
54. Canzani D, Laszlo KJ, Bush MF (2018) Ion mobility of proteins in nitrogen gas: effects of charge state, charge distribution, and structure. J Phys Chem A 122:5625–5634. https://doi.org/10.1021/acs.jpca.8b04474
55. Baker ES et al (2007) Ion mobility spectrometry-mass spectrometry performance using electrodynamic ion funnels and elevated drift gas pressures. J Am Soc Mass Spectrom 18:1176–1187. https://doi.org/10.1016/j.jasms.2007.03.031
56. Waraksa E et al (2016) Dopants and gas modifiers in ion mobility spectrometry. TrAC Trends Anal Chem 82:237–249. https://doi.org/10.1016/j.trac.2016.06.009
57. Fernandez-Maestre R, Wu C, Hill HH (2010) Using a buffer gas modifier to change separation selectivity in ion mobility spectrometry. Int J Mass Spectrom 298:2–9. https://doi.org/10.1016/j.ijms.2010.08.009
58. McMinn DG, Kinzer JA, Shumate CB, Siems WF, Hill HH, J. (1990) Ion mobility detection following liquid chromatographic separation. J Micrucol Sep 2:188–192
59. Jackson SN et al (2014) MALDI-ion mobility mass spectrometry of lipids in negative ion mode. Anal Methods 6:5001–5007. https://doi.org/10.1039/C4AY00320A
60. Borsdorf H, Nazarov EG, Eiceman GA (2002) Atmospheric pressure chemical ionization studies of non-polar isomeric hydrocarbons using ion mobility spectrometry and mass spectrometry with different ionization techniques. J Am Soc Mass Spectrom 13:1078–1087
61. Zhou Q, Hua L, Wang C, Li E, Li H (2015) Improved analytical performance of negative 63Ni ion mobility spectrometry for on-line measurement of propofol using dichloromethane as dopant. J Am Soc Mass Spectrom 26:190–193. https://doi.org/10.1007/s13361-014-0977-x
62. Sundarapandian S, May JC, McLean JA (2010) Sundarapandian, May, McLean - 2010 - Dual source ion mobility-mass spectrometer for direct comparison of electrospray ionization and mal-annotated. Anal. Chem 82:3247–3254. https://doi.org/10.1021/ac902980r
63. Sherrod SD, McLean JA (2016) Systems-wide high-dimensional data acquisition and informatics using structural mass spectrometry strategies. Clin Chem 62:77–83. https://doi.org/10.1373/clinchem.2015.238261
64. Buchberger AR, DeLaney K, Johnson J, Li L (2018) Mass spectrometry imaging: a review of emerging advancements and future insights. Anal Chem 90:240–265. https://doi.org/10.1021/acs.analchem.7b04733
65. Sans M, Feider CL, Eberlin LS (2018) Advances in mass spectrometry imaging coupled to ion mobility spectrometry for enhanced imaging of biological tissues. Curr Opin Chem Biol 42:138–146. https://doi.org/10.1016/j.cbpa.2017.12.005
66. Chouinard CD, Wei MS, Beekman CR, Kemperman RH, Yost RA (2016) Ion mobility in clinical analysis: current progress and future perspectives. Clin Chem 62:124–133. https://doi.org/10.1373/clinchem.2015.238840
67. Benigni P, Thompson CJ, Ridgeway ME, Park MA, Fernandez-Lima F (2015) Targeted high-resolution ion mobility separation coupled to ultrahigh-resolution mass spectrometry of endocrine disruptors in complex mixtures. Anal Chem 87:4321–4325. https://doi.org/10.1021/ac504866v
68. Kyle JE et al (2016) Uncovering biologically significant lipid isomers with liquid chromatography, ion mobility spectrometry and mass spectrometry. Analyst 141:1649–1659. https://doi.org/10.1039/c5an02062j
69. Zhang X et al (2016) SPE-IMS-MS: An automated platform for sub-sixty second surveillance of endogenous metabolites and xenobiotics in biofluids. Clin Mass Spectrom 2:1–10. https://doi.org/10.1016/j.clinms.2016.11.002
70. Sinclair E et al (2018) Mobilising ion mobility mass spectrometry for metabolomics. Analyst 143:4783–4788. https://doi.org/10.1039/c8an00902c

71. Poad BLJ et al (2018) Online ozonolysis combined with ion mobility-mass spectrometry provides a new platform for lipid isomer analyses. Anal Chem 90:1292–1300. https://doi.org/10.1021/acs.analchem.7b04091
72. Klein RJ, Welna DT, Weikel AL, Allcock HR, Runt J (2007) Counterion effects on ion mobility and mobile ion concentration of doped polyphosphazene and polyphosphazene ionomers. Macromolecules 40:3990. https://doi.org/10.1021/ma070357o
73. Wishart DS et al (2018) HMDB 4.0: the human metabolome database for 2018. Nucleic Acids Res 46:D608–D617. https://doi.org/10.1093/nar/gkx1089
74. Zheng X et al (2017) Structural elucidation of cis/trans dicaffeoylquinic acid photoisomerization using ion mobility spectrometry-mass spectrometry. J Phys Chem Lett 8:1381–1388. https://doi.org/10.1021/acs.jpclett.6b03015
75. Zhou Z, Xiong X, Zhu ZJ (2017) MetCCS predictor: a web server for predicting collision cross-section values of metabolites in ion mobility-mass spectrometry based metabolomics. Bioinformatics 33:2235–2237. https://doi.org/10.1093/bioinformatics/btx140
76. Soper-Hopper MT et al (2017) Collision cross section predictions using 2-dimensional molecular descriptors. Chem Commun (Camb) 53:7624–7627. https://doi.org/10.1039/c7cc04257d

Chapter 3

Untargeted Differential Metabolomics Analysis Using Drift Tube Ion Mobility-Mass Spectrometry

Rick Reisdorph, Cole Michel, Kevin Quinn, Katrina Doenges, and Nichole Reisdorph

Abstract

Mass spectrometry-based metabolomics is being increasingly applied to a number of applications, including the fields of clinical, industrial, plant, and nutritional science. Several improvements have advanced the field considerably over the past decade, including ultra-high performance liquid chromatography (uHPLC), column chemistries, instruments, software, and molecular databases. However, challenges remain, including how to separate small molecules that are part of highly complex samples; this can be accomplished using chromatographic techniques or through improved resolution in the gas phase. Ion mobility-mass spectrometry (IM-MS) provides an extra dimension of gas phase separation that can result in improvements to both quantitation and compound identification. Here we describe a typical drift tube IM-MS metabolomics workflow, which includes the following steps: (1) Data acquisition, (2) Data preprocessing, (3) Molecular feature finding, and (4) Differential analysis and Molecular annotation. Overall, these methods can help investigators from a variety of scientific fields use IM-MS metabolomics as part of their own workflow.

Key words Drift tube ion mobility, Metabolomics, Acquisition parameters, Four-dimensional feature finding, Differential analysis, Mass spectrometry

1 Introduction

Liquid chromatography mass spectrometry (LCMS) is widely used for untargeted differential analysis of molecules from a large array of materials. These include biological fluids and tissues [1, 2] (analysis of these sample types is typically referred to as "metabolomics"), foods and plants, and non-biological materials such as raw materials used in manufacturing, and drinking water. While these materials are diverse in origin and composition, they all share one characteristic: they are all complex matrices composed of hundreds to thousands of small molecule compounds. Despite the application of diverse liquid chromatography separation chemistries coupled to high-resolution mass spectrometers, achieving sufficient

Giuseppe Paglia and Giuseppe Astarita (eds.), *Ion Mobility-Mass Spectrometry: Methods and Protocols*, Methods in Molecular Biology, vol. 2084, https://doi.org/10.1007/978-1-0716-0030-6_3,

separation of molecules within these matrices is challenging and typically not possible using standard methodologies. Inadequate separation results in poor detection of low abundance species, molecular feature extraction artifacts, incorrect annotations from database searches, and inaccurate quantitation. New approaches are required to achieve sufficient separation for meaningful analysis of highly complex samples. Ion mobility-mass spectrometry (IM-MS) is such an approach that provides an extra dimension of separation to increase confidence in compound annotation while simultaneously improving detection and quantitation of molecules in complex matrices [3]. The focus of this chapter is on untargeted differential analysis of metabolomics samples using liquid chromatography coupled to a drift tube ion mobility (DTIM) quadrupole time-of-flight (Q-TOF) mass spectrometer (MS); however, the same general principles may apply to other types of IM-MS platforms. The methods detailed below can also be applied to the analysis of foods, plants, and non-biological samples.

1.1 Metabolomics Data Acquisition, Processing, Extraction, and Analysis Workflow

The protocols detailed below are broken into four primary categories that encompass steps in the metabolomics workflow on the Agilent Technologies drift tube ion mobility quadrupole time-of-flight mass spectrometry platform (IM-QTOF) (Fig. 1). They are (1) Data acquisition, (2) Data preprocessing, (3) 4D molecular feature finding, and (4) Differential analysis and Molecular annotation. While the data preprocessing steps are unique to the Agilent drift tube system, much of the workflow detailed in this chapter can be applied to other IM-MS platforms.

1.2 Sample Preparation

Although ion mobility combined with liquid chromatography and high-resolution mass spectrometry provides great improvements in separation and detection of molecules in complex matrices, it is vital to carefully consider sample preparation techniques and liquid chromatography column chemistries. As a starting point, it is recommended to consult the literature for protocols appropriate for your specific application [3, 4]. Generally, sample preparation protocols for

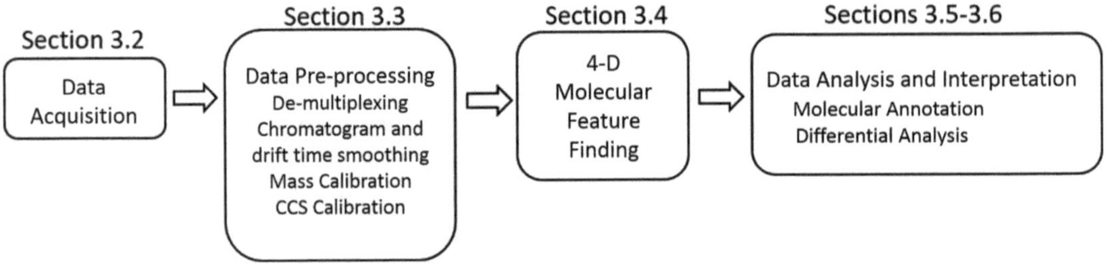

Fig. 1 DTIM-MS-based metabolomics experiments can be broken down into four general steps: (1) Data acquisition, (2) Data preprocessing, (3) 4D molecular feature finding, and (4) Differential analysis and Molecular annotation. Each step is labeled with the relevant protocol section numbers

standard metabolomics LCMS experiments are equally suited for LC-IM-MS. For highly complex samples such as plasma, urine, or tissues, care should be taken to select appropriate sample preparation techniques best suited for the class of molecules of interest. The goals of the experiment should also be taken into consideration. For example, a simple methanol crash may provide sufficient metabolite coverage required for you to achieve the goals of your experiment; conversely, you may need to separate lipids and hydrophilic molecules during sample preparation and analyze them separately (in both positive and negative ion modes) to provide greater metabolome coverage [3].

1.3 Liquid Chromatography

Liquid chromatography methods should be carefully considered. Ion mobility should be treated as an enhancement to chromatographic separation, not as a replacement. Although high-throughput methods that rely more heavily on ion mobility separation are being tested [5], when deep metabolome coverage is required, it is best to begin with standard LCMS methods and column chemistries before attempting to develop infusion methods that omit columns. Both infusion and extremely short chromatographic methods can result in unacceptable ion suppression.

2 Materials

Prepare all solutions using LCMS-grade solvents. Prepare and store all reagents at room temperature (unless indicated otherwise). Diligently follow all waste-disposal regulations when disposing of waste materials.

2.1 Data Acquisition

1. Agilent 6560 Drift Tube Ion Mobility Quadrupole Time-of-Flight (IM-QTOF) mass spectrometer equipped with an ion mobility gas supply kit and dual AJS ESI source.
2. High performance or ultra-high performance liquid chromatography system: UPLC pump, autosampler, and column heater. For example, an Agilent 1290 Infinity II High Speed Pump, 1290 Infinity II Multisampler, and 1290 Infinity II multicolumn thermostat is sufficient.
3. Tuning solution: Dilute 10 mL of the Agilent ESI-L Low Concentration Tuning Mix (100 mL) in 90 mL of 95% aqueous acetonitrile. Add 3 μL of Agilent 0.1 mM HP-0321 and mix.
4. Reference mass solution: To 1 L of 95% acetonitrile, 5% water, add 2.0 mL of 5 mM Purine and 0.8 mL of Agilent 2.5 mM HP-0921 and mix by inverting the bottle several times. Store at 4 °C
5. Agilent MassHunter Data Acquisition software version 8.0 or later version.

2.2 Data Preprocessing

1. Agilent De-multiplexing Tool (agtdemp.exe). This software is included in the Agilent 6560 software suite.
2. Pacific Northwest Laboratory's PNNL PreProcessor (https://omics.pnl.gov/software/pnnl-preprocessor).
3. Agilent IM-MS Data File Reprocessing Utility. This software is included in the Agilent 6560 software suite.
4. Agilent Mass Hunter IM-MS Browser B.08.00 or later version. This software is included in the Agilent 6560 software suite.

2.3 Four-Dimensional DTIM-MS Data Extraction

1. Agilent MassHunter Mass Profiler B.08.01 or later version. This software is included in the Agilent 6560 software suite.

2.4 Molecular Annotation

1. Agilent MassHunter ID Browser B.08.00 or later version. This software is included in the Agilent 6560 software suite.
2. CCS-AMRT databases (*see* **Note 23**).
3. Metlin AMRT database (*see* **Note 23**). This database is available in the Agilent software suite.

2.5 Differential Analysis

1. Agilent MassHunter Mass Profiler B.08.01 or later version. This software is included in the Agilent 6560 software suite.

3 Methods

3.1 Instrument Tuning and Calibration

1. Open MassHunter Data Acquisition software and enter the tune context.
2. If the instrument has not recently been tuned using System Tune, perform a positive or negative (or both if required) polarity QTOF System Tune in standard (3200 m/z) mass range and 2 GHz Extended Dynamic Range mode. For this tune, set the slicer mode to *High Resolution* (*see* **Note 1**). If the instrument does not need a system tune skip to **step 3**.
3. After the System Tune has completed, switch the instrument mass range to *low (1700 m/z)* and the slicer mode to *High Sensitivity* (*see* **Note 2**).
4. Perform a positive or negative TOF transmission tune in the 50—1700 m/z range or other range of your choice (*see* **Note 2**). Select the desired ion polarity boxes, the *TOF* radio button, the *Transmission tune* radio button, and the desired *m/z* range radio button. Then select *Start TOF Transmission Tune.*
5. Exit to the Acquisition context, and when prompted, save the tune file.

3.2 Data Acquisition Method Setup

Liquid chromatography methods are highly dependent on sample type, mobile phases, and column chemistries; therefore, liquid chromatography method details are not included in the acquisition method setup. Use LC methods most suited for your applications and sample types.

This section details setting up IM-QTOF parameters for metabolomics experiments. There are two methods to create. The first is the method used to analyze metabolomics samples. The second is used to acquire drift times of calibration standards for single-field collisional cross section (CCS) calibration. These drift times will be used during data preprocessing to allow for the calculation of CCS values of compounds.

1. In the Acquisition context of the IM-QTOF acquisition software, open the Method Editor.
2. Under *Acquisition Mode*, select *IM-QTOF.*
3. Under the *General* tab, select *Positive* Ion Polarity.
4. Ensure that *MS* is selected under *LC stream* and click *Apply Now.*
5. Under the Source tab, set the following source parameters: Gas temp 300 °C, Drying Gas flow to 12 L/min, nebulizer pressure to 35 psi, Sheath Gas temp to 275 °C, Sheath Gas flow to 12 L/min, VCap to 3500 V, Nozzle Voltage to 250 V, Fragmentor to 360 V, and the Oct1 RF Vpp to 750 V (*see* Fig. 2) (*see* **Notes 3** and **4**).
6. Under the *Acquisition, Spectral Parameters* tab, set the mass range to *50—1700 m/z* or other desired range (*see* **Note 2**).

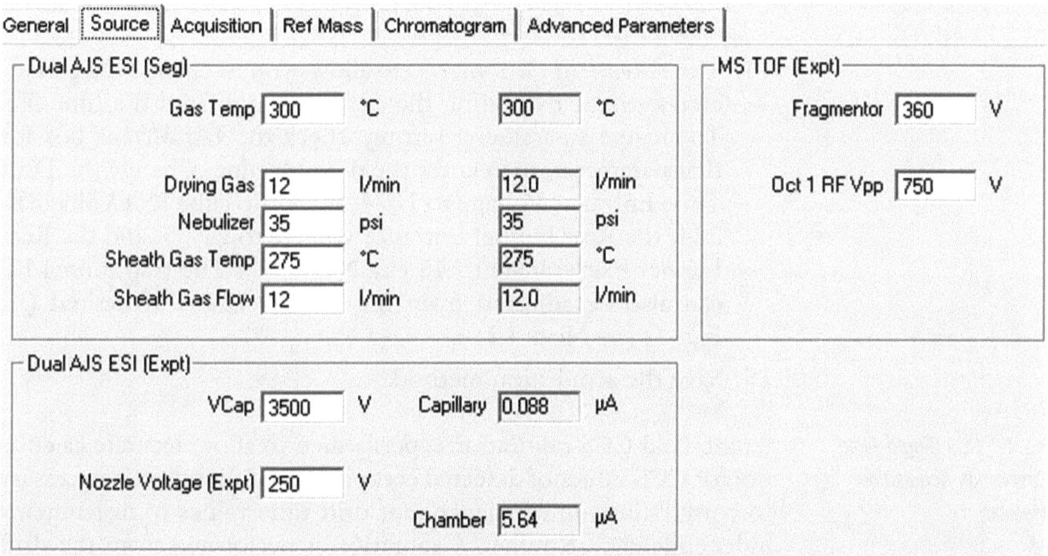

Fig. 2 Basic positive polarity dual Agilent Jet Stream source parameters for DTIM-MS metabolomics analysis

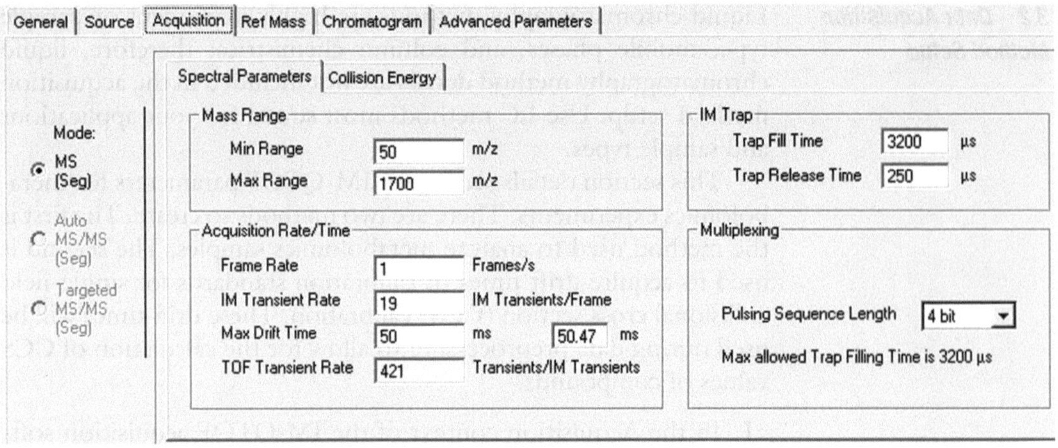

Fig. 3 IM-QTOF Acquisition parameter settings: Mass range, frame rate, max drift time, trap fill time, trap release time, and multiplexing (4 bit)

7. Set the *Frame Rate* to 1 frames/s (*see* Fig. 3) (*see* **Note 5**).
8. Set the *Max Drift Time* to *50 ms* (*see* Fig. 3) (*see* **Note 6**).
9. Set the *Trap Fill Time* to *3200 µs* (*see* Fig. 3) (*see* **Note 7**).
10. Set the *Trap Release Time* to *250 µs* (*see* Fig. 3) (*see* **Note 8**).
11. Set the *Multiplexing Pulse Sequence Length* to *4 bit* (*see* Fig. 3) (*see* **Note 9**).
12. Under the *Collision Energy* tab, make sure that *Fragmentation* is set to *Not Fragmented* under *Single Scheme* (*see* **Note 10**).
13. Under the *Reference Mass* tab, select *Use bottle A*. Click *Apply now* so that the calibrant delivery system is delivering reference mass solution to the reference nebulizer. Ensure that the dual AJS splitter valve is open.
14. The *Advanced Parameters* tab allows you to change IM parameter settings, overriding the default values from the tune file. To change a parameter setting, check the *Use Method* box for that parameter, then enter the desired value. Change the Drift Tube Entrance Voltage to 1574, the Drift Tube Exit Voltage to 224, the Rear Funnel Entrance voltage to 217.5, and the Rear Funnel Exit voltage to 45 (*see* **Note 11**). The trap funnel RF can also be adjusted from the default 150 V if desired (*see* Fig. 4) (*see* **Note 12**).
15. Save the acquisition method.

3.2.1 CCS Single-Field Calibration Acquisition Method

Single-field CCS calibration is performed to allow accurate calculation of CCS values of detected compounds. Calibration is necessary to convert instrument-dependent drift time values to instrument-independent CCS values. Calibration is performed from the drift times of the ions present in the tune ion mixture. The method

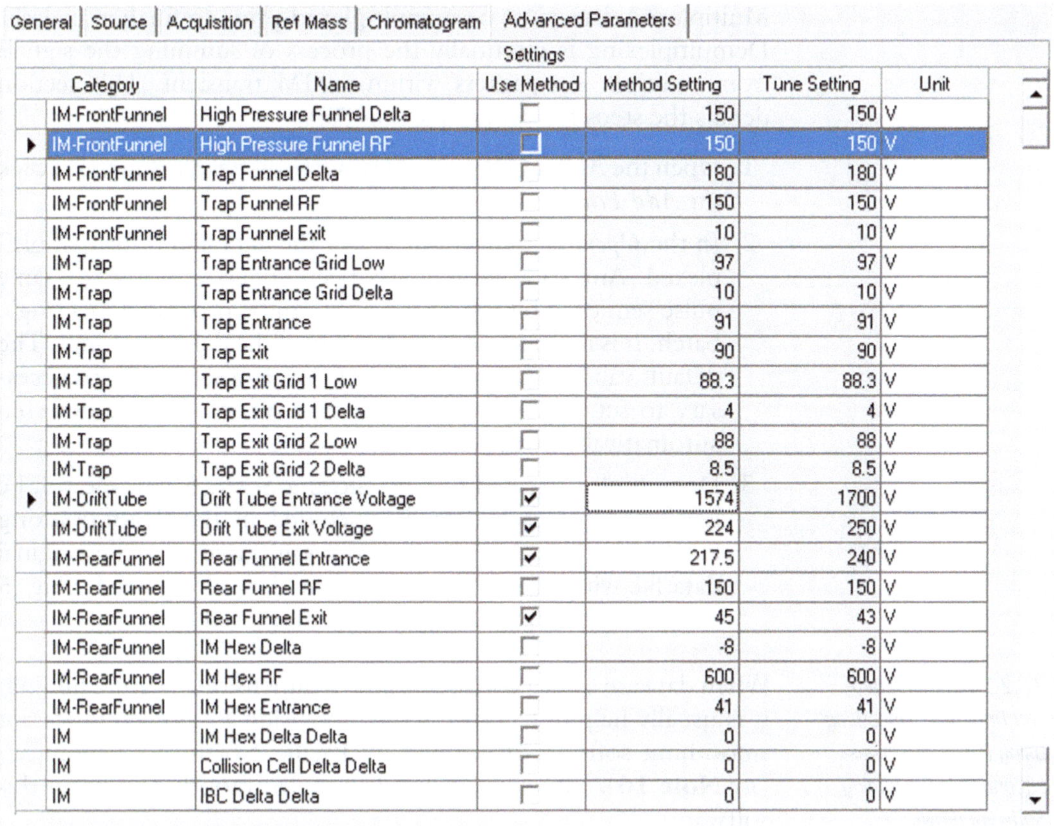

General | Source | Acquisition | Ref Mass | Chromatogram | Advanced Parameters

Settings

Category	Name	Use Method	Method Setting	Tune Setting	Unit
IM-FrontFunnel	High Pressure Funnel Delta	☐	150	150	V
IM-FrontFunnel	High Pressure Funnel RF	☐	150	150	V
IM-FrontFunnel	Trap Funnel Delta	☐	180	180	V
IM-FrontFunnel	Trap Funnel RF	☐	150	150	V
IM-FrontFunnel	Trap Funnel Exit	☐	10	10	V
IM-Trap	Trap Entrance Grid Low	☐	97	97	V
IM-Trap	Trap Entrance Grid Delta	☐	10	10	V
IM-Trap	Trap Entrance	☐	91	91	V
IM-Trap	Trap Exit	☐	90	90	V
IM-Trap	Trap Exit Grid 1 Low	☐	88.3	88.3	V
IM-Trap	Trap Exit Grid 1 Delta	☐	4	4	V
IM-Trap	Trap Exit Grid 2 Low	☐	88	88	V
IM-Trap	Trap Exit Grid 2 Delta	☐	8.5	8.5	V
IM-DriftTube	Drift Tube Entrance Voltage	☑	1574	1700	V
IM-DriftTube	Drift Tube Exit Voltage	☑	224	250	V
IM-RearFunnel	Rear Funnel Entrance	☑	217.5	240	V
IM-RearFunnel	Rear Funnel RF	☐	150	150	V
IM-RearFunnel	Rear Funnel Exit	☑	45	43	V
IM-RearFunnel	IM Hex Delta	☐	-8	-8	V
IM-RearFunnel	IM Hex RF	☐	600	600	V
IM-RearFunnel	IM Hex Entrance	☐	41	41	V
IM	IM Hex Delta Delta	☐	0	0	V
IM	Collision Cell Delta Delta	☐	0	0	V
IM	IBC Delta Delta	☐	0	0	V

Fig. 4 Drift tube advanced parameter settings for single-field experiments: drift tube entrance, drift tube exit, rear funnel entrance, and rear funnel exit voltages

detailed below allows acquisition of the calibrant mixture for this purpose.

1. Load the same method that will be used to acquire data for samples of interest (*see* **Note 13**). Go to the *Binary pump* tab and remove the gradient.
2. Set the *Stop time* to *1.0 min*.
3. Go to the *Properties* tab and select the *prerun script SCP_AcquireCalibrantData* (*see* **Note 14**).
4. Save the single-field calibration method.
5. At the beginning of every batch of samples, run the single-field calibration acquisition method (*see* **Note 15**).

3.3 Data Preprocessing

Before metabolomics data can be analyzed to determine differences between sample groups, it must be processed using the steps below.

3.3.1 Demultiplexing

The data acquisition method above utilized multiplexing to increase sensitivity and reduce the likelihood of detector saturation.

Multiplexed data must be demultiplexed prior to further analysis. Demultiplexing is essentially the process of summing the signals from multiple ion packets within an IM transient. This section details the steps for demultiplexing data.

1. Open the Agilent De-multiplexing Tool, *agtdemp.exe*, and click on *Add Files*.
2. In the *Open* dialog window, select the data files to be demultiplexed. Any data file that has been collected using the same pulse sequence bit rate (e.g., 3, 4) can be processed in a single batch. It is not necessary to alter any of the default settings. The default values will work well for most data sets. It is not necessary to set the bit rate. The correct bit rate will be detected automatically by the software.
3. Demultiplex selected files by clicking *Start*. A new file will be generated for each data file included in the demultiplexing batch. New files are saved in the same folder as the original data file with .DeMP.d added to the data file name. *See* Fig. 5 for examples of pre- and post-demultiplexed data.

3.3.2 Chromatogram and Drift Time Smoothing Using Pacific Northwest Laboratory's Smoothing Software (PNNL PreProcessor)

When data are to be processed using 4-dimensional feature finding, it is usually helpful to process the data using PNNL PreProcessor smoothing software developed by Pacific Northwest Laboratory (*see* **Note 16**). This section details how to smooth data using this software.

1. Transfer all demultiplexed files that require smoothing into a single folder. The PNNL PreProcessor software operates by using folders rather than individual files.
2. Select the folder containing the files to be smoothed by left clicking in the white box located at the top of the PreProcessor utility, *see* Fig. 6.
3. Using the drop down menus, set the drift smoothing value to 3 and the chromatography/infusion value to 3. *See* **Note 16** for additional information about DTIM-MS data smoothing.
4. Smooth the data files by clicking *Process* (*see* **Note 17**).

3.3.3 Mass Calibration

On standard Agilent Q-TOF instruments, reference mass correction is performed real time during acquisition. In IM-QTOF mode on the 6560, mass correction is performed post acquisition.

1. Transfer all demultiplexed smoothed files to a new folder.
2. Batch process multiple files using Agilent Technologies' IM-MS Reprocessing Utility by clicking on the box next to *only this file*. Select a .d file for mass correction, and then select *All file(s) in the same directory*.

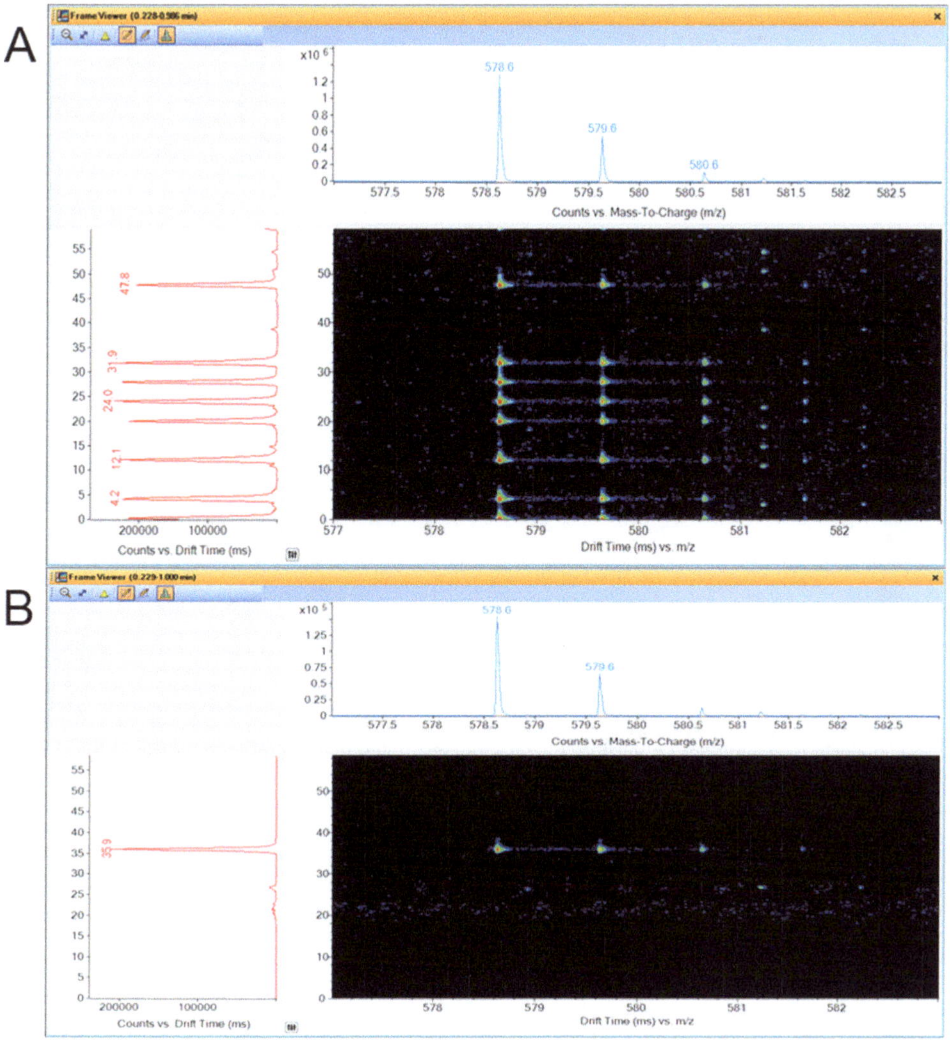

Fig. 5 Multiplexed DTIM-MS experiment before (**a**) and after demultiplexing (**b**)

3. Click on the *Recalibrate* tab. Select the reference masses that were used to lock masses during the run. Check the boxes next to 121.050873 (purine) and 922.009798 (HP-0921) (Fig. 7).
4. Click on the *Operations* tab, then click *Start*. The reprocessing utility modifies the original data files. It does not generate a new mass-corrected data file.

3.3.4 Single-Field Collisional Cross Section (CCS) Calibration

Single-field collisional cross section calibration is required for automated CCS calculation. This step must be performed if molecular features are to be annotated in database searches where CCS matching is required.

1. Open the ESI tune mix calibration .d file in MassHunter IM-MS Browser (this is the data file created using the *CCS single-field calibration acquisition method* in Subheading 3.2, **step 2**).

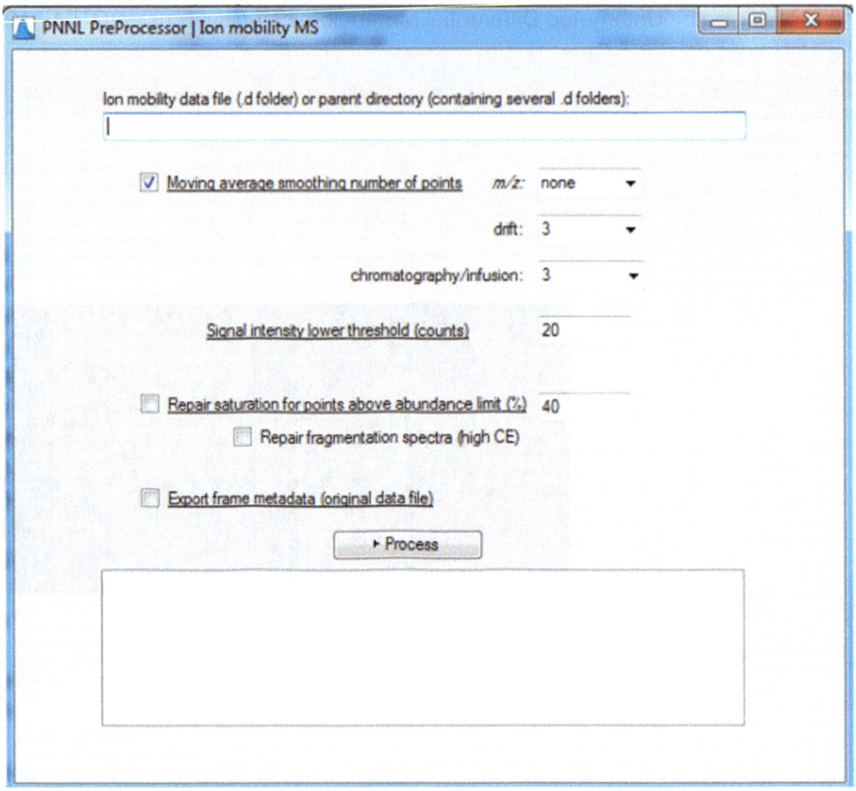

Fig. 6 Pacific Northwest National Laboratory's PreProcessor utility with suggested starting smoothing values for DTIM-MS experiments

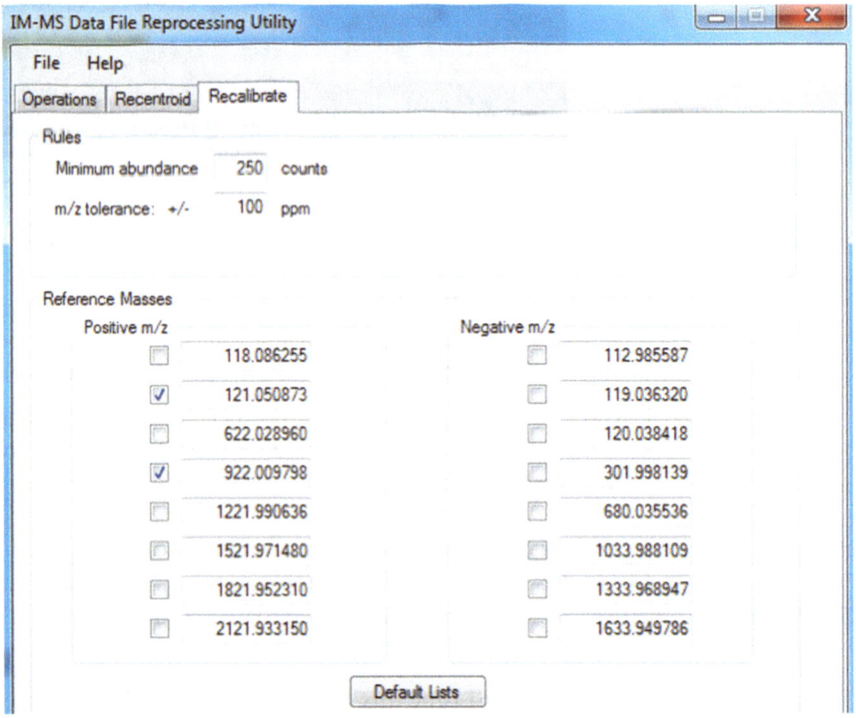

Fig. 7 Agilent IM-MS Data File Reprocessing Utility Recalibration tab with suggested Positive m/z reference correction parameters

2. From the *View menu*, select *CCS Calibration (Single-Field)*.
3. In the CCS Calibration (Single-Field) window, select the *Agilent ESI Tune Mix (POS) reference set*.
4. Extract frame data in the Total Ion Chromatogram (TIC) window that represents a region of sufficient signal associated with LC/MS Calibration standard ions. Frame extraction is performed by selecting the region of interest and double-clicking on that range.
5. Find drift times associated with ions in the selected reference set (**step 3**) by clicking on the *Find Drift Times* button in the CCS Calibration (Single-Field) window. Drift times will automatically be populated in the CCS Calibration (Single-Field) window. The coefficients associated with the linear-regression algorithm used to perform the single-field CCS calibration will also be populated (*see* **Note 18**).
6. Click on *Save or Restore*. A new window will open with options to save the CCS calibration. The Single-Field CCS calibration can now be applied to all data files that were run with the same drift parameters as the currently open CCS calibration file. Batch CCS calibration can be performed by selecting the *Save to Multiple Files* option and selecting the files to be calibrated in the *Save to Data File(s)* window. Linear regression parameters defined in **step 5** are now applied to all selected data files.

3.4 Four-Dimensional DTIM-MS Data Extraction Using Ion Mobility Feature Extractor (IMFE)

The ion mobility feature extraction algorithm mines data across four dimensions: mass, drift time, retention time, and abundance. Feature extraction results include drift time, mass, RT, abundance, and CCS for all extracted features; the latter is only done if single-field calibration has been performed. IMFE can also be used in IM-MS Browser software. However, for batch processing in metabolomics applications, extraction must be performed in MassHunter Mass Profiler (MP) as described below. MP can also perform basic, two-parameter differential comparisons with basic statistical tools.

1. Open Agilent MassHunter Mass Profiler and click on the *Create New Project* icon.
2. In the popup window, name the project and designate the experimental groups if applicable (*see* **Note 19**).
3. Add CCS-calibrated sample .d data files to their respective group(s) and click *OK*.
4. In the *Feature Finding* window, select *Chromatographic data*, set the desired retention time windows, set *ion intensity* > *=200.0* counts, use the drop down for *Isotope model* to select *Common organic (no Halogens)*, limit *charge state* to a range of *1—2*, and check the box next to *Report single-ion features with charge state z* = *1* (*see* **Note 20**).

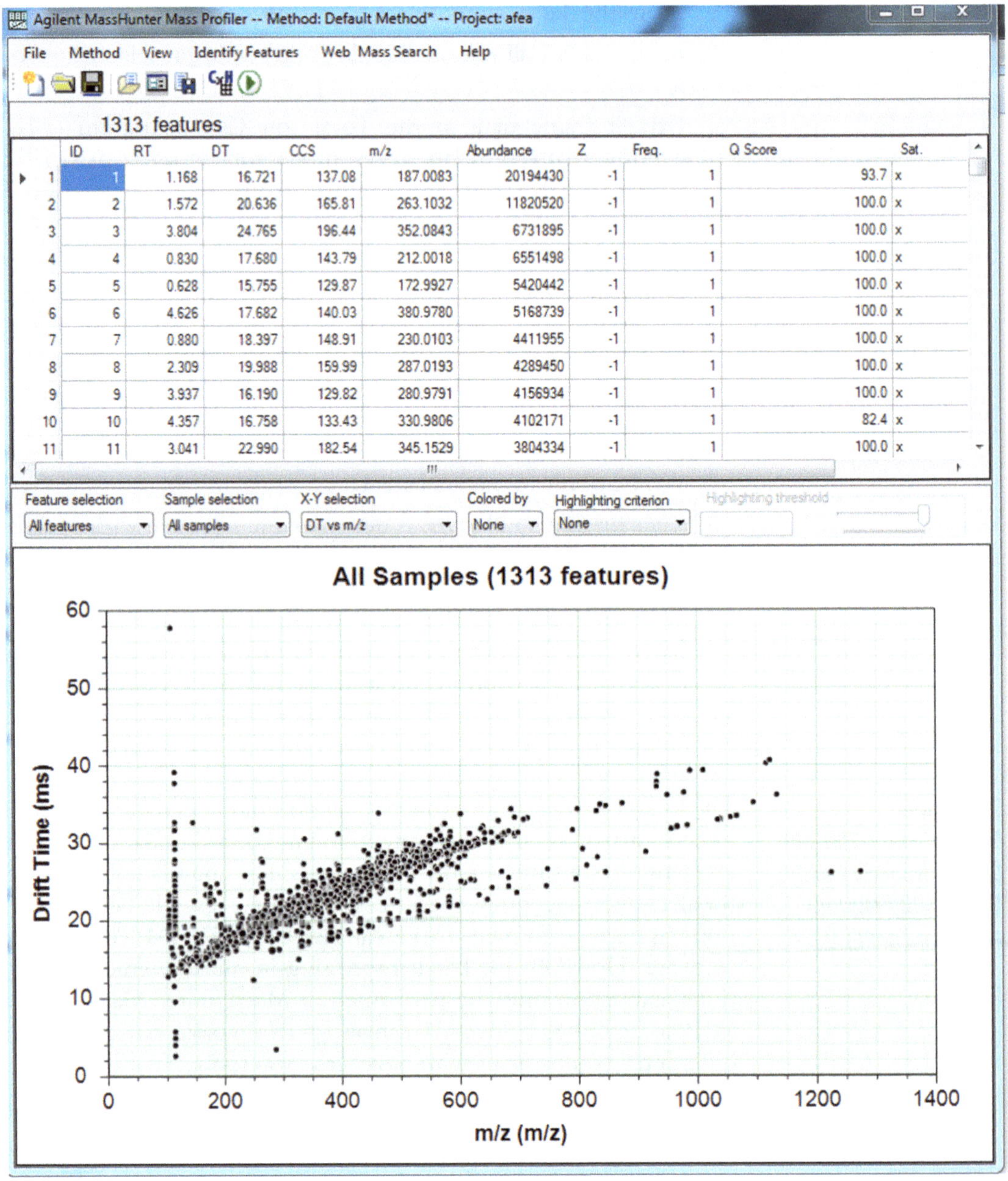

Fig. 8 Agilent MassHunter Mass Profiler visualizing a List of features extracted from the DTIM-MS data file(s), Top: Extracted DTIM-MS features with their respective numerical parameters; Bottom: Drift time vs. m/z 2D plot, each dot represents a features displayed in the top table

5. Click *Run* to start data extraction. Following feature finding, two windows will appear in the MassHunter Mass Profiler software. The top window displays extracted features with their respective numerical parameters (e.g., CCS, DT, RT, etc.). The bottom window provides a 2D plot with selectable *Z* and *Y* values, *see* Fig. 8.

3.5 Molecular Annotation: Collisional Cross Section Accurate Mass and Time (CCS-AMRT) Database Searching

Annotating metabolites and other small molecules by searching conventional metabolomics databases do not always result in accurate results. Not all metabolites will be matched to database entries. Many compounds will have several possible database hits, all with the same formula and/or database search score. And some database hits are simply incorrect. Requiring a retention time match when using accurate mass and time (AMRT) databases can improve confidence in annotations. However, matching chromatographic conditions in analytical methods to those used to create the AMRT databases is not always possible and is often difficult. Adding CCS to create CCS, or CCS-AMRT databases can greatly improve annotation confidence, even if retention time is not included in matching requirements. The protocol below provides details for annotating compounds using a CCS-AMRT database. In this protocol, retention time is not utilized. *See* **Note 23** for details about available CCS databases.

1. In Mass Profiler, click on the *Identify Features menu* and select *For All Features in Table*. This will open MassHunter ID Browser software (*see* **Note 21**).
2. In the Compound Identification Window select *Identify all compounds*. Check the box adjacent to *Database search (CSV, PCD/METLIN)* and click *Next*.
3. In the Search Criteria window, set the mass tolerance to ±5 ppm, check the box next to *CCS* and *Required*. Set the Tolerance to ±5% (Fig. 9) (*see* **Note 22**).
4. Select the desired CCS-AMRT database in the database tab by clicking the gray box next to *Database Path*, then navigate to the desired database (*see* **Note 23**).
5. In the *Positive Ions* table, check the box next to +H and click *Finish* (*see* **Note 24**).
6. A new window will open with Annotation results. Additional information about the compound annotation can be viewed by clicking on a compound in the bottom table.
7. Perform additional searches for unannotated compounds. Highlight all unannotated compounds. Click the green arrow next to *Run ID Wizard*.
8. In the Compound Identification Window select *Identify only highlighted compounds*. Check the box adjacent to *Database search (CSV, PCD/METLIN)* and click *Next*.
9. In the Search Criteria window, set the mass tolerance to ±5 ppm. *Uncheck* the box next to *CCS* and *Required*.
10. In the *Positive Ions* table, select the desired adducts and click *Finish*.
11. To save the results, click "Save and Return."

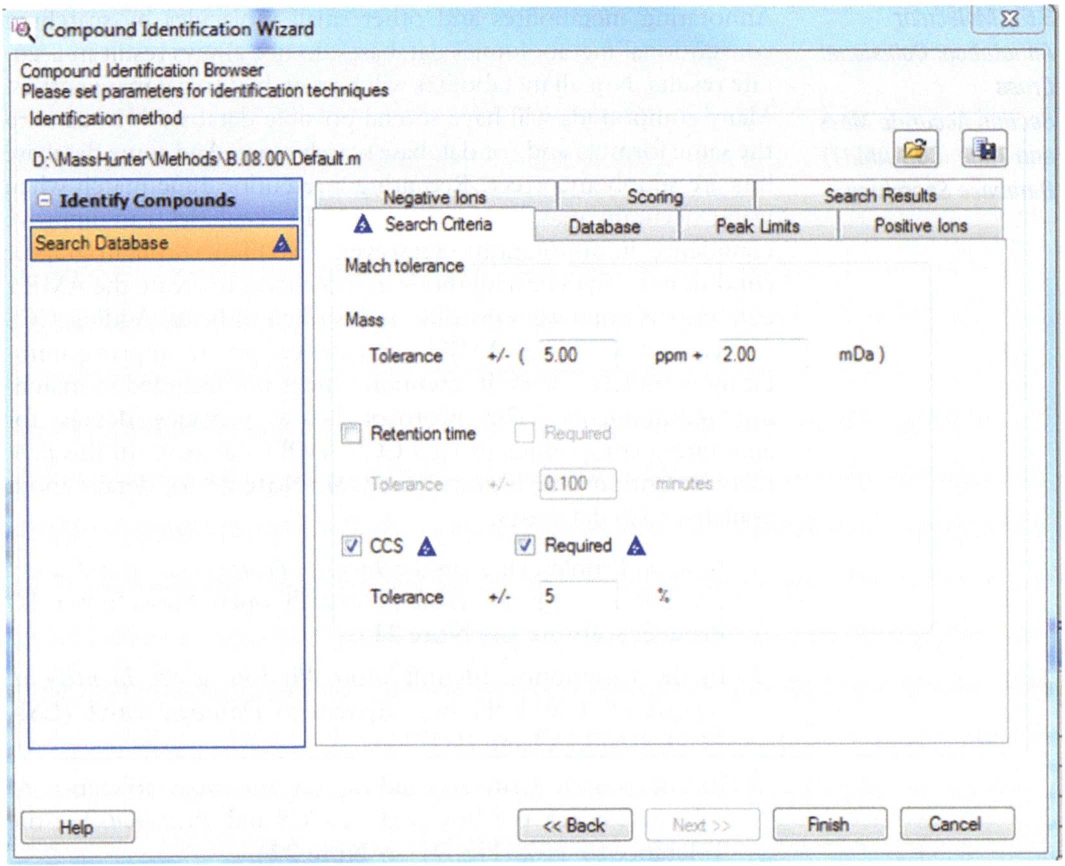

Fig. 9 Agilent MassHunter IDBrowser suggested compound annotation setting used for IM-MS database searching

3.6 Differential Analysis

Differential analysis of metabolomics data is often a complex process. Most data sets are composed of more than two parameters and may contain multiple parameter conditions. Not all data sets are created equal; when statistical analysis is performed, it is important to use appropriate statistical tests and other available tools such as clustering algorithms to obtain meaningful results. Thus, differential analysis is not a quick, one- or two-step process. Mass Profiler contains two basic analysis tools: a student's *t*-test and principal components analysis (PCA). Differential analysis in Mass Profiler can only be performed on two experimental conditions. The protocol below details steps for using these basic tools. For more complex analysis, data can be exported in .CEF format for analysis in Mass Profiler Professional, or in .csv format for analysis in third party software or freeware, e.g., MetaboAnalyst (https://www.metaboanalyst.ca/faces/home.xhtml).

3.6.1 Filter Data by Differential Score (t-Test) and Fold Change Filters

1. From the *Method menu*, select *Edit method*.
2. At the bottom of the method editor, select *All parameters* (Fig. 10).
3. Select the *Statistics and Filters* tab (Fig. 10).
4. For *Missing sample treatment*, select either *Assign 0 abundance* or *Exclude from analysis* (Fig. 10) (*see* **Note 25**).
5. In the *Group difference* section, to filter for compounds that have a higher abundance in group 2 (e.g., treated samples) compared to group 1 (e.g., control samples), click the radio button for *Up* (Fig. 10). Enter a *Fold change* value of your

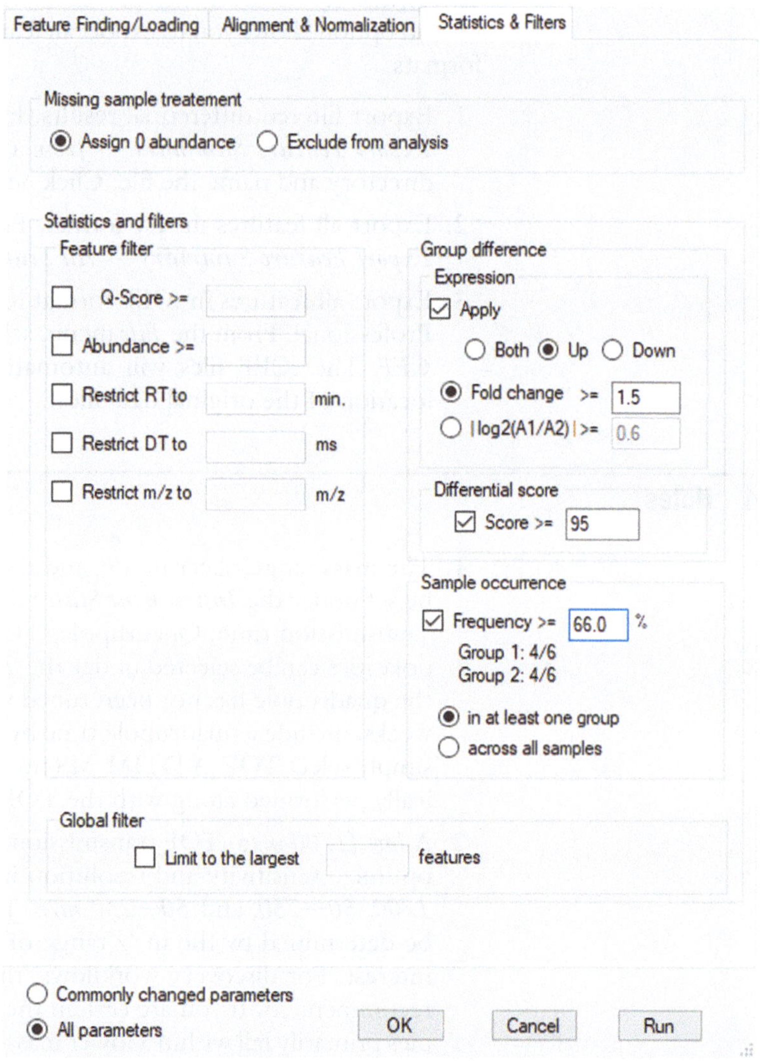

Fig. 10 MassHunter Mass Profiler differential score and fold change filter parameters for the comparison of two experimental conditions

choosing (*see* **Note 26**). Set the Differential score to 95 (*see* **Note 27**).

6. Enter a *Sample occurrence Frequency filter* percentage (Fig. 10) (*see* **Note 28**).
7. Click *Run* to apply the filters.
8. After the filter application has completed, the plot window will display only the features that pass the applied filters. Entities that were annotated in a database will be labeled with the annotation.

3.6.2 Export Results

There are a variety of export options to choose from. Two options are detailed below. The first option allows you to export the filtered differential results from the previous section in .csv format. The second option allows you to export all features in either .csv or .CEF formats.

1. Export filtered differential results. From the *File* menu, select *Export Feature Summary → from Graphed Features.* Select a directory and name the file. Click *Save.*
2. Export all features in .csv format. From the File menu, select *Export Feature Summary → All Features.*
3. Export all features in .CEF format for analysis in Mass Profiler Professional. From the *File* menu, select *Export Each Sample to CEF.* The .CEF files will automatically be exported to the location of the original data files.

4 Notes

1. The mass range, slicer mode, and the instrument mode can all be set under the *Instrument State* tab in the Tune context. The Transmission tune, Quadrupole/TOF/Both options, and the polarities can be selected under the *Tune & Calibration* tab. If the quadrupole has not been tuned within the last two to four weeks, include a quadrupole tune by selecting *Both.* Otherwise simply select TOF. A DTIM-MS transmission tune is automatically performed along with the TOF tune.
2. A *low (1700 m/z)* TOF transmission tune can be performed to optimize sensitivity and resolution in three m/z ranges: *50—1700*, *50—750*, and *50—250 m/z*. The appropriate range will be determined by the m/z range of the molecules of greatest interest. For discovery workflows, the *50—1700 m/z* range is recommended. If you are certain the compounds in your samples primarily fall within a lower mass range, e.g., *50—750 m/z*, selecting a smaller mass range option may provide an improvement in sensitivity and resolution. Regardless of which

transmission tune was performed, any m/z range between 20 and 1700 m/z can be selected for data collection.

A *Fragile Ions* tune option is also available. This option adjusts instrument parameters to reduce the likelihood of "in source" fragmentation. Be aware that using this option will result in a significant loss of signal. Unless you are confident that molecules of interest are likely to fragment under standard conditions, it may be best not to use this option to maintain instrument sensitivity. Setting the TOF slicer mode to High Sensitivity will improve detection of low abundance compounds. If TOF resolution is more critical for your purposes, use the High Resolution mode.

3. These source parameters are a good starting point for LC flow rates between 0.2 and 0.8 mL/min. Optimize as required for your applications. Higher flow rates can be used with uHPLC columns; this will result in narrower peaks, and potentially, better feature extraction results. Higher flow rates can result in reduced sensitivity. Low flow rates tend to produce wider, shorter peaks. The latter can reduce effective sensitivity, particularly for low abundance compounds. The key in choosing a chromatographic method is to match flow rates and column chemistries, to the sample type and complexity.
4. This method can be run in either positive or negative polarity. For positive ion polarity, the source Nozzle voltage should be set between 0 and 500 V. For negative ion polarity, set the source Nozzle voltage to 1000 V. Regardless of polarity, the Nozzle voltage is a compound-dependent parameter, and for certain molecules, sensitivity may be enhanced by voltage optimization.
5. The Frame Rate, in frames per second, is analogous to that of a scan rate measured in spectra per second on TOF instruments. Typical frame rates are set between 1 and 2 frames/s. Higher frame rates produce more data points across chromatographic peaks but at the cost of reduced signal intensity. When 4D feature finding will be used to mine data, it is recommended to use a frame rate between 1 and 2 frames/s to improve extraction of low abundance features.
6. If the Max Drift Time is set too short, a molecule with a large CCS may not exit; the drift tube before the next IM transient (ion packet release) is performed. The molecule will appear to have "wrapped around" into the second IM transient and be assigned a short drift time (i.e., small CCS). This will cause erroneous results in the calculated CCS value. Wrap-around compounds will be observed as a compound with a large m/z and a very small drift time. If the max drift time is set too long, the sensitivity of the instrument will be reduced because of the

relationship between IM and QTOF transients. The optimal drift time will vary based on the CCS of molecules analyzed and the strength of the electric field applied in the drift tube (*see* **Note 11** for additional details on the electric field).

7. If multiplexing is disabled, the optimal trap fill time will depend on quantity and m/z of molecules ionized before space charge, leaking and detector saturation issues can arise. Refer to **Note 9** for additional details.

8. Shorter Trap Release times generally result in lower signal intensity but better IM drift resolving power for molecules [6]. Longer release times generally result in higher signal intensity and slightly lower resolving power for molecules. The trap release time effects will be somewhat compound-dependent.

9. Multiplexing provides increased sensitivity and is recommended for virtually all applications [6, 7]. In DTIM-MS acquisition, packets of ions must be introduced into the drift tube. This is accomplished by using an ion trapping funnel at the entrance of the drift tube. Because delivery of ions into the DTIM-MS is not continuous, the overall sensitivity is reduced compared to a standard time-of-flight mass spectrometer [6]. In theory, ions can be trapped during the full time used for ion mobility separation thus achieving 100% duty cycle. In practice, however, the Trapping Funnel can overfill with ions and reach a space charge limit in a shorter time (microseconds scale) than ion mobility separation is accomplished (millisecond scale). Overfilling results in leakage of ions into the drift tube causing a "smearing" effect resulting in poor IMS resolution; it also increases the likelihood of detector saturation. Low m/z ions experience this space charge limit in a shorter time than larger m/z ions. Multiplexed ion packet release into the drift tube can overcome the space charge limit, ion leakage, and detector saturation [6]. It also increases sensitivity by effectively increasing the duty cycle [7].

10. When multiplexing mode is enabled, tandem MS/MS experiments cannot be performed. When multiplexing mode is disabled, either targeted tandem MS/MS or alternating frame experiments can be performed. Data-dependent MS/MS is not allowed when Ion Mobility mode is active.

11. The drift tube entrance voltage (DTEV) can be set to alter the strength of the electric field applied in the drift tube. This will alter the drift resolving power. The larger the difference between the drift tube entrance and exit voltages, the greater the electric field. Ions move faster through the drift tube in higher electric fields thus decreasing drift peak widths; this generally results in an increase in resolving power [6]. However, there will also be a decrease in the drift time difference between

ions and therefore peak to peak resolution may not improve. The maximum DTEV that can be set is 2000 V, and it is not recommended to go below 1200 V. The max drift time should be adjusted in relation to the strength of the electric field.

12. Low mass and high mass ions show different behavior with respect to the trap RF voltage. Low mass ion signal is optimized at low-mid RF values; ~80–130 V. Mid-High Mass ions are more stable at higher RF values near the default 150 V. Shorter trap fill times require higher trap funnel RF voltages for mid-high mass ions.
13. It is important to use the exact same IM-QTOF Source and Acquisition settings as those in the method used to acquire sample data to generate accurate CCS values for sample compounds.
14. The script, SCP_AcquireCalibrantData, is used to deliver calibrant mix from bottle B of the calibrant delivery system (CDS). The calibration standards are those in the Agilent TOF tune/calibration standard that have been diluted 1:10.
15. The drift time and ultimately the calculated CCS area of a molecule are directly affected by gas type, gas temperature, and gas pressure in the drift tube. In order to obtain reproducible inter-day and inter-laboratory CCS values, the gas temperature and pressure must be as tightly controlled as possible [8]. The instrument should be equipped with a gas kit for this reason. However, even with a gas kit, there still could be minor fluctuations in the gas pressure and temperature in the drift tube over time. Therefore, the single-field calibration method should be run at the beginning of every batch to ensure that CCS values will be accurate and precise throughout an experiment with multiple batches.
16. The PNNL PreProcessor software smooths chromatographic and drift time peaks using moving averages and is freely available on the PNNL website (https://omics.pnl.gov/software/pnnl-preprocessor). While not strictly necessary, smoothing generally reduces four-dimensional feature finding artifacts. Smoothing values must be specified for drift or LC smoothing. The value you enter determines the number of data points used to compute the new smoothed values. For example, a smoothing value of 5 would compute a value by averaging the original data point with the two points before and after it. Optimal smoothing values are dependent on chromatographic peak width and IM drift resolving power. Use the values provided in this protocol as a starting point and adjust as necessary. For poorly separated isomeric species, smoothing can result in the loss of one species.

17. The PNNL PreProcessor will generate a new data file in the same folder as the original data file. The smoothed data will have a name that includes _MA-dn-cn.d. "MA" (moving average) indicates the file was smoothed, "d" represents drift, "c" represents chromatogram, and "n" indicates the smoothing value applied. For example, a data file with the name "example.d" that is smoothed using 3 point smoothing for both the drift spectra and chromatograms will have a new data file created with the name "example.d_MA-d3-c3.d".
18. It may be necessary to manually validate CCS drift times in the single-field calibration. Low intensity or absent tune ions may be erroneously extracted creating errors in the CCS calibration. This can be verified by reviewing drift times (tD (ms)) populated in the CCS Calibration (Single-Field) window and comparing these values to peaks present in the m/z vs DT plot. If necessary, low-level tune ions can be removed from the calibration by right clicking on the selected tune ion and selecting delete. If the signal intensity of the tune ions is too low in general, it will be necessary to reacquire data for the tune ions as well as all samples in the batch.
19. Mass profiler is capable of performing basic statistical analysis on two experimental groups. Group designation is not necessary in Mass Profiler if more than three groups are going to be analyzed or if separate statistical software will be used for data analysis. In order to extract all sample data as a single group, select the option for one group.
20. All parameters can be modified based on the LC-IM-QTOF method being used and sample type being analyzed. Feature extraction performed in Mass Profiler can combine multiple isotopes of a compound into a single "molecular feature." A feature can contain one or more isotopes, but it does not contain multiple adducts or charge states; the algorithm does not collapse different charge carriers into a single compound. The *common organic (no halogens) isotope model* should not be used if a sample is expected to have halogenated compounds. If a halogenated compound is detected in the mass spectra, the common organic (no halogens) isotope model will define the compound as two separate species due to the increased M + 2 peak not traditionally associated with "small" compounds composed of C, H, N, or O. Using the *Report single-ion features with charge state* $z = 1$, option should only be used if known low-intensity compounds are not being extracted with at least two isotopes in the mass spectra. Using this option will result in a greater number of feature extraction artifacts. A *Q*-Score (feature quality) filter will filter out features below a specified quality score. It is recommended that the *Q*-Score filter is not used during feature extraction; it can be applied

after feature extraction has completed. This filter should only be applied at this step if the number of features being extracted by Mass Profiler is preventing the extraction from completing.

21. Database searching can be performed on user-filtered feature lists, rather than on all features in the data set. For example, *Highlighting criteria* can be used to filter features on the feature plot, so that only features that pass the applied filter are visible (colored black). If the option, *For highlighted features on graph*, is selected from the Identify compounds menu, only those features will be included in the database search. Alternatively you can apply differential (or other) filters by selecting *Edit method* from the Method menu.
22. The Agilent 6560 TOF typically operates with a mass error of less than 5 ppm when calibrant ions are detected during the sample run. It is important to perform postacquisition mass calibration on IM-QTOF data prior to feature extraction (Subheading 3.6). The mass tolerance window can be widened if calibrant ions are not detected in the analysis. Matrix effects and detector saturation can also shift mass; a subsequent database search for features not annotated in the first search using a wider ppm tolerance, e.g., 10 ppm, may provide a great number of annotated features. The CCS tolerance of $\pm 5\%$ is the suggested starting point for an instrument with a gas kit that has been calibrated using single-field CCS calibration. If you are using the same chromatographic method (pump configuration, column, flow rate, and mobile phases) used to build the CCS-AMRT database, requiring retention time matches will improve confidence in feature annotations, particularly when CCS values are not available for all features in your data set.
23. Several CCS accurate mass databases are now publicly available. For example, Zheng et al., characterized >500 metabolites and xenobiotics using DTIM-MS [9]. Ridenour characterized phospholipid and peptide standards using traveling-wave ion mobility [10]. Other databases are theoretical [11]. Most databases contain CCS values for separation using either helium or nitrogen gas, while a small portion of labs use argon. In order to use published CCS databases, it will likely be necessary to manually populate your own databases. Databases used for Agilent DTIM-MS compound annotation should be in Agilent Compound Database (.CDB) format. In-house .CDB CCS databases can be created and edited using Agilent PCDL Manager software, which is part of the Agilent software suite. For data collected on the Agilent 6560 IM-QTOF, only databases with CCS values calculated using nitrogen as a drift gas should be used for feature annotation. Not all of the compounds detected in your experiment will be included in CCS databases. Therefore, to add additional annotations to your data set, it is

recommended to follow CCS-AMRT database searches with standard AMRT database searches for compounds not annotated in the CCS-dependent search. Although annotation confidence will not be as high for these compounds, the annotations can provide a starting point for understanding your data using tools such as pathway analysis. In all cases, annotations that support your hypotheses, or help to generate a new hypothesis, should be bolstered using MS/MS spectral libraries and/or synthesized standards.

The Metlin metabolite database, which can be obtained as part of the Agilent software suite, can be used for standard metabolomics database searches, i.e., when CCS is not required. Additional non-CCS databases such as the Human Metabolome Database (HMDB, http://www.hmdb.ca/) and Lipid Maps (https://www.lipidmaps.org/) are freely available and can be converted to compatible formats in either .csv or . CDB.

24. Ion Mobility Feature Extractor does not condense multiple adducts. For common organic molecules, it is likely that +H or –H are the most common adducts for positive and negative ionization modes, respectively. Prior to performing annotation, plausible adducts for compound annotation can be determined by using the Molecular Feature Extractor (MFE) algorithm in Mass Hunter Qualitative Analysis. MFE operates off TOF only data (drift time is not utilized), thus it is able to condense multiple adducts for data review. Following annotation using +H or –H adducts, unannotated compounds can be annotated using additional adducts. If compounds that are known to form nonprotonated adducts are expected, for example, triglycerides primarily forming +NH4 adducts, other charge carriers can be used for the first pass annotation.

25. The Missing Values Treatment options allow you to choose how to handle feature abundance in cases where a particular feature is not present in one or more samples. *Assign 0 abundance* makes the assumption that the compound is actually absent from the sample and not missing due to a feature extraction artifact. *Exclude from analysis* assumes that the missing value is due to a feature extraction artifact. In that case, the "0" value will be excluded for average and *p*-value calculations.

26. Mass spectrometry data are often affected by compression of actual fold changes. For example, a fourfold biological change may be detected as a two- or threefold change in the mass spectrometry results. Therefore, if you are interested in compounds with a twofold change between treatment groups, you may want to apply a filter of 1.5-fold, for example. Keep in mind that when small fold change filters are applied, it is

particularly important to filter fold change results using appropriate statistical significance criteria.

27. In Mass Profiler, the Differential Score is based on a *p*-value. The score is calculated using the formula ($1 - p \times 100$), where p is the desired *p*-value. For example, to apply a significance filter using a *p*-value of 0.05, set the Differential Score to 95 ($1–0.05 \times 100$). To apply a filter using a *p*-value of 0.01, set the Differential Score to 99.

28. It is important to apply a Frequency filter prior to performing statistical analysis. Not all compounds extracted by feature finding algorithms will be extracted efficiently across the entire data set. This is often a result of matrix effects, or low abundance of features. By applying a frequency filter you can remove these compounds from the analysis. This has the benefit of reducing the number of statistical tests being performed which in turn reduces the effects of multiple testing. Data exported in .csv format can be analyzed in third party software. The CEF format is used for analysis in Mass Profiler Professional.

References

1. Heischmann S, Gano LB, Quinn K, Liang L-P, Klepacki J, Christians U, Reisdorph N, Patel M (2018) Regulation of kynurenine metabolism by a ketogenic diet. J Lipid Res 59(6):958–966
2. Cruickshank-Quinn CI, Mahaffey S, Justice MJ, Hughes G, Armstrong M, Bowler RP, Reisdorph R, Petrache I, Reisdorph N (2014) Transient and persistent metabolomic changes in plasma following chronic cigarette smoke exposure in a mouse model. PLoS One 9(7): e101855. https://doi.org/10.1371/journal.pone.0101855
3. Zhang X, Kew K, Reisdorph R, Sartain M, Powell R, Armstrong M, Quinn K, Cruickshank-Quinn C, Walmsley S, Bokatzian S, Darland E, Rain M, Imatani K, Reisdorph N (2017) Performance of a high-pressure liquid chromatography-ion mobility-mass spectrometry system for metabolic profiling. Anal Chem 89(12):6384–6391. https://doi.org/10.1021/acs.analchem.6b04628
4. Cruickshank-Quinn C, Quinn KD, Powell R, Yang Y, Armstrong M, Mahaffey S, Reisdorph R, Reisdorph N (2014) Multi-step preparation technique to recover multiple metabolite compound classes for in-depth and informative metabolomic analysis. J Vis Exp 89:51670. https://doi.org/10.3791/51670
5. Zhang X, Romm M, Zheng X, Zink EM, Kim Y-M, Burnum-Johnson KE, Orton DJ, Apffel A, Ibrahim YM, Monroe ME, Moore RJ, Smith JN, Ma J, Renslow RS, Thomas DG, Blackwell AE, Swinford G, Sausen J, Kurulugama RT, Eno N, Darland E, Stafford G, Fjeldsted J, Metz TO, Teeguarden JG, Smith RD, Baker ES (2016) SPE-IMS-MS: an automated platform for sub-sixty second surveillance of endogenous metabolites and xenobiotics in biofluids. Clin Mass Spectrom 2:1–10. https://doi.org/10.1016/j.clinms.2016.11.002
6. Belov ME, Clowers BH, Prior DC, Danielson Iii WF, Liyu AV, Petritis BO, Smith RD (2008) Dynamically multiplexed ion mobility time-of-flight mass spectrometry. Anal Chem 80(15):5873–5883. https://doi.org/10.1021/ac8003665
7. Liu W, Zhang X, Knochenmuss R, Siems WF, Hill HH (2016) Multidimensional separation of natural products using liquid chromatography coupled to hadamard transform ion mobility mass spectrometry. J Am Soc Mass Spectrom 27(5):810–821. https://doi.org/10.1007/s13361-016-1346-8
8. Stow SM, Causon TJ, Zheng X, Kurulugama RT, Mairinger T, May JC, Rennie EE, Baker ES, Smith RD, McLean JA, Hann S, Fjeldsted

JC (2017) An interlaboratory evaluation of drift tube ion mobility–mass spectrometry collision cross section measurements. Anal Chem 89(17):9048–9055. https://doi.org/10.1021/acs.analchem.7b01729

9. Zheng X, Aly NA, Zhou Y, Dupuis KT, Bilbao A, Paurus VL, Orton DJ, Wilson R, Payne SH, Smith RD, Baker ES (2017) A structural examination and collision cross section database for over 500 metabolites and xenobiotics using drift tube ion mobility spectrometry. Chem Sci 8(11):7724–7736. https://doi.org/10.1039/c7sc03464d
10. Ridenour WB, Kliman M, McLean JA, Caprioli RM (2010) Structural characterization of phospholipids and peptides directly from tissue sections by MALDI traveling-wave ion mobility-mass spectrometry. Anal Chem 82 (5):1881–1889. https://doi.org/10.1021/ac9026115
11. Zhou Z, Tu J, Xiong X, Shen X, Zhu Z-J (2017) LipidCCS: prediction of collision cross-section values for lipids with high precision to support ion mobility–mass spectrometry-based lipidomics. Anal Chem 89 (17):9559–9566. https://doi.org/10.1021/acs.analchem.7b02625

Chapter 4

Drift-Tube Ion Mobility-Mass Spectrometry for Nontargeted 'Omics

Tim J. Causon, Ruwan T. Kurulugama, and Stephan Hann

Abstract

This chapter describes the developments in drift-tube ion mobility-mass spectrometry (DTIM-MS) that have driven application development in 'omics analyses. Harnessing the additional, orthogonal separation that DTIM provides increased confidence in compound identifications as the mass spectral complexity can be reduced and mobility-derived parameters (most prominently the collision cross section, *CCS*) used to support identity confirmation goals for a variety of 'omics application areas. Presented within this contribution is a methodology for improving the transmission and maintaining accurate determination of drift time-derived *CCS* (^{DT}CCS) for low molecular weight compounds for a typical nontargeted 'omics (metabolomics) workflow using liquid chromatography in combination with DTIM-MS.

Key words Collision cross section, *CCS* databases, Drift-tube ion mobility-mass spectrometry, Ion mobility, Mass spectrometry, Metabolomics, Non-targeted analysis, Small molecules

1 Introduction

Ion mobility-mass spectrometry (IM-MS) has evolved from a research technique into a viable commercial option for analytical laboratories during the last 10 years. Key factors enabling this transformation have been research-driven improvements in instrumental design, better understanding of fundamental mobility behavior, and a continuing strong focus on validation of analytical methods using this technology within various application areas. One of the key areas where IM-MS has demonstrated substantial potential is within the scope of "'omics" analysis whereby full-scan MS information (typically in combination with a chromatographic separation) can be supported by mobility-related information in order to better annotate and finally identify molecules from within complex samples of biological origin.

This chapter focuses on the development and use of low-field drift-tube IM-MS (DTIM-MS) for 'omics applications where

Giuseppe Paglia and Giuseppe Astarita (eds.), *Ion Mobility-Mass Spectrometry: Methods and Protocols*, Methods in Molecular Biology, vol. 2084, https://doi.org/10.1007/978-1-0716-0030-6_4,

successful analytical campaigns are crucial for generation of high-value data sets suitable for statistical evaluation. Like other IM-MS technologies, major developments in instrument design have rendered low-field DTIM-MS as a platform with significant potential for ′omics applications. Fundamentals behind key instrumental settings for low-field DTIM-MS including maintaining mass accuracy, avoiding saturation events, improving transmission of small molecules, and employing quality control steps are introduced and methodology provided. Finally, the generation of ^{DT}CCS values traceable to a set of consensus calibrant set is highlighted to be of paramount importance for the development of library entries for different ion species in supporting ′omics applications.

1.1 Ion Mobility-Mass Spectrometry

1.1.1 Basic Characteristics of Atmospheric and Low-Pressure Drift Cells

Ion mobility spectrometry is collectively described as a gas phase ion separation technique that utilizes an electric field for ion transport through a buffer gas. The early experiments that describe the movement of ions through a gas under the influence of an electric field are first reported by Zeleny in 1898 [1]. However, ion mobility as an analytical technique was developed only after McDaniel and coworkers developed theoretical basis for this phenomenon in early 1970s [2, 3]. Coupling of ion mobility to mass spectrometry (IM-MS) was first reported by Cohen and Karasek [4], and the application of IM-MS to provide answers for analytical questions was first reported by Bowers and coworkers [5, 6]. After the introduction of soft ionization methods such as electrospray ionization (ESI) and matrix-assisted laser desorption ionization (MALDI), the use of IM-MS for bioanalytical applications progressed rapidly [7]. There are three major ion mobility techniques which depend on the nature of the electric field used and the nature of the drift gas used. The most common technique is the low-field ion mobility operated using a drift tube filled with a static inert buffer gas. The field-asymmetric waveform ion mobility (FAIMS or differential mobility spectrometry (DMS)) technique utilizes both low and high electric fields and also a device with flowing drift gas. The third technique, differential mobility analyzer (DMA), includes the use of a low electric field and a flowing drift gas orthogonal to the direction of electric field [8]. For conventional low-field IM instruments, the mechanism of ion separation is based on the collisions between the analyte ions and the neutral buffer gas molecules, where ions with larger collision cross sections experience higher number of collisions and traverse more slowly through the drift cell compared to the smaller ions. Several good reviews covering these topics in more detail have been published in recent years [9–12].

Ion mobility spectrometry has become a powerful analytical tool in complex sample analysis due to the introduction of several commercial IM-MS instruments. Currently there are three different variants of low-field ion mobility instruments in the market.

The conventional drift-tube ion mobility spectrometry (DTIMS) coupled to a TOF mass spectrometer, traveling wave ion mobility spectrometry (TWIMS) coupled to a TOFMS, and trapped ion mobility spectrometry (TIMS) coupled to a TOFMS.

1.2 Drift-Tube Ion Mobility-Mass Spectrometry

There are two primary categories of drift-tube ion mobility-mass spectrometry (DTIM-MS) instruments which operate at low pressures (<10 Torr) or with atmospheric pressure drift cells. Most of the low-pressure drift-tube instruments operate around 2–4 Torr drift gas pressure to provide stability and good transmission. Atmospheric pressure DTIM-MS instruments have low duty cycle and, therefore, low sensitivity, but provide higher drift resolving powers compared to low-pressure drift-tube instruments with similar drift cell lengths. The ion introduction mechanism into the drift tube is also different between these two techniques. DTIM-MS instruments operating at higher pressures use an ion-gating mechanism to introduce ion packets into the drift tube usually employing a Bradbury-Nielsen gate (BNG) [13], while low-pressure DTIM-MS instruments primarily use ion funnel technology developed by Smith and coworkers for ion accumulation and release to enhance duty cycle [14–18]. Atmospheric pressure DTIM-MS instruments can be equipped with either Hadamard- or Fourier Transform-based multiplexing techniques in conjunction with ion funnel traps to improve dynamic range and duty cycle up to 50% based on a pseudorandom sequence-based ion-pulsing mechanism [19]. Conventional, low-pressure DTIM-MS instruments with ion funnel technology provide several advantages related to the overall sensitivity of the instrument as well as for accurate collision cross section calculation for analyte ions.

1.2.1 Key Theoretical and Instrumental Design Aspects for DTIM-MS

The only commercially available low-pressure DTIM-MS instrument consists of a high-resolution uniform field drift cell coupled to a high-resolution quadrupole-TOF mass spectrometer (Fig. 1). This instrument utilizes three ion funnels to improve ion throughput, two prior to the drift cell and one immediately behind the drift cell. Ions generated via an atmospheric ESI source are carried in to the first ion funnel through a resistively coated glass capillary. The ESI source voltage is maintained at ground potential while the entrance to the capillary is maintained at high voltage to generate gas phase ions. The drift cell is composed of stacked ring electrodes connected with a resistor series to allow a uniform electric field across the drift cell, a coupling region with transmission ion funnel, hexapole, or quadrupole ion guides and a high-resolution quadrupole-TOF mass spectrometer. Coupling of ion mobility spectrometry with TOF mass spectrometry is possible due to the well-matched transient times for the two techniques. DTIM works on a millisecond time scale (usually <50 ms) while flight time for a single TOF experiment occurs in microsecond time scale (usually

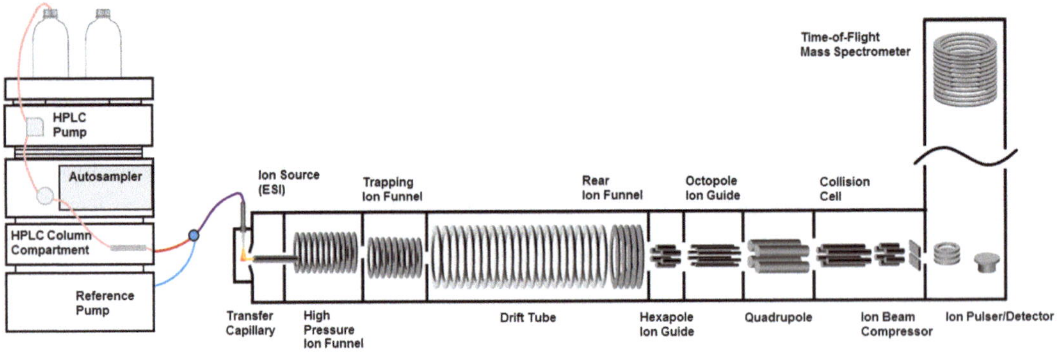

Fig. 1 A schematic representation of a commercially available DTIM-MS instrument (Agilent 6560 IM-QTOFMS) with combined HPLC modules suitable for 'omics applications. Reference ions can be infused for TOFMS mass axis calibration via an additional pump

about 100 μs). Therefore, hundreds of TOF transients can be collected for any given DTIM experiment. Typical drift peaks are in the order of 0.5–1.0 ms FWHM allowing a minimum of 5–10 data points across a given ion mobility peak. DTIM-MS instruments can be also coupled to liquid chromatography (LC) due to the relatively slow liquid phase separation, with peak widths of at least several seconds. This allows the sampling of an LC peak with sufficient frequency to obtain multiple data points including drift time and mass spectral information across the LC peak.

In a DTIM-MS instrument, the velocities of ions are proportional to the applied electric field, and the proportionality constant is the mobility constant (Eq. 1).

$$v_{\mathrm{d}} = KE, \quad (1)$$

where v_{d} is ion velocity, K is ion mobility, and E is applied electric field.

In this way, ion mobility can be used as a physical parameter to characterize analyte ions based on the experimental parameters as given in Eq. 2:

$$K = \frac{v_d}{E} = \frac{L^2}{t_d V} \quad (2)$$

where L is the length of the drift cell, t_{d} is the drift time of the analyte ion, and V is the voltage across the drift cell.

Ion mobilities are typically reported as the reduced mobility (K_0) values where temperature (T) and the pressure (p) of the buffer gas are normalized to standard temperature (273.15 K) and pressure (101,325 Pa or 760 Torr) to facilitate easy comparison of mobility values between experiments ($K_0 = K[p/101325 \times 273.15/T]$). Generally, DTIM-MS instruments are operated under low-field limits, where ion mobility (K) is independent of the electric field used. Under this low-field condition, average collision cross sections can be calculated directly from DTIM-MS measurements using the Mason–Schamp Eq. 3 [20].

$$\Omega = \frac{3}{16}\frac{ze}{N_0 K_0}\sqrt{\frac{2\pi}{\mu k_B T}} \tag{3}$$

where Ω is the diffusion collision integral (CCS), ze is the charge of the ion (z is the integer charge), N_0 is normalized number density, μ is reduced mass of the ion-neutral gas pair, k_B is the Boltzmann constant, and T is absolute temperature of the buffer gas.

To obtain highly accurate collision cross section values for analyte ions, it is very important to precisely control the pressure and the temperature of the buffer gas. Also, the length of the drift cell has to be known very accurately.

For DTIM-MS instruments, the drift resolving power is a function of three parameters: initial ion pulse width, space charge-related peak broadening, and the diffusion-limited peak broadening. This can be expressed by the following Eq. 4.

$$\Delta t = \sqrt{t_{\text{init}}^2 + \left(\frac{t_{\text{drift}}}{R_d}\right)^2 + t_{\text{sc}}^2} \tag{4}$$

where Δt is the time dispersion for a given peak at the end of the drift cell, t_{init} is the initial pulse width, R_d is the diffusion-limited resolving power, and t_{sc} is the peak broadening related to the coulombic repulsion.

Usually, the method parameters for a given DTIM experiments are selected to reduce the contributions from the initial pulse width and the space charge-related peak broadening. Under those conditions, the resolving power is primarily governed by the diffusion-limited resolving power (R_d) given in the following Eq. 5.

$$R_d \sqrt{\frac{LEze}{16 k_B T \ln 2}} \tag{5}$$

1.2.2 From Continuous Beam to Discrete Ion Packets

Dispersive analytical techniques such as ion mobility, time of flight mass spectrometry, and liquid or gas chromatography require discrete ion packets or sample "plug" introduction. This inherently reduces the duty cycle and throughput of these techniques. For atmospheric pressure DTIM-MS, the only feasible ion introduction method is based on the BNG ion gate. This results in <1% ion utilization efficiency, unless an ion multiplexing method is used as described previously. However, for low-pressure DTIM instruments, ion funnel trap technology can be used to accumulate ions prior to pulsing into the drift tube. This allows theoretical ion utilization efficiency approaching 100%. The current state-of-the-art ion funnel traps can store about 10^7 charges. However, when analyzing very complex biological samples, the total ion utilization for a given IM experiment much lower, especially if the avoidance of detector and data acquisition system saturation is desired (*see* Subheading 3.1). As described previously, ion multiplexing methods can also be utilized to overcome this low-throughput issue in conventional IM experiments where a single ion packet is introduced in to the drift tube for each mobility transient.

1.2.3 Assessing Dynamic Range and Sensitivity in MS Only and IM-MS Modes

All hybrid IM-MS instruments can (in theory) operate in IM-MS mode as well as in "MS only" modes. When the ion pulsing before the drift tube is disabled and the drift tube is operated in the ion transmission mode, the instrument essentially operates as a full scan mass spectrometer (typically QTOF or TOF mode). Due to the changes in ion current in IM mode versus QTOF/TOF mode, the dynamic range of the instrument in these two modes is different. Usually, the QTOF/TOF mode dynamic range is higher due to the larger number of transients summed for a given experiment and also due to relatively low ion currents per TOF transient. For IM-MS mode, since the ion signal is preconcentrated and pulsed into the drift tube for mobility separation, the instantaneous ion concentration at the detector is much higher and therefore leads to an earlier onset of detector and data acquisition system saturation. For QTOF mode operation, typically a few thousand TOF scans are summed per one second of data acquisition (i.e., 1 spectrum/s), while for IM-MS mode operation, only 10–20 TOF scans can be summed for any given m/z (due to additional ion mobility separation). This fewer number of summed TOF scans reduces the signal dynamic range. However, the sensitivity remains high due to the ion accumulation before IM-MS experiments.

1.2.4 DTIM-MS for 'Omics Studies

The pioneering work of the group of several groups focused on proteomics workflows with their implementations of coupling drift-tube ion mobility with time of flight mass spectrometry [21, 22]. The potential demonstrated in these early works was groundbreaking as the DTIM separation allows peptides to be resolved by charge state envelopes, and a subsequent increase in proteome coverage is apparent according to the number of successfully annotated proteins. Since these early works, a focus on other 'omics fields including lipidomics [23–26], glycomics [27], exposomics [28, 29], and metabolomics [30–34] has emerged demonstrating the interest in applying this technology for other challenging application areas. Particular focus in these works has been placed on the accurate derivation and addition of ^{DT}CCS values for identity confirmation based on library building. This aspect remains a critical focus for use of DTIM-MS in 'omics applications as the chromatographic separation remains superior to current DTIM-MS technology for realization of isomer separations [35, 36].

2 Materials

2.1 ^{DT}CCS and Mass Calibration Standards

1. *Compounds for single-field ^{DT}CCS and m/z calibration.*
   As the stepped-field approach of DTIM-MS under low-field conditions is considered as a primary method of measurement, calibrant ions with *CCS* values measured using this approach should be employed for further single-field measurements. Moreover, chosen calibrant ions should be measured within round robin studies to provide best estimates of their true values conditional to the type of measurement employed. An example of this is found in [33] where a commercially available tune mix covering a *m/z* range of 113–2834 have been established as conditional reference standards in nitrogen drift gas ($^{DT}CCS_{N2}$) in both positive and negative ionization modes. Analysis of additional "in-house" standards may also be used for quality control purposes during the analytical sequences (*see* Subheading 3.5).
2. *Online m/z recalibration solution.*
   A suitable set of mass axis calibrant ions should be concomitantly infused with the chromatographic effluent to allow post hoc recalibration of the mass axis. The choice of "lock masses" for this set must suit the algorithm used for the mass recalibration, which often requires a low and high mass to allow determination of two mathematical terms for TOF recalibration (for example). The infusion of this solution should not substantially dilute the chromatographic eluent, and the abundances must remain below the saturation threshold of the TOFMS. An example of a suitable infusion setup is included in Fig. 1.
3. *Chemicals for mobile phase and sample preparation.*
   All solvents used for the liquid phase separation (typically reversed-phase liquid chromatography) should be MS grade or equivalent. For example, LC-MS grade acetonitrile, methanol and water, ammonium formate, and formic acid should be used to prepare mobile phases and for sample dilutions. Care should be taken with cleaning practices, and exchanging of solvent bottles as contamination of the mobile phase can severely impair interpretation of full-scan DTIM-MS data in typical metabolomics applications.

3 Methods

3.1 Tuning for Low Molecular Mass Ions

1. The majority of analytes studied in metabolomics, exposomics and related environmental applications fall in the mass range of 50–300 *m/z*. Therefore, such applications require IM-MS instruments to be optimized for the transmission of and

sensitivity for low mass ions. Mass spectrometers coupled with ion mobility spectrometry and operated in IM-MS mode need careful optimization of ion accumulation/trapping and releasing mechanisms. The only commercially available low-pressure DTIM-MS instrument utilizes an ion funnel for ion accumulation and release into the drift tube for ion mobility separation. This instrument can also operate in QTOF-only mode where the ion funnel trap and drift tube are essentially operated as an ion guides with ion trap and release mechanism disabled. The relative ion signal for a given ion in QTOF-only mode and IM-MS mode depends on several factors. For IM-MS mode, the instantaneous ion signal at the detector is much higher compared to the QTOF-only mode due to the ion accumulation/concentration in the ion funnel trap prior to ion release into the drift tube for ion mobility separation. However, due to limited dynamic range of the microchannel plate-based TOF detectors and analog data acquisition systems, the apparent absolute signal is lower for IM-MS mode. In general, for QTOF-only mode, the detector dynamic range is about four orders of magnitude while, for IM-MS mode, it is about three orders of magnitude under best conditions. Therefore, due to this dynamic range disadvantage as well as space charge-related ion signal discrimination for low mass ions in the ion funnel trap, metabolomics experiments require special ion funnel trap optimization [37]. In addition to optimizing the transmission of small molecules, the sample dilution level must also be adjusted in order to avoid overfilling of the ion funnel trap and/or acquisition system saturation.

2. Optimization of the ion funnel conditions for improving transmission of small molecules is shown in Fig. 2. Transmission of

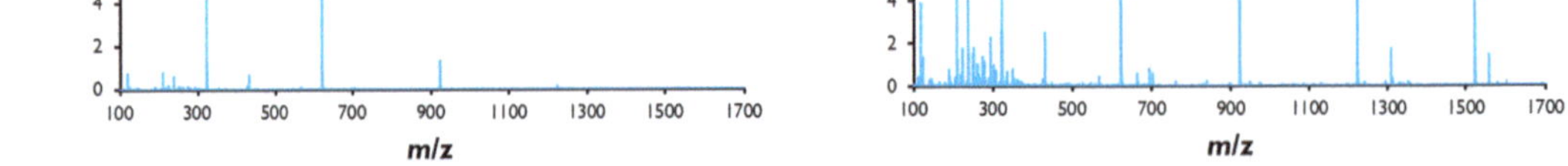

Fig. 2 Improvement of transmission of small molecules using DTIM-MS. (**a**) Result of tune ion infusion following standard tuning using ion funnel trap RF of 80 V, trap fill time of 20 ms, and trap release time of 150 μs. (**b**) Result of tune ion infusion using optimized ion funnel trap RF of 120 V, trap fill time of 15 ms, and trap release time of 300 μs

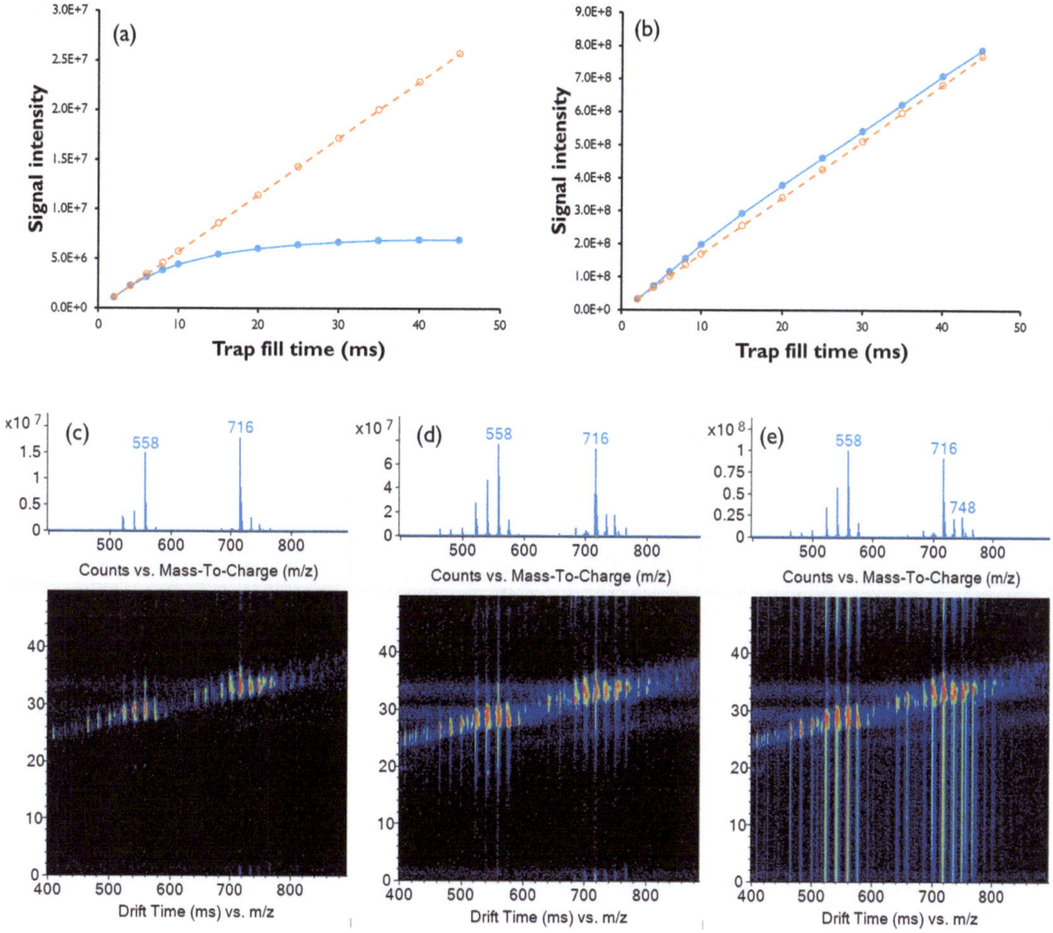

Fig. 3 Examples of DTIM-MS system saturation: (**a**) diagram of peak abundance vs. trap fill time for tetraethylammonium bromide ion (*m/z* 130) indicating overfilling of ion funnel trap (**b**) total ion current data for the same experiment indicate no concomitant saturation of the detection system. Dashed lines represent theoretically optimum transmission; solid lines are experimental data. (**c–e**) Show the deleterious effect of trap overfilling using 1 mg/mL erythromycin with 1.00, 20.0, and 40.0 ms trapping times, respectively

small molecules is improved by optimization of the ion funnel trap RF and increasing the trap release time.

3. Following optimization, exemplary samples or compounds should be evaluated using the chosen DTIM-MS conditions. In Fig. 3, the deleterious effects of ion funnel trap and TOFMS detector saturation are shown, which must be addressed during method development for 'omics applications by optimization of sample dilution, reduction of injection volume, and/or reduction of the trapping time.
4. Finally, when options associated with MS tune parameters are exhausted, implementation of modified Hadamard transform multiplexing with hardcoded pseudorandom sequences can

also be used to improve ion utilization efficiency and reduce trapping bias for samples characterized by a high dynamic range [39]. In such implementations, a maximum of ~50% ion utilization efficiency can be realized with trapping times up to a maximum of 1/15th of the IM maximum drift time (i.e., for a 4-bit pseudorandom sequence resulting in eight ion pulsing events). The methodology described for all data acquisition steps is exactly the same with the exception that all multiplexed data files (including ^{DT}CCS calibration files) must be demultiplexed prior to data processing steps (*see* Subheading 3.6).

3.2 Single-Field ^{DT}CCS Calibration

1. Calibration for single-field operation involves measurement of a mixture of ions covering the desired m/z and mobility range to be measured. Calibration ions should be well-characterized consensual ^{DT}CCS standards that are supported by round robin testing under specified conditions of measurement, which includes the type of drift gas, adherence to low-field conditions, and knowledge of voltage settings that influence zones of the instrument where ion mobility processes occur (*see* **Note 1**).
2. A short (<1 min) infusion is performed prior to the start of sample measurements and the linear regression results applied to all datafiles within the same sequence. An exemplary single-field calibration function is shown in Fig. 4.

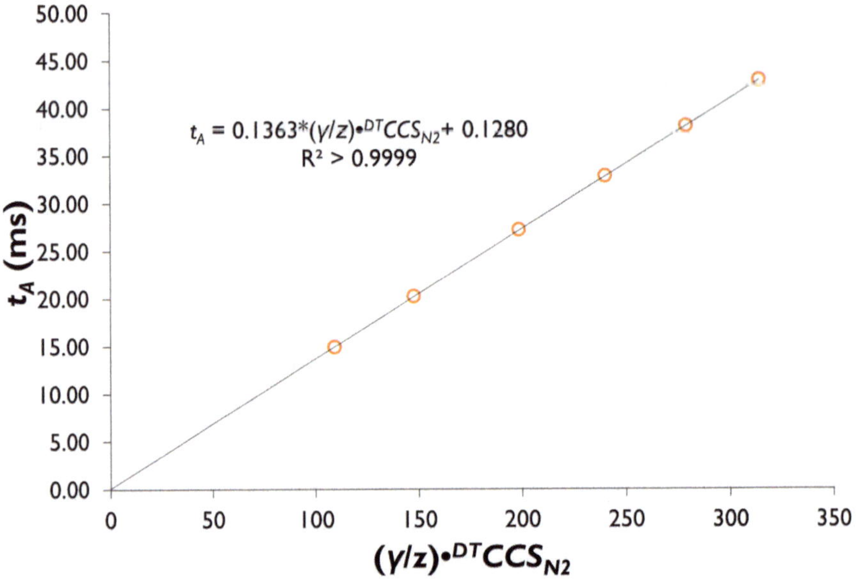

Fig. 4 Example of single-field calibration function used for derivation of ^{DT}CCS values via DTIM-MS. The reduced mass term (γ) is given by $[m_i/(m_g + m_i)]^{1/2}$, where m_i is the mass of the calibrant ion and m_g is the mass of the drift gas

3.3 System Suitability Check: LC-DTIM-MS with Well-Characterized Standards

1. In addition to the preliminary calibration, a mixture of well-characterized independent standards should be selected and employed as a quality control sample for the single-field calibration. Samples should only be prepared following successful check-out with this sample including the parameters use for identity confirmation: retention time, ^{DT}CCS, and accurate mass (including chemical formula annotation). This sample should be run using the full analytical method employed to ensure that system performance meets expectations prior to the commencement of an analytical sequence.

3.4 Preparation of Samples and Pooled Quality Controls

1. Sample preparation for ′omics studies must follow established protocols particularly where temperature or other environmental changes may change the sample significantly (e.g., analytes with fast metabolic turnover rates). Samples must finally be prepared in high-quality solvents amenable to the mode of chromatography employed (e.g., primarily aqueous for reversed-phase HPLC).
2. Additional quality control samples should also be prepared, which are typically a matrix blank (if available) and one or more pools of samples representing the investigated population prepared simply by aliquoting equal amounts of each sample into a single vial.
3. Finally, samples should be transferred to a cooled sample tray while awaiting injection and analysis to minimize potential temperature-induced changes to the samples.

3.5 Performing Nontargeted Metabolomics Using LC-DTIM-MS

1. Within most nontargeted metabolomics measurement campaigns, the sample order should be randomized or semirandomized according to knowledge of the samples under investigation.
2. In addition to periodic measurements of procedural blanks and various types of QC and reference samples, a pooled sample is often measured multiple times at the end of a sequence with the goal of annotation of high abundance metabolites using high-resolution fragment (QTOF) data and application of suitable databases.
3. Like mass accuracy and retention time stability, ^{DT}CCS stability should be included as part of the QC assessment to ensure that relevant instrument parameters (e.g., drift-tube pressure and temperature) are suitably stable for accurate determination of ^{DT}CCS for all samples across the entire sequence. A typical analytical sequence will comprise up to 20% QC samples with a minimum of three replicates per QC required for statistical considerations.

4. For retention time and m/z abundance comparability, batch effects may also need to be considered for measurements made at wide intervals or following instrument modification or service. By following the appropriate single-field calibration and system suitability check, the ion mobility parameters should remain within expected tolerances in such cases. In any case, suitable QC and data evaluation standards relevant to the type of investigation should be sought in literature and adhered to both publication requirements and community acceptance of data (e.g., [38–41]) (*see* **Note 2**).

3.6 Data Preprocessing

1. Some important measurement parameters should be checked for collected data to ensure the accuracy of IM-MS information. The stability of key parameters including drift-tube temperature and pressure should be inspected to ensure that the single-field calibration is not biased for any measurement files. Temperature and pressure are key parameters for the accurate measurement of arrival time and maximum deviations of <0.5% and <0.2% (respectively) should be observed in order to maintain the required accuracy of ^{DT}CCS determination (Fig. 5).
2. Following this step, all datafiles should be made trackable regarding the calibrations applied for m/z and ^{DT}CCS accuracy. Datafiles are typically postprocessed to recalibrate the mass axis using the reference masses of calibrant ions infused

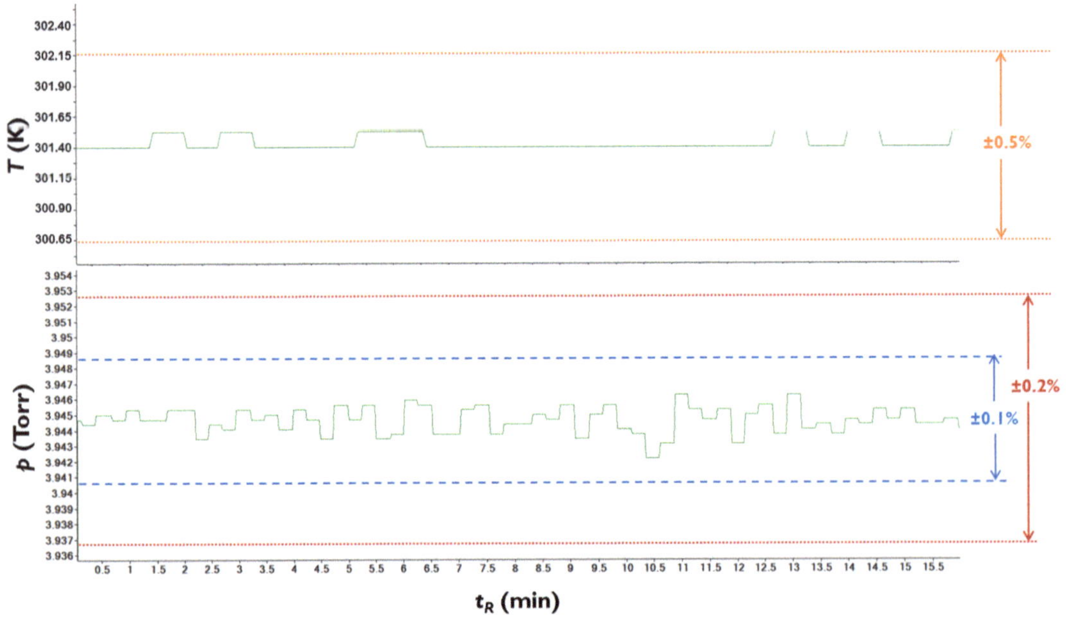

Fig. 5 Experimental stability of measured drift gas temperature and pressure during 16 min acquisition from LC × DTIM-MS analysis. Blue, green, and orange intervals indicate approximate deviations of ±0.1%, ±0.2%, and ± 0.5%, respectively

during the measurements prior to the ^{DT}CCS calibration. In this regard, it is important that the abundances of these ions do not exceed detector saturation limits during the LC gradient as the post hoc recalibration will be biased according to the determination of the centroid of reference peaks.

3. The $^{DT}CCS_{N2}$ values for metabolites are determined using a single-field calculation following determination of calibration parameters (slope, intercept) from the measurement of a series of calibrant ions (Fig. 4) with conditional reference $^{DT}CCS_{N2}$ values (e.g., [33]). A plot of the arrival time (t_A) against a parameter derived from the reduced mass and reference $^{DT}CCS_{N2}$ values yields a linear function which is applied to all datafiles to allow direct determination of all $^{DT}CCS_{N2}$ values (*see* **Notes 1** and **3**).
4. Further data processing and statistical evaluation can then be undertaken (Fig. 6). Vendor tools available for peak picking and annotation of molecular features are available, but other options capable of treating such multidimensional datafiles are also available (or emerging) that are also suitable for such tasks [42, 43].

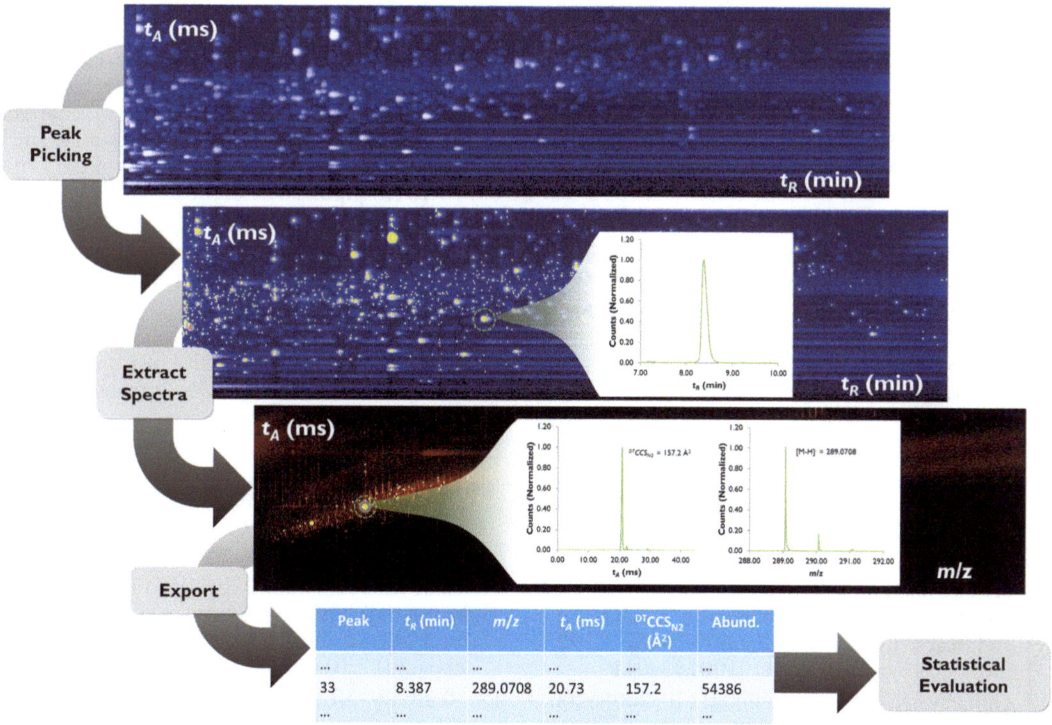

Fig. 6 Schematic representation of molecular feature finding (peak picking) for three-dimensional LC × DTIM-MS data and exporting for multivariate statistical analysis

4 Notes

1. Treatment of compounds with very large calculated ^{DT}CCS values with respect to their m/z should be evaluated carefully. Fragmentation after the drift-tube separation can occur and will hinder annotation.
2. Collection of additional data without DTIM (i.e., TOF/ or QTOF only) can be valuable to address sample sets with a large dynamic range.
3. General recommendations for reporting of data from IM-MS experiments have been recently published to promote development and standardization [44].

Acknowledgements

Vienna Business Agency and EQ BOKU VIBT GmbH are acknowledged for providing mass spectrometry instrumentation.

References

1. Zeleny J (1898) VI. On the ratio of the velocities of the two ions produced in gases by Röntgen radiation; and on some related phenomena. Lond Edinb Dublin Phil Mag J Sci 46:120–154. https://doi.org/10.1080/14786449808621173
2. McDaniel EW, Mason EA (1973) The mobility and diffusion of ions in gases. Wiley, Hoboken, NJ
3. Mason EA, McDaniel EW (1988) Transport properties of ions in gases. Wiley, Hoboken, NJ
4. Cohen MJ, Karasek FW (1970) Plasma chromatography™—a new dimension for gas chromatography and mass spectrometry. J Chromatogr Sci 8:330–337. https://doi.org/10.1093/chromsci/8.6.330
5. Bowers MT, Kemper PR, von Helden G, van Koppen PAM (1993) Gas-phase ion chromatography: transition metal state selection and carbon cluster formation. Science 260:1446–1451
6. von Helden G, Hsu MT, Gotts N, Bowers MT (1993) Carbon cluster cations with up to 84 atoms: structures, formation mechanism, and reactivity. J Phys Chem 97:8182–8192. https://doi.org/10.1021/j100133a011
7. Guharay SK, Dwivedi P, Hill HH Jr (2008) Ion mobility spectrometry: Ion source development and applications in physical and biological sciences. IEEE Trans Plasma Sci 36:1458–1470. https://doi.org/10.1109/TPS.2008.927290
8. Rus J, Moro D, Sillero JA, Royuela J, Casado A, Estevez-Molinero F, Fernández de la Mora J (2010) IMS–MS studies based on coupling a differential mobility analyzer (DMA) to commercial API–MS systems. Int J Mass Spectrom 298:30–40. https://doi.org/10.1016/j.ijms.2010.05.008
9. Kanu AB, Dwivedi P, Tam M, Matz L, Hill HH Jr (2008) Ion mobility-mass spectrometry. J Mass Spectrom 43:1–22. https://doi.org/10.1002/jms.1383
10. May JC, McLean JA (2015) Ion mobility-mass spectrometry: time-dispersive instrumentation. Anal Chem 87:1422–1436. https://doi.org/10.1021/ac504720m
11. Ewing MA, Glover MS, Clemmer DE (2016) Hybrid ion mobility and mass spectrometry as a separation tool. J Chromatogr A 1439:3–25. https://doi.org/10.1016/j.chroma.2015.10.080
12. Gabelica V, Marklund E (2018) Fundamentals of ion mobility spectrometry. Curr Opin Chem Biol 42:51–59. https://doi.org/10.1016/j.cbpa.2017.10.022
13. Bradbury NE, Nielsen RA (1936) Absolute values of the electron mobility in hydrogen. Phys Rev 49:388–393. https://doi.org/10.1103/PhysRev.49.388

14. Shaffer SA, Tang K, Anderson GA, Prior DC, Udseth HR, Smith RD (1997) A novel ion funnel for focusing ions at elevated pressure using electrospray ionization mass spectrometry. Rapid Commun Mass Spectrom 11:1813–1817. https://doi.org/10.1002/(SICI)1097-0231(19971030)11:16<1813::AID-RCM87>3.0.CO;2-D
15. Tang K, Shvarisburg AA, Lee H-N, Prior DC, Buschbach MA, Li F, Tolmachev AV, Anderson GA, Smith RD (2005) High-sensitivity ion mobility spectrometry/mass spectrometry using electrodynamic ion funnel interfaces. Anal Chem 77:3330–3339. https://doi.org/10.1021/ac048315a
16. Page JS, Tolmachev AV, Tang K, Smith RD (2006) Theoretical and experimental evaluation of the low m/z transmission of an electrodynamic ion funnel. J Am Soc Mass Spectrom 17:586–592. https://doi.org/10.1016/j.jasms.2005.12.013
17. Baker ES, Clowers BH, Li F, Tang K, Tolmachev AV, Prior DC, Belov ME, Smith RD (2007) Ion mobility spectrometry—mass spectrometry performance using electrodynamic ion funnels and elevated drift gas pressures. J Am Soc Mass Spectrom 18:1176–1187. https://doi.org/10.1016/j.jasms.2007.03.031
18. Clowers BH, Ibrahim YM, Prior DC, Danielson WF, Belov ME, Smith RD (2008) Enhanced ion utilization efficiency using an electrodynamic ion funnel trap as an injection mechanism for ion mobility spectrometry. Anal Chem 80:612–623. https://doi.org/10.1021/ac701648p
19. Clowers BH, Siems WF, Hill HH, Massick SM (2006) Hadamard transform ion mobility spectrometry. Anal Chem 78:44–51. https://doi.org/10.1021/ac050615k
20. Mason EA, Schamp HW (1958) Mobility of gaseous ions in weak electric fields. Ann Phys 4:233–270. https://doi.org/10.1016/0003-4916(58)90049-6
21. Valentine SJ, Plasencia MD, Liu X, Krishnan M, Naylor S, Udseth HR, Smith RD, Clemmer DE (2006) Toward plasma proteome profiling with ion mobility-mass spectrometry. J Proteome Res 5:2977–2984. https://doi.org/10.1021/pr060232i
22. Taraszka JA, Kurulugama R, Sowell RA, Valentine SJ, Koeniger SL, Arnold RJ, Miller DF, Kaufman TC, Clemmer DE (2005) Mapping the proteome of Drosophila melanogaster: analysis of embryos and adult heads by LC–IMS–MS methods. J Proteome Res 4:1223–1237. https://doi.org/10.1021/pr050038g
23. Kyle JE, Zhang X, Weitz KK, Monroe ME, Ibrahim YM, Moore RJ, Cha J, Sun X, Lovelace ES, Wagoner J, Polyak SJ, Metz TO, Dey SK, Smith RD, Burnum-Johnson KE, Baker ES (2016) Uncovering biologically significant lipid isomers with liquid chromatography, ion mobility spectrometry and mass spectrometry. Analyst 141:1649–1659. https://doi.org/10.1039/C5AN02062J
24. Kyle JE, Casey CP, Stratton KG, Zink EM, Kim Y-M, Zheng X, Monroe ME, Weitz KK, Bloodsworth KJ, Orton DJ, Ibrahim YM, Moore RJ, Lee CG, Pedersen C, Orwoll E, Smith RD, Burnum-Johnson KE, Baker ES (2017) Comparing identified and statistically significant lipids and polar metabolites in 15-year old serum and dried blood spot samples for longitudinal studies. Rapid Commun Mass Spectrom 31:447–456. https://doi.org/10.1002/rcm.7808
25. Burnum-Johnson KE, Baker ES, Metz TO (2017) Characterizing the lipid and metabolite changes associated with placental function and pregnancy complications using ion mobility spectrometry-mass spectrometry and mass spectrometry imaging. Placenta. https://doi.org/10.1016/j.placenta.2017.03.016
26. Groessl M, Graf S, Knochenmuss R (2015) High resolution ion mobility-mass spectrometry for separation and identification of isomeric lipids. Analyst 140:6904–6911. https://doi.org/10.1039/C5AN00838G
27. Fenn LS, McLean JA (2013) Structural separations by ion mobility-MS for glycomics and glycoproteomics. Methods Mol Biol 951:171–194. https://doi.org/10.1007/978-1-62703-146-2_12
28. Metz TO, Baker ES, Schymanski EL, Renslow RS, Thomas DG, Causon TJ, Webb IK, Hann S, Smith RD, Teeguarden J (2017) Integrating ion mobility spectrometry into MS-based exposome measurements: what can it add and how far can it go? Bioanalysis 9:81–98. https://doi.org/10.4155/bio-2016-0244
29. Zheng X, Dupuis KT, Aly NA, Zhou Y, Smith FB, Tang K, Smith RD, Baker ES (2018) Utilizing ion mobility spectrometry and mass spectrometry for the analysis of polycyclic aromatic hydrocarbons, polychlorinated biphenyls, polybrominated diphenyl ethers and their metabolites. Anal Chim Acta 1037:265–273. https://doi.org/10.1016/j.aca.2018.02.054
30. T. Causon, V. Ivanova-Petropulos, D. Petrusheva, E. Bogeva, S. Hann, Fingerprinting of traditionally produced red wines using liquid chromatography combined with

drift tube ion mobility-mass spectrometry, Anal Chim Acta (n.d.) in press

31. Mairinger T, Causon TJ, Hann S (2018) The potential of ion mobility–mass spectrometry for non-targeted metabolomics. Curr Opin Chem Biol 42:9–15. https://doi.org/10.1016/j.cbpa.2017.10.015
32. Zheng X, Aly NA, Zhou Y, Dupuis KT, Bilbao A, Paurus VL, Orton DJ, Wilson R, Payne SH, Smith RD, Baker ES (2017) A structural examination and collision cross section database for over 500 metabolites and xenobiotics using drift tube ion mobility spectrometry. Chem Sci 8:7724–7736. https://doi.org/10.1039/C7SC03464D
33. Stow SM, Causon TJ, Zheng X, Kurulugama RT, Mairinger T, May JC, Rennie EE, Baker ES, Smith RD, McLean JA, Hann S, Fjeldsted JC (2017) An Interlaboratory evaluation of drift tube ion mobility–mass spectrometry collision cross section measurements. Anal Chem 89:9048–9055. https://doi.org/10.1021/acs.analchem.7b01729
34. Nichols CM, Dodds JN, Rose B, Picache JA, Morris CB, Codreanu SG, May JC, Sherrod SD, McLean JA (2018) Untargeted molecular discovery in primary metabolism: collision cross section as a molecular descriptor in ion mobility-mass spectrometry. Anal Chem. https://doi.org/10.1021/acs.analchem.8b04322
35. Causon TJ, Hann S (2015) Theoretical evaluation of peak capacity improvements by use of liquid chromatography combined with drift tube ion mobility-mass spectrometry. J Chromatogr A 1416:47–56. https://doi.org/10.1016/j.chroma.2015.09.009
36. Dodds JN, May JC, McLean JA (2017) Investigation of the complete suite of the leucine and isoleucine isomers: toward prediction of ion mobility separation capabilities. Anal Chem 89:952–959. https://doi.org/10.1021/acs.analchem.6b04171
37. Tim J. Causon, Le Si-Hung, Kenneth Newton, Ruwan T. Kurulugama, John Fjeldsted, Stephan Hann, (2019) Fundamental study of ion trapping and multiplexing using drift tube-ion mobility time-of-flight mass spectrometry for non-targeted metabolomics. Analytical and Bioanalytical Chemistry 411 (24):6265–6274
38. Taylor CF, Paton NW, Lilley KS, Binz P-A, Julian RK Jr, Jones AR, Zhu W, Apweiler R, Aebersold R, Deutsch EW, Dunn MJ, Heck AJR, Leitner A, Macht M, Mann M, Martens L, Neubert TA, Patterson SD, Ping P, Seymour SL, Souda P, Tsugita A, Vandekerckhove J, Vondriska TM, Whitelegge JP, Wilkins MR, Xenarios I, Yates Iii JR, Hermjakob H (2007) The minimum information about a proteomics experiment (MIAPE). Nat Biotechnol 25:887–893. https://doi.org/10.1038/nbt1329
39. Liebisch G, Ekroos K, Hermansson M, Ejsing CS (2017) Reporting of lipidomics data should be standardized. Biochim Biophys Acta 1862:747–751. https://doi.org/10.1016/j.bbalip.2017.02.013
40. Burla B, Arita M, Arita M, Bendt AK, Cazenave-Gassiot A, Dennis EA, Ekroos K, Han X, Ikeda K, Liebisch G, Lin MK, Loh TP, Meikle PJ, Orešič M, Quehenberger O, Shevchenko A, Torta F, Wakelam MJO, Wheelock CE, Wenk MR (2018) MS-based lipidomics of human blood plasma: a community-initiated position paper to develop accepted guidelines. J Lipid Res 59:2001–2017. https://doi.org/10.1194/jlr.S087163
41. Broadhurst D, Goodacre R, Reinke SN, Kuligowski J, Wilson ID, Lewis MR, Dunn WB (2018) Guidelines and considerations for the use of system suitability and quality control samples in mass spectrometry assays applied in untargeted clinical metabolomic studies. Metabolomics 14:72. https://doi.org/10.1007/s11306-018-1367-3
42. Blaženović I, Shen T, Mehta SS, Kind T, Ji J, Piparo M, Cacciola F, Mondello L, Fiehn O (2018) Increasing compound identification rates in untargeted lipidomics research with liquid chromatography drift time–ion mobility mass spectrometry. Anal Chem 90:10758–10764. https://doi.org/10.1021/acs.analchem.8b01527
43. MacLean BX, Pratt BS, Egertson JD, MacCoss MJ, Smith RD, Baker ES (2018) Using skyline to analyze data-containing liquid chromatography, ion mobility spectrometry, and mass spectrometry dimensions. J Am Soc Mass Spectrom 29:2182–2188. https://doi.org/10.1007/s13361-018-2028-5
44. Gabelica V, Shvartsburg AA, Afonso C, Barran P, Benesch JLP, Bleiholder C, Bowers MT, Bilbao A, Bush MF, Campbell JL, Campuzano IDG, Causon T, Clowers BH, Creaser CS, Pauw ED, Far J, Fernandez-Lima F, Fjeldsted JC, Giles K, Groessl M, Hogan CJ, Hann S, Kim HI, Kurulugama RT, May JC, McLean JA, Pagel K, Richardson K, Ridgeway ME, Rosu F, Sobott F, Thalassinos K, Valentine SJ, Wyttenbach T Recommendations for reporting ion mobility mass spectrometry measurements. Mass Spectrom Rev. 38:291–320. https://doi.org/10.1002/mas.21585

Chapter 5

The Use of DMS-MS for the Quantitative Analysis of Acylcarnitines

Nicholas B. Vera, Michelle Clasquin, Stephen L. Coy, and Paul Vouros

Abstract

Differential mobility spectrometry (DMS) is capable of separating molecules based on their size and shape. When coupled to mass spectrometry (MS), DMS reduces chemical background and enhances signal-to-noise (S/N) ratio. Flow injection analysis (FIA) is a technique used to introduce samples into the source of the DMS-MS platform. Here we describe the application of FIA-DMS-MS/MS for the analysis of urinary acylcarnitine species. More than 20 acylcarnitine species can be detected and quantified during a single FIA-DMS-MS/MS acquisition.

Key words Differential mobility spectrometry, Mass spectrometry, Acylcarnitine, Solid-phase extraction

1 Introduction

FIA-DMS-MS/MS is an extremely high-throughput and quantitative analytical technique that can be utilized for the detection and quantitation of urinary acylcarnitines. The preparation of urine samples requires the use of solid-phase extraction (SPE) in order to remove nonspecific metabolites and salts. Following elution of the acylcarnitines, the extracts are dried down and resuspended in the FIA mobile phase.

DMS separates ions based on their differential mobility through an asymmetric field [1]. DMS separation occurs at atmospheric pressure prior to mass analysis. An alternating radio frequency voltage known as Separation Voltage (SV) is required for DMS, along with a direct current voltage known as Compensation Voltage (COV), which transmits ions through the electric field (Fig. 1).

Giuseppe Paglia and Giuseppe Astarita (eds.), *Ion Mobility-Mass Spectrometry: Methods and Protocols*, Methods in Molecular Biology, vol. 2084, https://doi.org/10.1007/978-1-0716-0030-6_5,

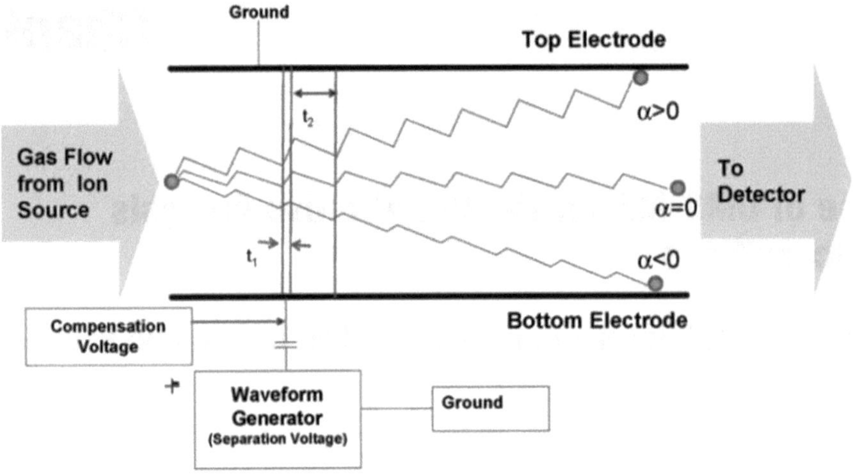

Fig. 1 Schematic of Sciex SelexIon DMS analyzer illustrating the process of ion mobility separation [10]. α refers to the differential mobility function, which is generally expressed as the normalized difference between the high- and low-field mobilities. t1 and t2 refer to the length of time that ions are exposed to low-field and high-field conditions from a radio frequency waveform applied to the parallel electrodes

Mass spectrometry distinguishes between analytes based on their mass and charge (m/z). In this case, a triple quadrupole (QQQ) mass spectrometer was used for acylcarnitine analysis, where two degrees of selectivity are imposed. In the first quadrupole (Q1), the precursor ion is selected, followed by fragmentation in the collision cell (Q2), and ultimately selection of the product ion in the third quadrupole (Q3). Acylcarnitines ionize very well under electrospray conditions, and they fragment fairly well with collision-induced dissociation (CID). Acylcarnitines generate a common product ion of 85.0, which is then used as the Q3 m/z in the multiple reaction monitoring (MRM) method. An MRM experiment is capable of measuring multiple analytes at extremely fast scan speeds. In addition, MRM experiments are extremely specific, selective, and sensitive, thus allowing for optimal quantitation.

Hence, we have coupled DMS to QQQ-MS in order to profile acylcarnitine species from nonhuman primate (NHP) and human urine samples [2, 3] in a high-throughput and quantitative manner [4–7].

2 Materials

2.1 Sample Preparation

1. Internal standard mix stock solution: prepare 2 mM of the following acylcarnitine internal standards in acetonitrile:water (1:1, v/v): deuterated acetylcarnitine (AC d3–2:0); deuterated

propionylcarnitine (AC d3–3:0); deuterated hexanoylcarnitine (AC d3–6:0); deuterated octanoylcarnitine (AC d3–8:0); deuterated decanoylcarnitine (AC d3–10:0); and deuterated palmitoylcarnitine (AC d3–16:0). Then dilute the internal standard mix stock solution with acetonitrile:water (1:1, v/v) to achieve a final concentration of 120 μM.

2. SPE elution buffer: methanol:water:acetic acid (50:45:5, v/v/v). Add 500 mL of methanol to a graduated cylinder and transfer to 1 L bottle. Add 450 mL of water to a graduated cylinder and transfer to 1 L bottle. Add 5 mL of acetic acid and transfer to 1 L bottle.
3. Resuspension buffer: methanol:water:formic acid (70:30:0.1, v/v/v). Add 70 mL of methanol to a graduated cylinder and transfer to 100 mL bottle. Add 30 mL of water to a graduated cylinder and transfer to 100 mL bottle. Add 100 μL of formic acid directly to 100 mL bottle.
4. FIA mobile phase: methanol:water:formic acid (70:30:0.1, v/v/v). Add 700 mL of methanol to a graduated cylinder and transfer to 1 L bottle. Add 300 mL of water to a graduated cylinder and transfer to 1 L bottle. Add 1 mL of formic acid directly to 1 L bottle.
5. Weak wash solvent: water:acetonitrile (1:1, v/v). Add 500 mL of water to a graduated cylinder and transfer to 1 L bottle. Add 500 mL of acetonitrile to a graduated cylinder and transfer to 1 L bottle.
6. Strong wash solvent: water (100%). Add 1000 mL of water to a graduated cylinder and transfer to 1 L bottle.
7. DMS modifier: ethyl acetate (100%). Add 1000 mL of ethyl acetate to a graduated cylinder and transfer to 1 L bottle.

3 Methods

3.1 Urinary Acylcarnitine Sample Preparation

1. Add 10 μL of the diluted internal standard mix solution into 190 μL of urine and dry down under nitrogen (N_2) at 60 °C. Resuspend in 200 μL of methanol, vortex, and centrifuge at 14,000 rpm (18845 × *g*) for 5 min.
2. Perform acylcarnitine extraction by solid-phase extraction (SPE) using SEP-PAK VAC 6 cc silica gel cartridges (Waters) [8]. Precondition SPE cartridges with 1.5 mL methanol. Add 200 μL of sample to SPE cartridge. Wash SPE cartridge with

2 mL methanol. Elute acylcarnitines with methanol:water: acetic acid (50:45:5, v/v/v) [9] (*see* **Note 1**).

3. Dry down acylcarnitine extract under N_2 at 50 °C. Resuspend in 1000 μL of methanol:water:formic acid (70:30:0.1, v/v/v). Vortex and centrifuge sample at 14,000 rpm (18845 × *g*) for 5 min. Transfer 200 μL to LC-MS vial.

3.2 FIA

1. FIA enabled sample introduction into the source. Prepare the following FIA mobile phase: methanol:water:formic acid (70:30:0.1, v/v/v). Set the flow rate at 0.050 mL/min. Use acetonitrile:water (1:1, v/v) as weak washing solvent and water (100%) as strong washing solvent. Inject a volume of 10 μL.

3.3 DMS

1. Use the Sciex SelexIon for all DMS analyses. Set the DMS temperature at low (150 °C). Use ethyl acetate as the DMS modifier (MOD) at low modifier composition (1.5% v/v). Set the DMS offset at −3.0 V. Set the DMS resolution enhancement (DR) off. Set the separation voltage (SV) at 3500 V. Optimize the compensation voltage (COV) daily since it needs to be varied for each acylcarnitine (Table 1) (*see* **Note 2**). Acquire DMS data with Analyst 1.6.2.

3.4 Mass Spectrometry (MS)

1. Use the Sciex QTRAP 5500 for the acquisition of all MS and tandem mass spectrometry (MS/MS) data. Analyze acylcarnitines using positive-ion electrospray ionization (ESI+) in the multiple reaction monitoring mode (MRM). Set the collision gas (CAD) at medium and the curtain gas (CUR) at 20 psi. Set the ion source gas at 60 psi for both ion source gas 1 (GS1) and ion source gas 2 (GS2). Set the ion spray (IS) voltage at 4500 V and the temperature (TEM) at 600 °C. Set the compound conditions for each acylcarnitine as follows: declustering potential (DP), 60 V; the entrance potential (EP), 10; the collision energy (CE), 21; and the collision exit potential (CXP), 7.0. Set the MRM mass transitions for each acylcarnitine as described in Table 1. The common acylcarnitine product ion is m/z = 85 ($^+$CH2-CH=CH-CO-OH), and the fragmentation mechanism is highlighted in Fig. 2.
2. Use Sciex Analyst 1.6.2 to acquire all MS and MS/MS data. Perform all FIA peak integration with Sciex MultiQuant software version 2.1. Use Microsoft Excel for data reduction and quantitation.

Table 1
Acylcarnitine MRM transitions with COV range

AC ID	Q1	Q3	COV (V)
AC 2:0	204.2	85.1	−50 ± 5
AC d3-2:0	207.2	85.1	−50 ± 5
AC 3:0	218.2	85.1	−45 ± 5
AC d3-3:0	221.2	85.1	−45 ± 5
AC 4:0	232.2	85.1	−40 ± 5
AC 5:0	246.2	85.1	−40 ± 5
AC 6:0	260.2	85.1	−35 ± 5
AC d3-6:0	263.2	85.1	−35 ± 5
AC 7:0	274.2	85.1	−35 ± 5
AC 8:0	288.2	85.1	−30 ± 5
AC d3-8:0	291.2	85.1	−30 ± 5
AC 8:1	286.2	85.1	−30 ± 5
AC 9:0	302.2	85.1	−25 ± 5
AC 10:0	316.2	85.1	−25 ± 5
AC d3-10:0	319.2	85.1	−25 ± 5
AC 10:1	314.2	85.1	−25 ± 5
AC 12:0	344.2	85.1	−20 ± 5
AC 14:0	372.2	85.1	−15 ± 5
AC 16:0	400.2	85.1	−15 ± 5
AC d3-16:0	403.2	85.1	−15 ± 5

List includes all labeled acylcarnitine internal standards

4 Notes

1. All SPE was performed using a traditional SPE manifold. As a result, the SPE process was low throughput. It is certainly possible to transfer this assay to a 96-well SPE plate.
2. The COV can change from day to day. Consequently, COV optimization is required daily for each analyte.

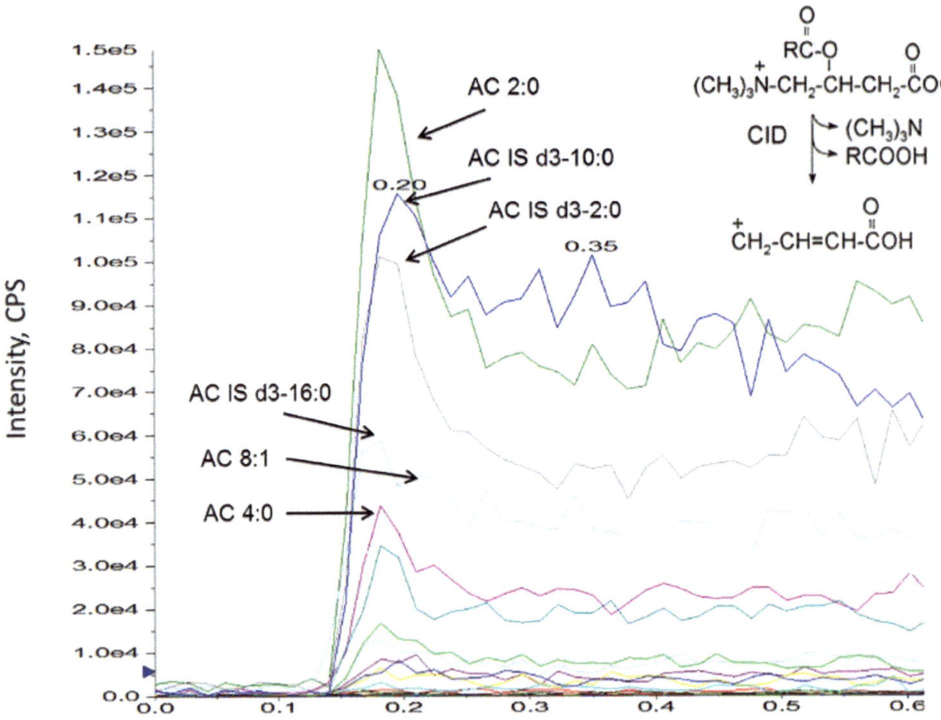

Fig. 2 FIA-DMS-MS/MS analysis of human urinary acylcarnitines. The acylcarnitine fragmentation mechanism is given in the top right

Acknowledgments

This work was funded by the Internal Medicine Research Unit (IMRU), within the Pfizer Global Research and Development Business Unit, at Pfizer.

References

1. Lapthorn C, Pullen F, Chowdhry BZ (2013) Ion mobility spectrometry-mass spectrometry (IMS-MS) of small molecules: separating and assigning structures to ions. Mass Spectrom Rev 32(1):43–71
2. Azzam EI, Jay-Gerin JP, Pain D (2012) Ionizing radiation-induced metabolic oxidative stress and prolonged cell injury. Cancer Lett 327(1–2):48–60
3. Barjaktarovic Z, Schmaltz D, Shyla A, Azimzadeh O, Schulz S, Haagen J, Dorr W, Sarioglu H, Schafer A, Atkinson MJ, Zischka H, Tapio S (2011) Radiation-induced signaling results in mitochondrial impairment in mouse heart at 4 weeks after exposure to X-rays. PLoS One 6(12):e27811
4. Pannkuk EL, Laiakis EC, Authier S, Wong K, Fornace AJ Jr (2015) Global metabolomic identification of long-term dose-dependent urinary biomarkers in nonhuman primates exposed to ionizing radiation. Radiat Res 184 (2):121–133
5. Laiakis EC, Pannkuk EL, Chauthe SK, Wang YW, Lian M, Mak TD, Barker CA, Astarita G, Fornace AJ Jr (2017) A serum small molecule biosignature of radiation exposure from total body irradiated patients. J Proteome Res 16 (10):3805–3815
6. Vera NB, Chen Z, Pannkuk E, Laiakis EC, Fornace AJ Jr, Erion DM, Coy SL, Pfefferkorn JA, Vouros P (2018) Differential mobility spectrometry (DMS) reveals the elevation of

urinary acetylcarnitine in non-human primates (NHPs) exposed to radiation. J Mass Spectrom 53(7):548–559

7. Chen Z, Coy SL, Pannkuk EL, Laiakis EC, Fornace AJ Jr, Vouros P (2018) Differential mobility spectrometry-mass spectrometry (DMS-MS) in radiation biodosimetry: rapid and high-throughput quantitation of multiple radiation biomarkers in nonhuman primate urine. J Am Soc Mass Spectrom 29 (8):1650–1664
8. Kafle A, Coy SL, Wong BM, Fornace AJ Jr, Glick JJ, Vouros P (2014) Understanding gas phase modifier interactions in rapid analysis by differential mobility-tandem mass spectrometry. J Am Soc Mass Spectrom 25 (7):1098–1113
9. Heinig K, Henion J (1999) Determination of carnitine and acylcarnitines in biological samples by capillary electrophoresis-mass spectrometry. J Chromatogr B Biomed Sci Appl 735(2):171–188
10. Schneider BB, Nazarov EG, Londry F, Vouros P, Covey TR (2016) Differential mobility spectrometry/mass spectrometry history, theory, design optimization, simulations, and applications. Mass Spectrom Rev 35 (6):687–737

Chapter 6

Traveling Wave Ion Mobility-Mass Spectrometry to Enhance the Detection of Low Abundance Features in Untargeted Lipidomics

Zeeshan Hamid and Andrea Armirotti

Abstract

Ion mobility has become a valuable tool in mass spectrometry-based lipidomics workflows, thanks to its ability to separate ions in the gas phase based on their size and conformation. Over the last years, it was demonstrated that the comparative analysis of ion mobility data has the potential to highlight the presence of low abundance species in untargeted lipidomics. The present chapter illustrates the background of the topic and guides the reader from the sample preparation to the data analysis of an untargeted lipidomics experiment performed using ion mobility.

Key words Lipidomics, Ion mobility-mass spectrometry, High-resolution LC-MS/MS

1 Introduction

In the most generic terms, lipidomics is the large-scale study of lipids as chemical entities involved in many and diverse physiological functions, from structural building blocks to signaling molecules [1]. With thousands of different possible lipid structures (43,137 lipids reported in LipidMaps as of June 2018) [2], the global analysis of lipids becomes a huge challenge, even for the most advanced analytical tools and skilled scientists. Furthermore, when addressing internal diversity related to isomeric species (like, for example, Sn1 or Sn2 positional isomers in glycerophospholipids), this task becomes even more challenging. However, over the last years, the field of lipidomics has witnessed a number of impressive technical advancements, resulting in the development of very sophisticated mass spectrometers (MS). These top-end, high-resolution, platforms are able to acquire accurate mass data (frequently below 1 ppm mass accuracy) and provide the fast acquisition of high-quality tandem mass spectra (MS/MS). Nowadays, the identification and quantification of hundreds of lipids can be done in a

Giuseppe Paglia and Giuseppe Astarita (eds.), *Ion Mobility-Mass Spectrometry: Methods and Protocols*,
Methods in Molecular Biology, vol. 2084, https://doi.org/10.1007/978-1-0716-0030-6_6,

Table 1
Some major advantages and disadvantages of the three main lipidomics workflows

	Dimensions of separation	Pros	Cons
Direct infusion	*m/z*	Faster	No isomer separation. Limits in dynamic range. Low sensitivity toward low abundance species. Needs very high-resolution MS
LC-MS/MS	Retention time, *m/z*	Separation by Log*P* (lipid classes). Two dimension in separation. Broader dynamic range	Slower. Requires efficient chromatographic setups
LC-TWIMS/MS	Retention time, drift time, *m/z*	Separation by LogP (Lipid classes). Three dimension in separation. Broader dynamic range. Allows measurent of CCS	Slower. Requires efficient chromatographic setups. Requires He gas for correct operation. Requires calibration for CCS measurement

matter of minutes [3]. Lipid annotation is normally done by searching the observed accurate mass value (following deconvolution of different adduct species) and by confirming the identity by investigating MS/MS data. Different workflows for MS-based lipid analysis have been developed (Table 1). One of the earliest known is shotgun lipidomics [3], where lipid extracts are directly infused into the MS, generating information about many lipids in very short timeframe. Shotgun lipidomics is relatively easy and fast to perform. However, the infusion of all lipids simultaneously makes isomeric species often undistinguishable, and it frequently masks the detection of low abundant species, given the limited sensitivity and dynamic range of the MS platforms. A more efficient, but longer, workflow consists in the use of liquid chromatography (LC) for lipid separation before MS analysis. This additional step greatly decreases sample complexity and allows the detection of tens of lipids for each molecular class. The elution of lipid species, in reversed phase chromatography, is sequential and Log*P*-related, with fatty acyls, for example, eluting earlier than glycerophospholipids and much earlier than triacylglycerols or cholesterylesters [4]. LC-MS/MS currently represents the state-of-the-art for lipidomics investigation. More recently, however, Waters Inc. developed and implemented in quadrupole-time of flight (QToF) MS, the Traveling Wave Ion Mobility (TWIMS) technology [5, 6]. In very few words, with this device, the gas phase ions are packed in a well-defined region filled with nitrogen gas and then moved forward by the sequential application of low/high potential pulses, similarly to surfers floating back to the beach by incoming sea waves. This gas phase ion transport adds an extra separation dimension based on size and shape of the lipid ions. This process generates an additional descriptor for the ion, named drift time. Based on this number, and with an adequate instrumental calibration, the gas phase collisional cross section of the ion (CCS) can be calculated and it can be used as additional parameter for lipid (and metabolite) identification and characterization [7]. Since the TWIMS process works in the millisecond scale, it is fully compatible with the upstream chromatographic separation (with peaks eluting in 3–5 s) and with downstream ToF analysis (microseconds). Compared to the three dimensions of exploration allowed by conventional LC-MS/MS, with TWIMS, the lipidome is represented by a 4-dimension box, where each molecule is distinguished by four coordinates: LC retention time (RT, in minutes), accurate mass (as *m/z*), intensity (as ion counts), and drift time (in milliseconds) or CCS (in $Å^2$). The additional separation dimension in the gas phase frequently allows the detection of isomeric species that would otherwise be undistinguishable [8–10]. Furthermore, TWIMS does not interfere with the continuous acquisition of MS/MS data. Indeed, on Waters Synapt® platforms, the user might decide to apply ion mobility separation also to MS/MS fragments, thus greatly enhancing the possibility for advanced structural characterization [11, 12].

Recently, it has also been shown that TWIMS is also useful at data analysis levels, thanks to its capability to enhance the detection of low abundant lipid species [13]. Indeed, the most widely used protocols for untargeted lipidomics (and metabolomics in general) make use of multivariate data analysis techniques (MVA), such as Principal Component Analysis (PCA) and others [14, 15]. These algorithms, particularly those publicly available [16, 17] normally handle 2D LC-MS data, where each feature is characterized by a (RT, *m/z*) pair. In this context, the benefits of TWIMS are thus lost, together with the drift time information. Furthermore, MVA often tends to overlook those features that represent a very minor part of the overall ion intensity observed in the whole LC-MS/MS run. Very often, this happens at data preprocessing stage, where a compromise threshold between absolute signal intensity and the need to filter out background signals has to be found. In other words, during classical MVA workflows, it happens that signals too close to the background are simply ignored, although their regulation might be very significantly altered between the different experimental groups. In this context, the visual inspection of TWIMS data is useful to detect these features, particularly when applied to mobilograms that are 2D plots reporting *m/z* vs. drift time values (Fig. 1).

By using dedicated Waters software (HDMS Compare®), mobilograms from two different experimental groups can be overlapped similarly to 2D gel electrophoresis of proteins. Instead of protein spots stained by Coomassie or colloidal-Ag, lipids (and

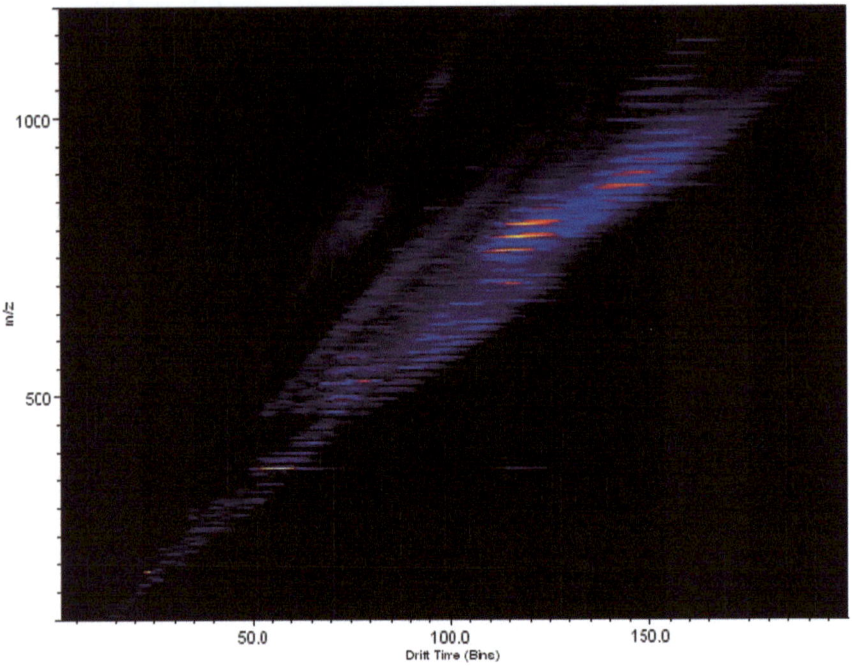

Fig. 1 Example of mobilogram generated from a human plasma sample analyzed by LC-TWIMS/MS. The plot refers to the total ion intensity observed during a full 30 min run. *X* axis reports the drift time values (in milliconds), *Y* axis reports high-resolution *m/z* values

other metabolites) being differently expressed between the groups are reported (and thus detected) by using colors of different intensity. Since the constant background present in both groups is summed with opposite signs and thus removed, this strategy enables the detection of features purely based on their different regulation between the groups, irrespectively from the signal-to-noise ratio they showed at the time of acquisition. This approach is called Comparative Mobilogram Analysis (CMA). CMA is a powerful tool for data analysis in lipidomics and metabolomics in general. The additional gas phase separation enabled by the TWIM technologies greatly enhances the analytical possibilities in untargeted LC-MS/MS experiments. Human plasma, for example, is perhaps the most complex lipidome an analyst might be facing. All the lipid classes are present and their abundance spans over the broadest concentration range. In ESI+ mode, at almost any retention time value between 3 and 18 min, the spectrum is dominated by high abundant glycerophospholipids, the rest of the signals is mostly made of tryacylglicerols and cholesterylesters. In this extremely complex mixture, still, the CMA approach enables the identification of other lipids whose abundance is much lower. Figure 2 shows the overlap between the total ion current generated by the plasma lipidome and the extracted ion current for ion

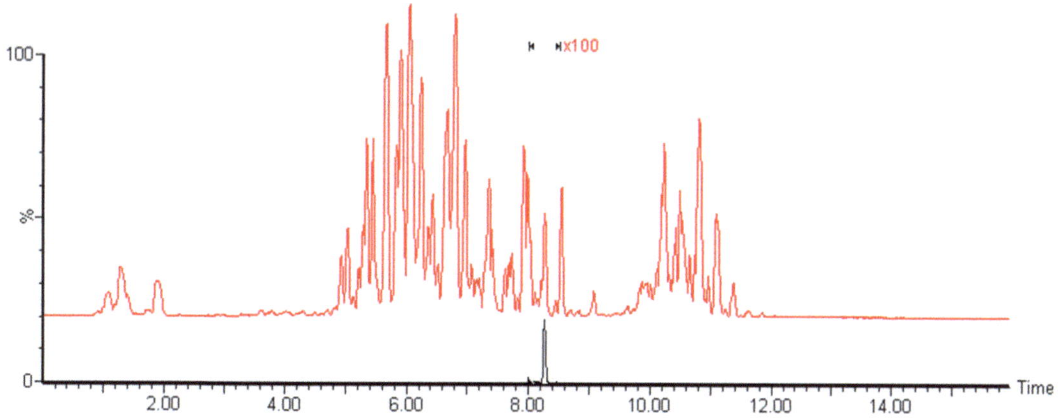

Fig. 2 Overlapped chromatograms of total ion current and 1224.50 *m/z* ion current for human plasma lipid extract. The extracted ion current has been magnified by 100. The two plots have a 20% offset on the *Y* axis for clarity purposes

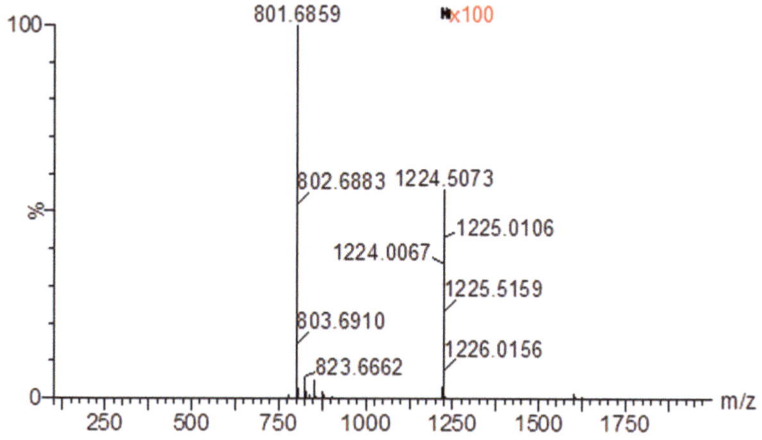

Fig. 3 Mass spectrum at 8.2–8.3 min retention time. The doubly charged GC1c (d18:1/20:0) is magnified 100× to become visible compared to coeluting very abundant glycerophospholipids

1224.50 *m/z*, a signal highlighted by CMA as differently expressed between mice and humans.

This ion is eluting at 8.2 min, together with many other abundant glycerophospholipids. The signal corresponds to the large ganglioside GC1c (d18:1/20:0) that appears in the corresponding mass spectrum as a doubly charged species (Fig. 3).

Multiply charged species are normally rarely observed in conventional lipidomics setups. Furthermore, this kind of huge glycosphingolipids would normally require dedicated sample preparation and extraction protocols and optimized targeted analysis to be efficiently detected and quantified. With the use of TWIMS technology and comparative mobilogram analysis, the presence of these species can be detected (and quantified) in extremely complex biological matrices with conventional untargeted lipidomics experiments. The following sections show how to prepare and acquire a generic lipidomic data set and guide the reader in the workflow.

2 Materials

- Standard laboratory pipettes, high-quality 4- or 8-ml glass vials with Teflon caps.
- Solvents of the highest LC-MS available quality (chloroform, methanol, isopropanol (IPA), and acetonitrile (ACN)).
- Ammonium formate (AmFor) and formic acid (FA) of the highest quality.
- Leucine enkephalin reference standard.
- Polyalanine reference standard.
- MilliQ water.
- A vortex mixer.
- A centrifuge.
- A nitrogen evaporator.
- A Synapt G2® Mass Spectrometer coupled to an UPLC chromatographic system.
- A reversed-phase C18 column 1.8 μm particle size, 2.1× 100 mm format.
- A data analysis workstation equipped with MassLynx® and HDMSCompare® software.

3 Methods

The present protocol shows the CMA workflow applied to brain tissue lipidomics. The workflow illustrates the extraction of lipids from the biological matrix, the setup of the chromatographic separation, the optimization of the ion source and TWIMS parameters, the data acquisition, and the data analysis process.

3.1 Extraction of Lipids

A number of extraction procedures have been described, but the choice of extraction method largely depends on the target lipids to be studied [18, 19]. The here reported protocol illustrates the widely used method described by Bligh and Dyer in 1959 [20], which exploits the sequential addition of a 2:1 methanol:chloroform mixture, water and chloroform to obtain a three-phase separation. At the end of this procedure, a solid layer of proteins separates the lower organic phase from the upper aqueous one.

The organic phase is highly enriched in most of the known lipid classes, like fatty acyls, triacylglycerols, sterols, sphingolipids, and glycerophospholipids.

1. Weigh frozen brain tissue in glass vials. Keep the vials on ice (*see* **Note 1**).
2. Add 2 ml of cold extraction solution (methanol:chloroform 2:1).
3. Homogenize the tissue keeping the vials in ice bath.
4. Add 0.6 ml of chloroform.
5. Vortex for 30 s.
6. Add 0.6 ml of water.
7. Vortex again for 30 s.
8. Centrifuge at 15 °C (2100 × *g* for 10 min).
9. You will observe two liquid phases separated by a protein disk. With a glass pipette tips, carefully collect the lower organic phase and move it to a new glass vial.
10. Evaporate the solvent under gentle nitrogen stream.
11. Reconstitute the dried sample in 100 μl of 9:1 methanol:chloroform and move it to a LC-MS vial. The sample is now ready for analysis.

3.2 Setup of the Chromatographic Conditions

The following conditions were optimized for a Waters CSH® column, but similar profiles should be expected when using other C18 columns from other vendors. An isopropanol gradient (IPA) is normally able to elute even the most apolar lipids from the column. Particular attention should be dedicated to column temperature, to be maintained above 50 °C to decrease IPA viscosity (thus system backpressure) and to enhance peak shape.

1. Prepare a 1 M AmFor solution in water. Adjust the pH to 5 with FA.
2. Prepare the mobile phases as follows (0.5 l for roughly 200 samples): Phase A: add 195 ml of water, 5 ml of the AmFor solution, and 300 ml of ACN. Phase B: add 445 ml of isopropanol, 5 ml of AmFor solution, and 50 ml of ACN.
3. Cap and shake the bottles.
4. Install the column and set the following gradient conditions. Make sure you gradually increase column flow up to the final 0.4 ml/min. Warm the column to 55 °C.
5. Set the gradient profile as described in Table 2.

This procedure refers to the setup of data acquisition process on Synapt G2 mass spectrometers. The protocol indicates generic tune conditions that might need to be checked and adjusted.

Table 2
Gradient profile for chromatographic separation (Subheading 3.2, step 4)

Time (min)	Flow rate (ml/min)	Percent A	Percent B
0	0.4	85	15
1	0.4	85	15
10	0.4	40	60
18	0.4	25	75
21.5	0.4	0	100
23	0.4	0	100
23.1	0.4	85	15
25	0.4	85	15

3.3 Setup, MS Tuning for Ion Mobility Experiments, and Data Acquisition

1. Prepare sodium formate reference solution (5 mM in 50:50 isopropanol:water).
2. Prepare leucine encephalin reference solution (2 μg/ml in 50:50 IPA:water with 0.1% FA).
3. Prepare polyalanine reference solution (0.1 mg/ml in 50:50 ACN:water with 0.1% FA).
4. Startup the system as indicated by the Vendor and allow it to reach the optimum operating conditions (warmup).
5. Open the He gas line and allow the system to equilibrate for 10 min.
6. Set the initial instrumental parameters as described in Table 3, checking both ESI+ and ESI− conditions.
7. Set the chromatographic flow at 0.4 ml/min, 50% Phase B and infuse leucine encephalin solution in the flow with a tee union (*combined* mode on the fluidics pane). Use the signal at ESI+ 556 *m/z* (554 *m/z* for ESI−) to optimize the source conditions and achieve the highest sensitivity.
8. Switch the acquisition to ion mobility mode and allow the system 5 minutes to equilibrate.
9. Pay particular attention to the two major parameters for TWIMS efficiency: wave velocity (WV) and wave height (VH) (*see* **Note 2**). Waters recommends setting WH to 40 V on the Synapt G2. Different settings might be needed on different models. Carefully change the WV parameter, so that the leucine encephalin peak in the mobilogram is close to the middle of the drift time scale (*X* axis). Values between 500 and 800 m/s are normally adequate for the 50–1200 *m/z* scan range. *See* Subheading 4 for further details. Perform standard instrument calibration in the 50–1200 *m/z* mass range by

Table 3
MS instrumental parameters (Subheading 3.3)

MS parameters	
Polarity	ESI-positive and ESI-negative
Acquisition mode	MS^e
Capillary voltage	3 kV (positive), 2.2 kV (negative)
Source temperature	90 °C
Desolvation gas flow	800 L/h
Cone gas flow	20 L/h
Desolvation temperature	400 °C
Wave velocity (WV)	*600 m/s*
Wave height (WH)	*40 V*
He cell gas flow	180 mL/min
IMS cell gas flow	90 L/min
Scan range	50–1200 *m/z*
Scan rate	0.3 s per spectrum

using sodium formate solution, as recommended by the Vendor. Perform calibration in both polarities, resolution mode.

10. When all source parameters are set, infuse polyalanine solution and acquire its mobility spectrum for two minutes, in both ESI+ and ESI− modes. This is required for CCS calibration (but not strictly mandatory for comparative mobilogram analysis).
11. The system is now ready to acquire data. Prepare a method for HDMSe acquisition with ion mobility (Mobility ToF function). Set the acquisition range to 50–1200 *m/z* and scan time to 0.5 s. Important: in the "Mobility" panel, make sure "Add Drift Time Function" and "Maintain Mobility Separation" options are checked. IMS data for CMA are acquired in *Profile* mode.
12. In the Transfer CE pane, set the low energy to 4 eV and ramp the high energy from 25 to 45 eV. *See* Subheading 4 for further details.
13. Set the Lock Mass acquisition to one spectrum every 30 s. DO NOT apply automatic recalibration.
14. Run the samples. Injection volumes of 3–5 μl are normally adequate for lipidomics. When preparing the acquisition batch, apply the general guidelines for metabolomics experiments: (a) add a consistent number of QC samples (usually

pools of samples from all the experimental groups), (b) run reference samples, (c) carefully check system carryover by running blank samples, and (d) fully randomize the batch acquisition to avoid any systematic acquisition bias.

3.4 Comparative Mobilogram Analysis Using HDMS Compare Software

The plots and screenshots here reported refer to CMA analysis of a brain lipidomics experiment. Two experimental groups were analyzed: animals with a genetically abolished enzyme (Knock-out, KO) versus the corresponding wild-type (WT) controls. These data are only used to illustrate the workflow. At present, HDMA Compare *only allows dual comparisons.*

1. In the "File" pane, start a "New analysis".
2. Load the RAW datafiles for each of the two groups in the *Sample 1 Data* and *Sample 2 Data* panes. In the *Advanced Settings*, choose *BPI* for chromatographic data and narrow down the LC retention time extraction window according to the part of the chromatogram to be investigated. For example, one might wish to focus the CMA analysis to triacylglycerols and sterols only (from around 13 min to the end of run). In the current example, the retention time range is set from 1 to 19 min (whole run). Make sure to apply the same range to all the samples of a group. HDMS compare allows the comparison of both low (function 1) and high (function 2) energy traces acquired by the MS. This allows, for example, to selectively extract CMA data on a given diagnostic fragment ion (for example, ion 264.26 *m/z* for d18:1 ceramides or 369.35 *m/z* for sterols). Figure 4 reports the screenshot for the analysis setup. Click OK to start the analysis.
3. After a few minutes, the analysis is complete, and a result window similar to that reported in Fig. 5 appears.
4. The bottom left pane of Fig. 5 represents the "Fusion Map" of the analysis. As described above, the constant signals have been summed with opposite signs and only differently expressed features are represented as colored dots (yellow for Sample 1 and blues for Sample 2). The color intensity is related to the observed differences in signal intensity.
5. The bottom right pane of Fig. 5 shows a zoomed region of the Fusion Map highly upregulated in Sample 1 (region A). This pane also allows to measure the CCS of the observed lipids, by using a CCS calibrated data (see above for polyalanine). Make sure the right molecular mass for the drift time gas is used (28 amu for N_2).
6. The upper left region of Fig. 5 represents the overlapped chromatograms (Sample 1 upward, Sample 2 downward). By means of dedicated selection tools, the software allows the manual selection of Regions of Interest in the Fusion Map.

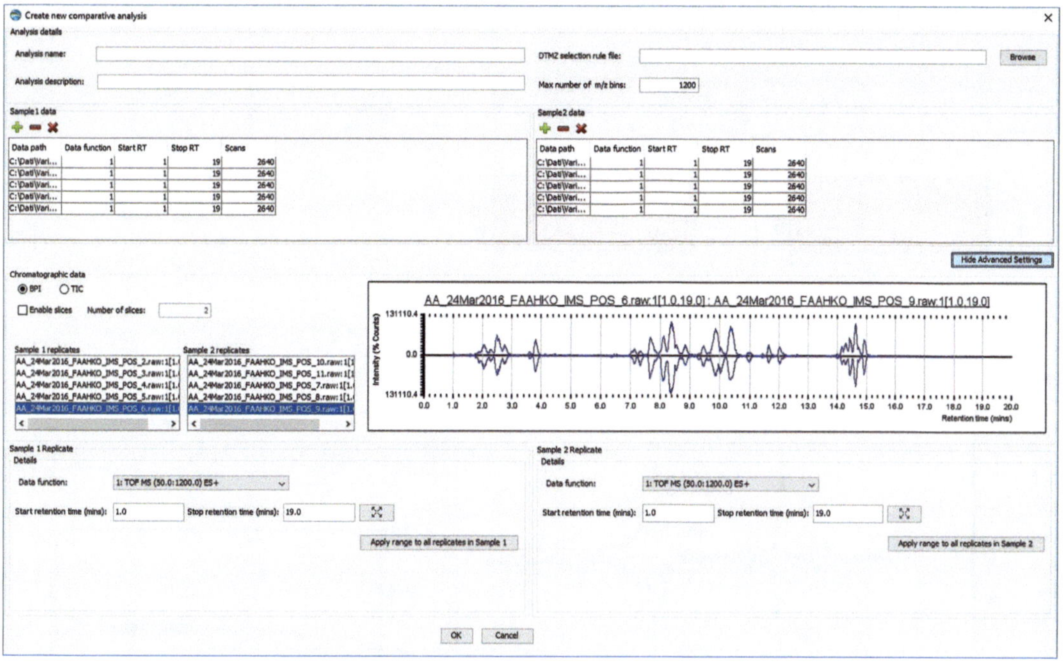

Fig. 4 Screenshot for HDMS Compare analysis setup window

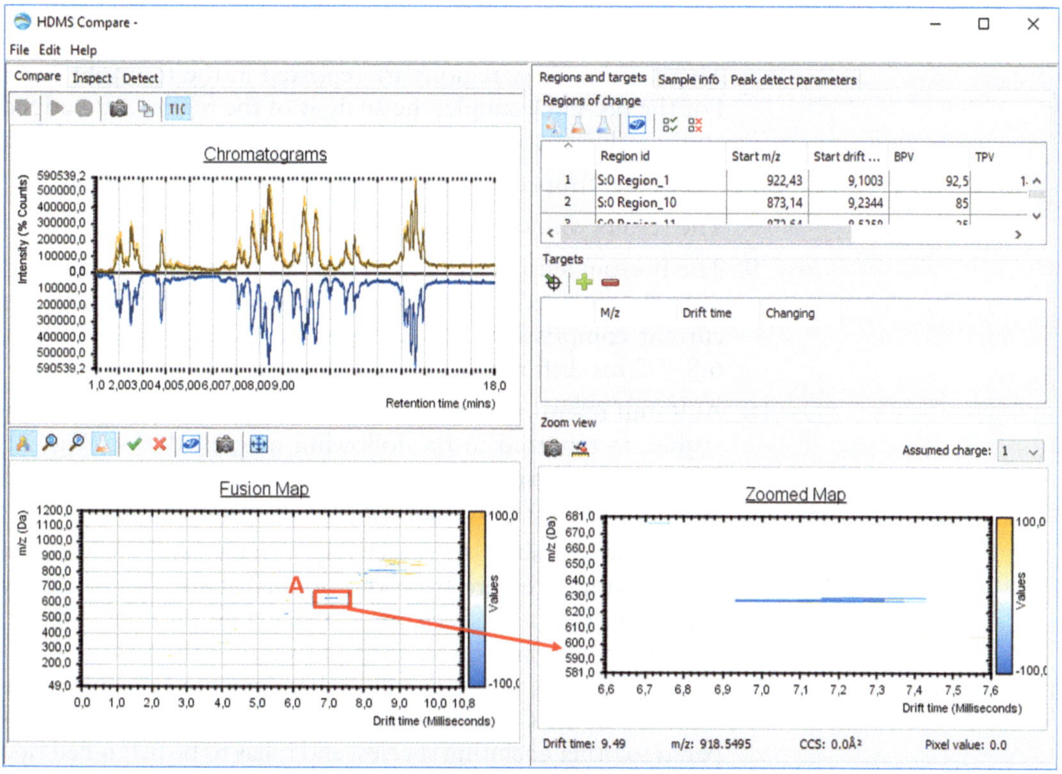

Fig. 5 Screenshot for HDMS Compare analysis results window

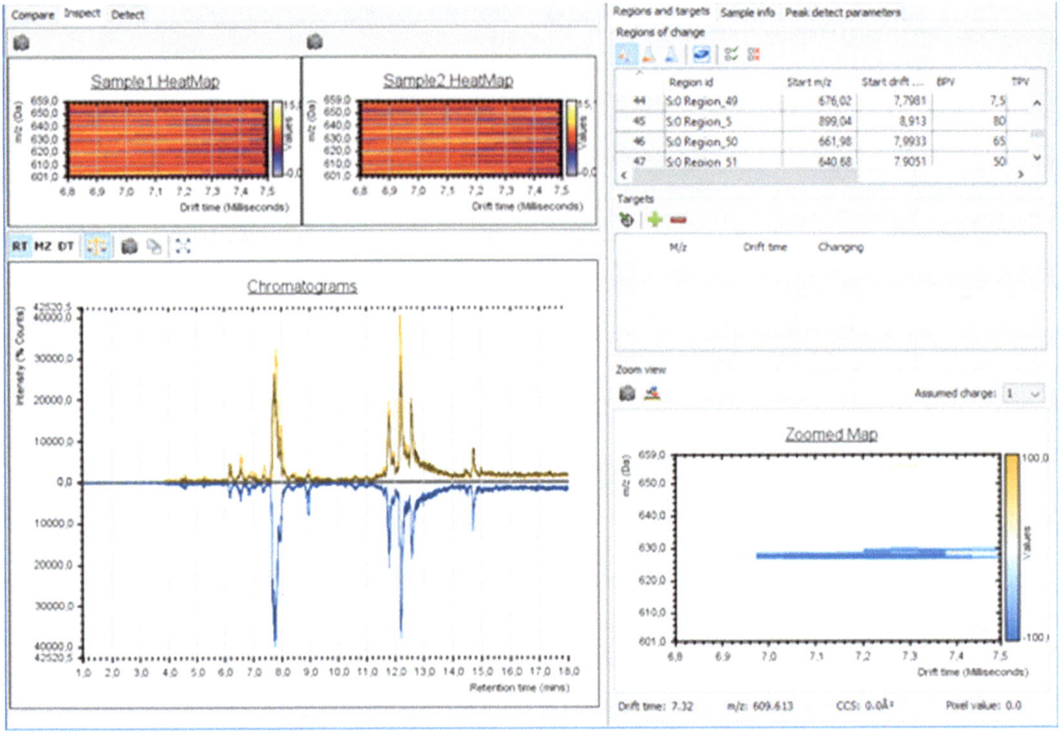

Fig. 6 Screenshot for HDMS Compare Inspect results window

The data for these regions are reported in the top right pane. For the present example, the analysis of the region A described above is illustrated.

7. Select the "Inspect" pane in the upper left of the window.
8. The results windows appear as reported in Fig. 6.
9. The bottom left pane of Fig. 6 shows the overlapped chromatograms obtained by extracting from all the runs the total ion current comprised in the A region of Fig. 2 (600–660 *m/z*, 6.8–7.5 ms drift time).
10. At 9 min retention time, a differently expressed peak is clearly visible, as reported in the following magnified plot (Fig. 7). This peak corresponds to a lipid significantly upregulated in *Sample 2 Data* (the KO group).
11. It is now possible to selectively extract and analyze the corresponding feature from the original RAW file, by simply inspecting the mass spectrum at that retention time (Fig. 8). The spectrum can be recalibrated by using the leucine encephalin lock mass value. The lipid that generated the difference reported in the CMA analysis shows a very low intensity compared to other coeluting species, and it has to be magnified by a factor of 15 to become visible. Indeed, the intensity of the 629.5532 *m/z* ion accounts for only 0.17% of the total ion

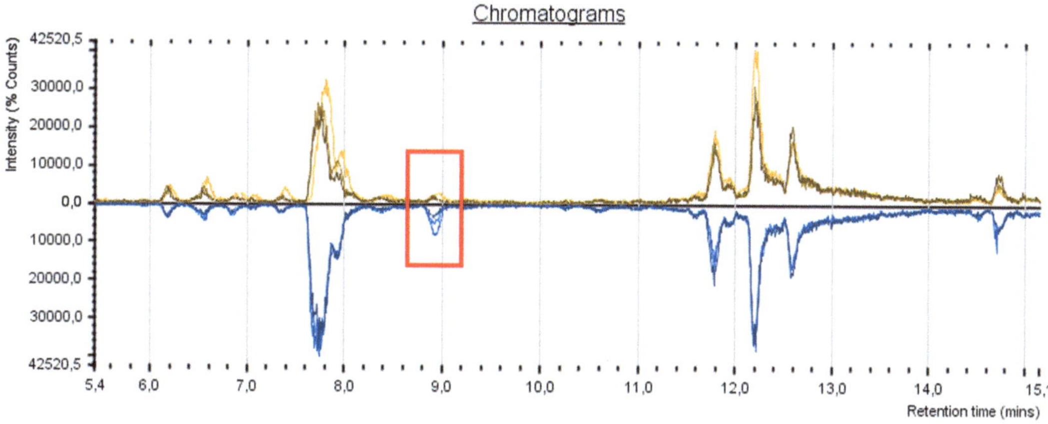

Fig. 7 Magnified view of chromatogram pane reported in Fig. 4. The peak corresponding to a lipid differentially expressed between the two groups is indicated by the red box

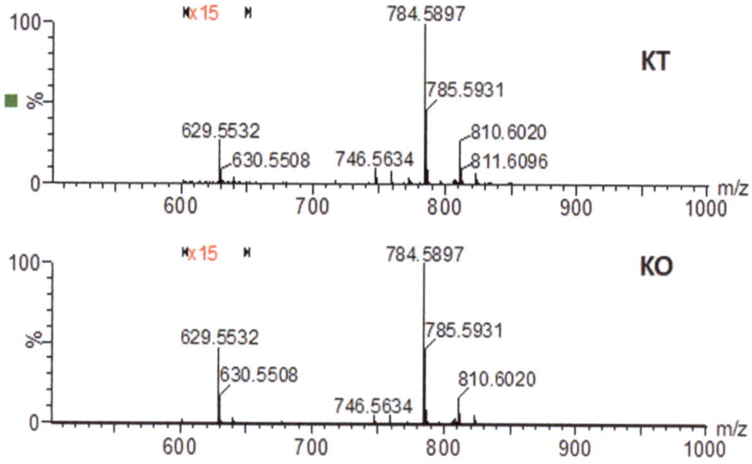

Fig. 8 Overlapped (WT top, KO bottom) mass spectrum at retention time 8.9–9.0 min retention time. The intensity of the lipid at 629.5532 *m*/*z* is magnified by a factor of 15 in both spectra

current at that retention time in the WT, and it raises to 0.32% in the KO group, thus making the comparative difference observed in the CMA. With a threshold set as low as 1% of the total ion current for the feature extraction process in an ordinary MVA analysis, this lipid would have go completely unnoticed.

12. Following a database search in LipidMaps [2], this lipid is identified as a diacylglycerol with 38 carbons and 3 double bonds: DG(38:3), brute formula C41H73O4, adduct $[M\text{-}H_2O + H]^+$, observed 629.5532, calculated 629.5503, mass accuracy 4.6 ppm. Further MS/MS analysis in ESI mode identified this lipid as DG (18:0/20:3) (*see* **Note 3**).

4 Notes

1. Use only glass vials since chloroform can extract many plasticizers from plastics causing unnecessary contamination to the lipid fraction [21].
2. Wave velocity is a very important parameter to tune when setting up the acquisition. It is highly recommended to run preliminary LC-MS runs of the samples of interest to check that (a) ions at higher *m/z* (es. triacylglycerols) are correctly reported at high drift time values. If the WV is too low, these ions do not come out of the IMS tube at the right time (delayed drift), and they are sampled at the beginning of the following ToF push, thus resulting in unrealistically low drift time values and (b) ions at low *m/z* (es. fatty acids) are correctly represented in the mobilograms. If the WV is too high, these ions might come out of the IMS tube too fast to be sampled by the ToF push (anticipated drift). This results in an unrealistically low abundance of these species. As a general rule, the mobilogram in Fig. 1 should not report significant signals at the upper left of the plot (high *m/z* but very low drift time). Furthermore, a TWIMS LC-MS run should look very similar to a conventional LC-MS one, with the same features showing similar abundances between the two approaches.
3. Tandem mass analysis can be performed in either the Trap region of the MS (before the TWIMS separation) or in the Transfer region (after the mobility separation). The first option is helpful in the case of advanced structural investigation of the ions, since the MS/MS fragments will undergo mobility separation (*see* for example Basit et al. [13] supplementary information). For standard applications, including comparative mobilogram analysis, it is suggested to perform MSe in the Transfer region of the MS. This way, precursor ions and their fragments will always share the same drift time values.

References

1. Wenk MR (2005) The emerging field of lipidomics. Nat Rev Drug Discov 4(7):594–610
2. Fahy E et al (2007) LIPID MAPS online tools for lipid research. Nucleic Acids Res 35(Web Server issue):W606–W612
3. Han X, Gross RW (2005) Shotgun lipidomics: electrospray ionization mass spectrometric analysis and quantitation of cellular lipidomes directly from crude extracts of biological samples. Mass Spectrom Rev 24(3):367–412
4. Cajka T, Fiehn O (2014) Comprehensive analysis of lipids in biological systems by liquid chromatography-mass spectrometry. Trends Analyt Chem 61:192–206
5. Hines KM, Herron J, Xu L (2017) Assessment of altered lipid homeostasis by HILIC-ion mobility-mass spectrometry-based lipidomics. J Lipid Res 58(4):809–819
6. Paglia G, Astarita G (2017) Metabolomics and lipidomics using traveling-wave ion mobility mass spectrometry. Nat Protoc 12 (4):797–813
7. Paglia G et al (2015) Ion mobility-derived collision cross section as an additional measure for

lipid fingerprinting and identification. Anal Chem 87(2):1137–1144
8. Lanucara F et al (2014) The power of ion mobility-mass spectrometry for structural characterization and the study of conformational dynamics. Nat Chem 6(4):281–294
9. Lapthorn C, Pullen F, Chowdhry BZ (2013) Ion mobility spectrometry-mass spectrometry (IMS-MS) of small molecules: separating and assigning structures to ions. Mass Spectrom Rev 32(1):43–71
10. May JC, McLean JA (2015) Ion mobility-mass spectrometry: time-dispersive instrumentation. Anal Chem 87(3):1422–1436
11. Groessl M, Graf S, Knochenmuss R (2015) High resolution ion mobility-mass spectrometry for separation and identification of isomeric lipids. Analyst 140(20):6904–6911
12. Kyle JE et al (2016) Uncovering biologically significant lipid isomers with liquid chromatography, ion mobility spectrometry and mass spectrometry. Analyst 141(5):1649–1659
13. Basit A et al (2016) Ion mobility mass spectrometry enhances low-abundance species detection in untargeted lipidomics. Metabolomics 12:50
14. Witting M (2018) Bio- and chemoinformatics approaches for metabolomics data analysis. Methods Mol Biol 1738:41–61
15. Marco-Ramell A et al (2018) Evaluation and comparison of bioinformatic tools for the enrichment analysis of metabolomics data. BMC Bioinformatics 19(1):1
16. Gowda H et al (2014) Interactive XCMS online: simplifying advanced metabolomic data processing and subsequent statistical analyses. Anal Chem 86(14):6931–6939
17. Chong J, Xia J (2018) MetaboAnalystR: an R package for flexible and reproducible analysis of metabolomics data. Bioinformatics 34 (24):4313–4314
18. Reis A et al (2013) A comparison of five lipid extraction solvent systems for lipidomic studies of human LDL. J Lipid Res 54 (7):1812–1824
19. Armirotti A et al (2014) Sample preparation and orthogonal chromatography for broad polarity range plasma metabolomics: application to human subjects with neurodegenerative dementia. Anal Biochem 455:48–54
20. Bligh EG, Dyer WJ (1959) A rapid method of total lipid extraction and purification. Can J Biochem Physiol 37(8):911–917
21. Yao CH et al (2016) Inaccurate quantitation of palmitate in metabolomics and isotope tracer studies due to plastics. Metabolomics 12:143

Chapter 7

Lipidomics by HILIC-Ion Mobility-Mass Spectrometry

Amy Li, Kelly M. Hines, and Libin Xu

Abstract

Lipidomics is a rapidly growing field that enables the characterization of the entire lipidome in cells, tissues, or an organism. Changes in lipid metabolism and homeostasis caused by different disease states or drug treatments can be probed by lipidomics experiments, which can aid our understanding of normal physiology and disease pathology at the molecular level. While current technologies using liquid chromatography coupled with high-resolution mass spectrometry have greatly increased coverage of the lipidome, there are still limitations in resolving the large number of lipid species with similar masses in a narrow mass window. We recently reported that two orthogonal separation techniques, hydrophilic interaction liquid chromatography (HILIC) and ion mobility (IM), enhance the resolution of lipid species based on headgroup polarity and gas-phase size and shape, respectively, of various classes of glycerolipids, glycolipids, phospholipids, and sphingolipids. Here we describe the application of our HILIC-IM-MS lipidomics protocol to the analysis of lipid extracts derived from either tissues or cells, to identify significant changes in the lipidome in response to an internal or external stimulus, such as exposure to environmental chemicals.

Key words Hydrophilic interaction liquid chromatography, Ion mobility, Mass spectrometry, Lipidomics, Collision cross section

1 Introduction

In recent years, the advancement of singular and multidisciplinary 'omics technologies have quickly taken storm as a systems biology approach is applied to biomedical research. They are highly informative experiments that can reveal molecular changes within some altered state of a system, such as a disease or treatment condition [1]. The central biological processes of DNA transcription to mRNA, subsequent translation to protein, and eventual production of metabolites correspond to the studies of genomics, transcriptomics, proteomics, and metabolomics, respectively. Lipidomics falls under the category of metabolomics as the lipidome represents a major fraction of the human metabolome [2]. Specifically, lipidomics aims to systematically identify and quantify individual lipid species within cells, tissues, or an organism, as well as map out the

Giuseppe Paglia and Giuseppe Astarita (eds.), *Ion Mobility-Mass Spectrometry: Methods and Protocols*,
Methods in Molecular Biology, vol. 2084, https://doi.org/10.1007/978-1-0716-0030-6_7,

complex interactions between lipids and other macromolecules. There is growing evidence that lipids are closely related to the onset and progression of several diseases, including neurodegenerative, neurodevelopmental, cardiovascular, and metabolic disorders. Thus, lipidomics represents a unique niche to address specific questions about the role of altered lipid metabolism within disease etiologies, as well as advance biomarker identifications to aid diagnosis [3].

Lipids cover a diverse range of functions within the cell as structural components of membranes, energy reservoirs, and signaling molecules, which reflects their large chemical diversity. There are eight main lipid classes (FA, fatty acyls; GL, glycerolipids; GP, glycerophospholipids; SP, sphingolipids; ST, sterol lipids; PR, prenol lipids; SL, saccharolipids; PK, polyketides) and multiple subclasses [4, 5]. Within each subclass, further variations are possible from multiple types of head groups, backbones, fatty acyl chains, linkages (ester, ether, or vinyl ether), and other accessory modifications, like sugar molecules [4, 5]. This vast combinatorial space is estimated (by the LIPID MAPS database) to contain over 40,000 distinct lipids, which presents a major analytical challenge for lipidomics because most lipid species occupy a narrow mass range of 600–900 *m/z*. Modern mass spectrometry has greatly advanced the field of lipidomics and increased coverage of the lipidome over the past two decades, but some limitations remain.

Mass spectrometry-based lipidomics can be divided into two categories: shotgun [6, 7] and liquid chromatography (LC)-based experiments [8–10]. In shotgun lipidomics, samples are directly infused into the mass spectrometer without any prior chromatographic separation, aiming to maximize the throughput of analysis. However, this approach is unable to resolve isomeric and isobaric lipids and limits the detection of lower abundance lipids due to ion suppression. LC-based lipidomics involves LC separation, such as normal phase (NP), reverse phase (RP), or hydrophilic interaction liquid chromatography (HILIC), prior to ionization. RP separation utilizes nonpolar C18 columns with polar solvents, where lipids are separated based on hydrophobicity, resulting in separation by chain length and degree of unsaturation of fatty acyl chains. HILIC separation utilizes columns with polar stationary phase in combination with typical reverse phase solvents, such as acetonitrile and water, where lipids partition at the water/silica interface. HILIC separation is mostly dependent on the polarity of lipid head groups [11, 12]. Further separation, although to a lesser extent, occurs within each subclass by chain length and degree of unsaturation of fatty acyl chains.

Ion mobility (IM) is another technology that has gained popularity and shown advantages in lipidomics studies as it contributes to both separation and structural elucidation of analytes and can be coupled to LC separation without affecting throughput

[13–16]. Ions are separated based on their size, shape, and charge as they are pushed through a drift tube by an electric field through an opposing flow of inert gas [16–18]. Drift times are related to a physicochemical property, called collisional cross section (CCS), which can be measured either directly on a drift-tube IM (DTIM) instrument using the Mason–Schamp equation [19, 20] or on a traveling wave IM (TWIM) instrument through calibration [21–23]. Previous papers from our group have shown that in TWIM experiments, matched lipid calibrants produce more accurate and precise CCS values of lipid standards compared to calibration by calibrants from other classes of molecules, such as poly-DL-alanine, tetraalkylammonium compounds, and hexakis(fluoroalkoxy)phosphazines [21]. IM is a gas phase technique that is orthogonal to LC and enables separation of lipid ions from other classes of macromolecules due to their distinct IM profiles [24, 25]. Even within total lipids, individual lipid classes can be separated from each other, although complete resolution of certain species is still limited by instrumentation [15, 26–28]. Coupled with retention times (RT) and mass-to-charge (*m/z*) ratios, CCS values greatly increase confidence in lipid identifications. Recently, the Zhu Lab reported the first platform for lipid identification using multidimensional RT-*m/z*-CCS data, LipidIMMS, which can utilize both RP-IM-MS and HILIC-IM-MS data [29].

Herein, we discuss the use of our HILIC-IM-MS protocol [14, 15] for the analysis of lipid species in the lipid extracts of biological samples using brain tissues and cells as examples. The workflow involves (1) sample preparation including tissue homogenization, lysis, and liquid phase extraction, (2) data acquisition on a Waters SYNAPT HDMS G2-Si Q-TOF ion mobility-mass spectrometer, and (3) data analysis using Progenesis QI and EZ Info software. The data included in Fig. 1 come from a paper by members of our research group, investigating the lipidomic changes in a neuroblastoma cell line exposed to an environmental chemical, benzylalkonium chlorides (BACs), and related compounds known to disrupt cholesterol and lipid homeostasis [14]. Readers are encouraged to consult the original paper for biological context.

2 Materials

2.1 General

1. Ammonium hydroxide (certified ACS reagent).
2. Ammonium acetate (Optima LC/MS grade).
3. Sodium chloride (certified ACS crystalline).
4. Acetonitrile LC/MS grade.
5. Water LC/MS grade.
6. Chloroform Optima LC/MS grade.

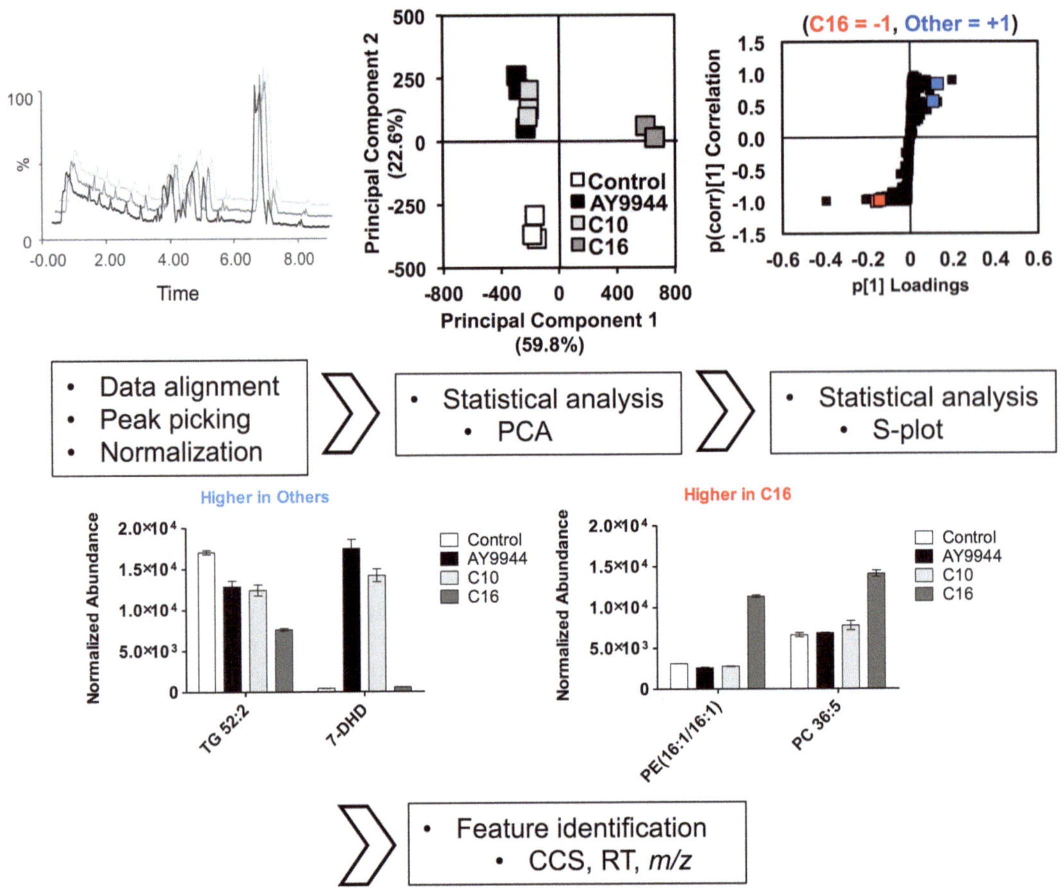

Fig. 1 Schematic of HILIC-IM-MS data analysis workflow from MS data processing to lipid identification

7. Methanol Optima LC/MS grade.
8. Methylene chloride Optima LC/MS grade.
9. Phosphate buffer saline (PBS).

2.2 Sample Preparation

1. 1× phosphate buffer saline.
2. *DC*™ Protein Assay Kit II.

2.3 Folch Lipid Extraction

1. Folch solution: 2:1 (v/v) chloroform/methanol.
2. Sodium chloride solution: 0.9% (w/v) NaCl aqueous solution.

2.4 HILIC-MS

1. HILIC A mobile phase: 95% ACN, 5% H_2O, 5 mM ammonium acetate.
2. HILIC B mobile phase: 50% ACN, 50% H_2O, 5 mM ammonium acetate.

Table 1
Lipid standards included in discrete lipid internal standards mixture

Lipid common name	Systematic name	Catalog no.
DG 13:0/13:0	Ditridecanoin glyceride	D-136 (Nu-Chek)
TG 15:0/15:0/15:0	Tripentadecanoin glyceride	T-145 (Nu-Chek)
Cer d18:1/17:0	*N*-Heptadecanoyl-D-erythro-sphingosine	860517
PG 15:0/15:0	1,2-Dipentadecanoyl-sn-glycero-3-phospho-(1′-rac-glycerol)	840446
PE 15:0/15:0	1,2-Dipentadecanoyl-sn-glycero-3-phosphoethanolamine	850704
PC 15:0/15:0	1,2-Dipentadecanoyl-sn-glycero-3-phosphocholine	850350
PA 12:0/12:0	1,2-Dilauroyl-sn-glycero-3-phosphate	840635
PS 12:0/12:0	1,2-Dilauroyl-sn-glycero-3-phospho-L-serine	840038
SM d18:1/17:0	*N*-Heptadecanoyl-D-erythro-sphingosylphosphorylcholine	860585
C12 Sphingosyl PE	*N*-Lauroyl-D-erythro-sphingosyl phosphoethanolamine	860529
Lyso PC 15:0/0:0	1-Pentadecanoyl-2-hydroxy-sn-glycero-3-phosphocholine	855576
Lyso PE 13:0/0:0	1-Tridecanoyl-sn-glycero-3-phosphoethanolamine	856706

2.5 Lipid Standards

1. Lipid internal standard solution: prepare 1 mM stock solutions of lipid standards in chloroform. Prepare a 5 μM mixture in HILIC A (*see* Table 1 for the list of lipid standards). All lipids were purchased from Avanti Polar Lipids (Alabaster, AL) unless otherwise noted as from Nu-Chek Prep (Elysian, MN).
2. Lipid mixture solution: extracts purchased from Avanti Polar Lipids were prepared at 1 mM concentration in chloroform. Prepare a 10 μM mixture in chloroform. For analysis, make a 1:2 dilution of the mixture in HILIC A (*see* Table 2 for the list of lipid extracts).
3. CCS calibrants: phosphatidylcholine (PC) and phosphatidylethanolamine (PE) lipid standards (Avanti Polar Lipids). Prepare 1 mM stock solutions of lipid standards. Prepare a mixture of PC 6:0–24:0 in methanol with 0.1% formic acid for positive mode analysis and PE 6:0–24:0 at 5–10 μM in methanol with 50 μM ammonium hydroxide for negative mode analysis. Prior to calibration, make a 1:4 dilution of the PC mixture and 1:2 dilution of the PE mixture in their respective solvents (*see* Table 3).

2.6 Supplies and Equipment

1. Analytical Column: HILIC Phenomenex Kinetex®, 2.1 × 100 mm, 1.7 μm.
2. Guard Cartridge: Phenomenex SecurityGuard® ULTRA HILIC Cartridge for 2.1 mm ID UHPLC columns.

Table 2
Lipid extracts included in lipid mixture

Lipid class	Catalog no.
Ceramide	860052P
Cerebroside	131303P
Sphingomyelin	860062C
L-α-Phosphatidic acid	840101C
L-α-Phosphatidylcholine	840051C
L-α-Lysophosphatidylcholine	830071P
L-α-Phosphatidylethanolamine	840022C
L-α-Phosphatidylglycerol	841138C
L-α-Phosphatidylinositol	840042C
L-α-Phosphatidylserine	840032C
Diacylglyceride	Nu-Chek Prep

3. Guard Cartridge Holder: Phenomenex SecurityGuard® ULTRA Holder for 2.1–4.6 mm ID UHPLC columns.
4. Instrument: Waters Synapt® G2-Si HDMS equipped with an ESI source (Waters Corporation, Milford, MA, USA).
5. Inlet System: Waters® Acquity I-Class FTN UPLC with Autosampler (Waters Corporation, Milford, MA, USA).

3 Methods

Carry out procedures at room temperature.

3.1 Sample Preparation: Cultured Cells

1. Pellet a minimum of 10^6 cells in Pyrex glass centrifuge tube at 104 × *g* (750 rpm) for 5 min at 4 °C (Sorvall® Legend® X1R Centrifuge) and keep samples on ice (*see* **Note 1**).
2. Lyse in 300 μL chilled 1× PBS using cold sonication for 30 min.
3. Conduct protein quantitation assay in a 96-well microplate (BioTek® Synergy® HTX Microplate Reader) (*see* **Note 2**).
4. Store at −80 °C or perform lipid extraction immediately.

3.2 Sample Preparation: Tissues

1. Measure tissue weight (*see* **Note 2**).
2. Add 4 mL Folch solution.

Table 3
PE and PC lipid species included in the lipid CCS calibration mixture

Lipid common name	Systematic name	Catalog no.	Conc. (μM)
PE			
PE 6:0/6:0	1,2-Dihexanoyl-sn-glycero-3-phosphoethanolamine	850697C	5
PE 8:0/8:0	1,2-Dioctanoyl-sn-glycero-3-phosphoethanolamine	850699C	5
PE 10:0/10:0	1,2-Didecanoyl-sn-glycero-3-phosphoethanolamine	850700C	5
PE 12:0/12:0	1,2-Dilauroyl-sn-glycero-3-phosphoethanolamine	850702X	5
PE 14:0/14:0	1,2-Dimyristoyl-sn-glycero-3-phosphoethanolamine	850745X	5
PE 15:0/15:0	1,2-Dipentadecanoyl-sn-glycero-3-phosphoethanolamine	850704X	10
PE 16:1/16:1	1,2-Dipalmitoleoyl-sn-glycero-3-phosphoethanolamine	850706C	10
PE 16:0/16:0	1,2-Dipalmitoyl-sn-glycero-3-phosphoethanolamine	850705X	10
PE 16:0/18:1	1-Palmitoyl-2-oleoyl-sn-glycero-3-phosphoethanolamine	850757C	10
PE 17:0/17:0	1,2-Diheptadecanoyl-sn-glycero-3-phosphoethanolamine	830756X	10
PE 18:1/18:1	1,2-Dioleoyl-sn-glycero-3-phosphoethanolamine	850725C	10
PE 18:0/18:1	1-Stearoyl-2-oleoyl-sn-glycero-3-phosphoethanolamine	850758C	10
PE 18:0/18:0	1,2-Distearoyl-sn-glycero-3-phosphoethanolamine	850715X	10
PE 20:4/20:4	1,2-Diarachidonoyl-sn-glycero-3-phosphoethanolamine	850800C	10
PC			
PC 6:0/6:0	1,2-Dihexanoyl-sn-glycero-3-phosphocholine	850305C	5
PC 8:0/8:0	1,2-Dioctanoyl-sn-glycero-3-phosphocholine	850315C	5
PC 10:0/10:0	1,2-Didecanoyl-sn-glycero-3-phosphocholine	850325C	5
PC 12:0/12:0	1,2-Dilauroyl-sn-glycero-3-phosphocholine	850335C	5
PC 14:0/14:0	1,2-Dimyristoyl-sn-glycero-3-phosphocholine	850345C	5
PC 16:0/16:0	1,2-Dipalmitoyl-sn-glycero-3-phosphocholine	850355C	10
PC 18:0/18:0	1,2-Distearoyl-sn-glycero-3-phosphocholine	850365C	10
PC 20:0/20:0	1,2-Diarachidoyl-sn-glycero-3-phosphocholine	850368C	10
PC 22:0/22:0	1,2-Dibehenoyl-sn-glycero-3-phosphocholine	850371C	10
PC 24:0/24:0	1,2-Dilignoceroyl-sn-glycero-3-phosphocholine	850373C	10

3. Disrupt and homogenize tissue for 30 s using a tissue homogenizer (Polytron® PT 1200 Kinematica).
4. Add 1 mL NaCl solution.
5. Continue to **step 2** of "Folch Lipid Extraction".

3.3 Folch Lipid Extraction

1. Add 4 mL Folch solution and 1 mL NaCl solution to cell lysates.
2. If including lipid internal standard, add to sample (*see* **Note 3**).
3. Vortex for 30 s.
4. Centrifuge at 1660 × *g* (3000 rpm) for 5 min at 4 °C.
5. Using a 9″ glass Pasteur pipet with a mechanical pipet pump, pipette through the upper, aqueous phase and protein layer and transfer approximately 2 mL of the lower organic layer to a 10 mL glass tube.
6. Dry lipid extracts with a speed vacuum concentrator (Thermo Fisher Savant SpeedVac® SC210A).
7. Reconstitute in 300 μL CH_2Cl_2 for cells and tissue samples with weights below 100 mg. Reconstitute heavier tissue samples in 1 mL CH_2Cl_2 and transfer to HPLC vials with screw caps. Store at −80 °C until HILIC-MS analysis.

3.4 Data Acquisition

1. To prepare samples for analysis, make a 1:4 dilution of lipid extracts in an LC glass autosampler vial. Adjust dilutions according to instrument signal intensity (*see* **Note 4**).
2. Prepare a pooled QC sample, combining an equal volume from each final sample into a separate LC autosampler vial.
3. Equilibrate LC column and instrument (*see* Tables 4, 5, and 6). The pressure should be around 4000 psi when the 0.5 mL/min flow rate is established.

Table 4
HILIC-IM-MS parameters: LC conditions

LC method	
Injection volume	5 μL for positive mode, 10 μL for negative mode
Flow rate	0.5 mL/min
Column temperature	40 °C
Sample temperature	6 °C
Mobile phase gradient	0–1 min: Hold at 100% HILIC A 4 min: Hold at 90% HILIC A 7–8 min: Linear gradient to 70% HILIC A 9–12 min: Equilibrate at 100% HILIC A

Table 5
MS source conditions

Source conditions	
Mode	**Resolution**
Mass range	50–1200 m/z
Capillary voltage	+2.5 kV for positive, −2.0 kV for negative
Sampling cone	40 V
Source offset	80 V
Source temperature	150 °C
Desolvation temperature	500 °C
Desolvation gas flow	1000 L/h
Cone gas flow	100 L/h
Nebulizer gas flow	5.0 bar

Table 6
Ion mobility parameters

IM/MS conditions	
IMS wave velocity	500 m/s
IMS wave height	40.0 V
Trap wave velocity	311 m/z
Trap wave height	4.0 V
Trap collision energy	4.0 eV
Transfer wave velocity	380 m/s
Transfer wave height	4.0 V
Transfer collision energy, function 1 (MS)	2.0 eV
Transfer collision energy, function 2 (MS^E)	Low, 35 eV High, 45 eV
Trap gas flow	2 mL/min
Helium cell gas flow	180 mL/min
IMS gas flow	90 mL/min

4. Perform detector setup through Intellistart interface (*see* **Note 5**).
5. Mass calibration. Infuse sodium formate at 20 μL/min and perform mass calibration through Intellistart interface for both positive and negative modes.
6. Switch instrument to Mobility TOF mode and allow the IM cell to equilibrate for an hour.
7. CCS calibration. Direct infuse PC and PE calibrant mixtures at 15 μL/min and calibrate through Intellistart interface (*see* Fig. 2).
8. LockSpray Source Setup to check calibration curve quality (checks that it gives the right CCS for LeuEnk based on the observed DT).
9. Inject the lipid extract mix (*see* Figs. 3 and 4 for example chromatograms), standard mix and a blank sample.
10. Randomize the sample list and increment a QC sample every ten injections.

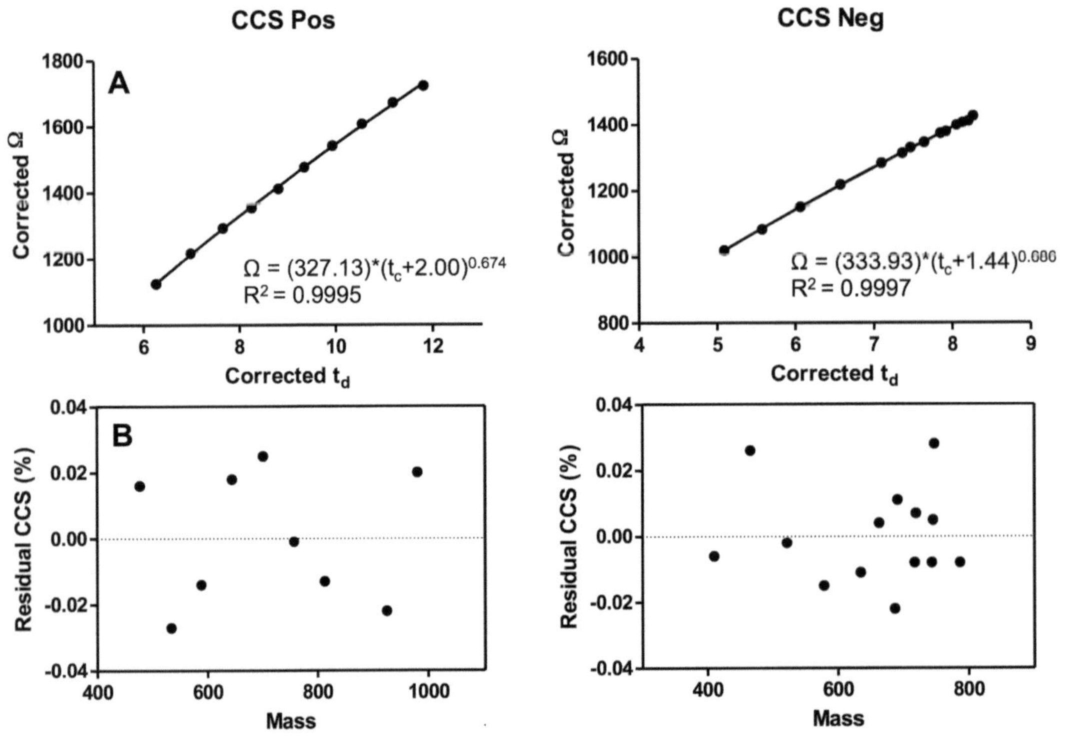

Fig. 2 (**a**) Example calibration plots and (**b**) corresponding residual plots for CCS calibration in both positive and negative ionization modes using PC and PE calibrant mixtures, respectively

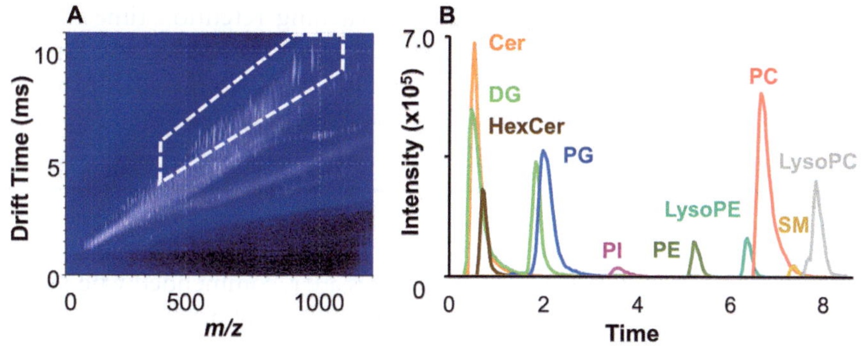

Fig. 3 (**a**) Drift time versus *m*/*z* spectrum showing IM profile of lipids. (**b**) Chromatogram of the most intense lipids from individual lipid subclasses that are present in the mammalian lipid extract mixture. Ceramide (Cer) d36:1; Diacylglycerol (DG) 36:4; Hexosylceramide (HexCer) d42:2; Phosphatidylglycerol (PG) 34:1; Phosphatidylinositol (PI) 38:4; Phosphatidylethanolamine (PE) 36:1; LysoPE 18:0; Phosphatidylcholine (PC) 34:1; Sphingomyelin (SM) d36:1; LysoPC 16:0

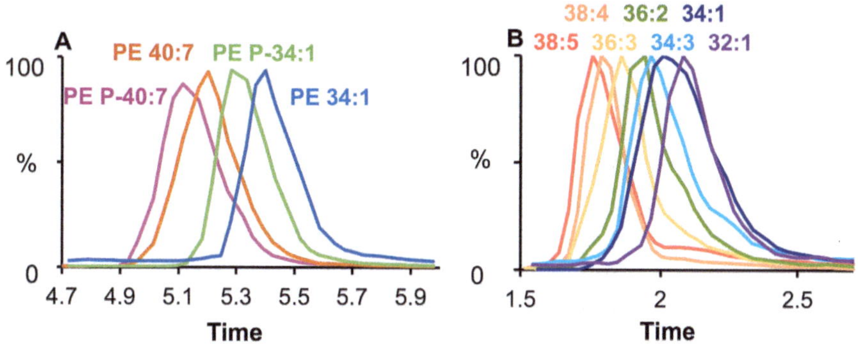

Fig. 4 (**a**) Chromatogram showing separation of phosphatidyl and plasmenyl PEs. M + H adducts shown. (**b**) Chromatogram showing separation of PGs by fatty acyl chain composition. From left to right: PG 38:5; PG 38:4; PG 36:3; PG 36:2; PG 34:3; PG 34:1; PG 32:1. M + Na adducts shown

3.5 Data Processing and Analysis (See Fig. 1)

1. Import raw files into Progenesis® QI (Nonlinear Dynamics) and process data in the chromatographic region between 0.4 and 8.5 min: data alignment to a reference QC sample, automatic peak detection of profile (continuum) data, and normalization to the external standard (such as tissue weight or protein weight).
2. Perform multivariate statistical analysis of the data (EZInfo).
3. If an internal standard was used, normalize the signals to the signal intensity of the appropriate internal standard in addition to external normalization.
4. Determine features that are significantly different between groups.

5. Identify features by matching retention time to specific lipid classes, *m/z* to METLIN, LIPID MAPS, and LipidBlast databases, and CCS values to LipidCCS/LipidIMMS [28, 29] and our own lipid CCS database [14, 15].

4 Notes

1. Glass tubes and pipettes are recommended when working with chloroform to prevent leaching of plasticizers, typically eluting between 2–3 min, which will interfere with data analysis.
2. Data should be normalized to protein weight or tissue weight, and internal standard (if used) post-data acquisition.
3. An example internal standard is d17:1/12:0 sphingo PE, N-lauroyl-D-erythro-sphingosyl phosphoethanolamine, catalog no. 860529. Please refer to the original paper (*see* ref. [12]) for more information on quantitative analysis using a single internal lipid standard.
4. Signal intensity of TIC should be on the order of 10^7 to ensure detection of lower concentration lipids. Preparing samples at two concentrations may be necessary to get the best results for both high and low abundance lipids.
5. LockSpray should be utilized (leucine enkephalin; positive mode: *m/z* 566.2771, CCS 226.0; negative mode: *m/z* 564.2615, CCS 224.0) to correct shifts in masses and drift times throughout the run.

Acknowledgements

This work was supported by grants from National Institutes of Health (R00HD073270 and R01HD092659). A.L. is an appointed trainee of the Pharmacological Sciences Training Program funded by the National Institutes of Health (T32GM007750).

References

1. Jurowski K, Kochan K, Walczak J, Baranska M, Piekoszewski W, Buszewski B (2017) Analytical techniques in lipidomics: state of the art. Crit Rev Anal Chem 47:418–437
2. Wishart DS, Knox C, Guo AC, Eisner R, Young N, Gautam B, Hau DD, Psychogios N, Dong E, Bouatra S, Mandal R, Sinelnikov I, Xia J, Jia L, Cruz JA, Lim E, Sobsey CA, Shrivastava S, Huang P, Liu P, Fang L, Peng J, Fradette R, Cheng D, Tzur D, Clements M, Lewis A, De Souza A, Zuniga A, Dawe M, Xiong Y, Clive D, Greiner R, Nazyrova A, Shaykhutdinov R, Li L, Vogel HJ, Forsythe I (2009) HMDB: a knowledgebase for the human metabolome. Nucleic Acids Res 37:D603–D610
3. Rustam YH, Reid GE (2018) Analytical challenges and recent advances in mass spectrometry based lipidomics. Anal Chem 90:374–397

4. Fahy E, Subramaniam S, Brown HA, Glass CK, Merrill AH Jr, Murphy RC, Raetz CR, Russell DW, Seyama Y, Shaw W, Shimizu T, Spener F, van Meer G, VanNieuwenhze MS, White SH, Witztum JL, Dennis EA (2005) A comprehensive classification system for lipids. J Lipid Res 46:839–861
5. Fahy E, Subramaniam S, Murphy RC, Nishijima M, Raetz CR, Shimizu T, Spener F, van Meer G, Wakelam MJ, Dennis EA (2009) Update of the LIPID MAPS comprehensive classification system for lipids. J Lipid Res 50 (Suppl):S9–S14
6. Han X, Gross RW (2005) Shotgun lipidomics: electrospray ionization mass spectrometric analysis and quantitation of cellular lipidomes directly from crude extracts of biological samples. Mass Spectrom Rev 24:367–412
7. Schwudke D, Liebisch G, Herzog R, Schmitz G, Shevchenko A (2007) Shotgun lipidomics by tandem mass spectrometry under data-dependent acquisition control. Methods Enzymol 433:175–191
8. Quehenberger O et al (2010) Lipidomics reveals a remarkable diversity of lipids in human plasma. J Lipid Res 51:3299–3305
9. Merrill AH Jr, Sullards MC, Allegood JC, Kelly S, Wang E (2005) Sphingolipidomics: high-throughput, structure-specific, and quantitative analysis of sphingolipids by liquid chromatography tandem mass spectrometry. Methods 36:207–224
10. Ivanova PT, Milne SB, Byrne MO, Xiang Y, Brown HA (2007) Glycerophospholipid identification and quantitation by electrospray ionization mass spectrometry. Methods Enzymol 432:21–57
11. Baker PR, Armando AM, Campbell JL, Quehenberger O, Dennis EA (2014) Three-dimensional enhanced lipidomics analysis combining UPLC, differential ion mobility spectrometry, and mass spectrometric separation strategies. J Lipid Res 55:2432–2442. https://doi.org/10.1194/jlr.D051581
12. Cifkova E, Holcapek M, Lisa M, Ovcacikova M, Lycka A, Lynen F, Sandra P (2012) Nontargeted quantitation of lipid classes using hydrophilic interaction liquid chromatography-electrospray ionization mass spectrometry with single internal standard and response factor approach. Anal Chem 84:10064–10070
13. Hinz C, Liggi S, Griffin JL (2018) The potential of ion mobility mass spectrometry for high-throughput and high-resolution lipidomics. Curr Opin Chem Biol 42:42–50
14. Hines K, Herron J, Xu L (2017) Assessment of altered lipid homeostasis by HILIC-ion mobility-mass spectrometry-based lipidomics. J Lipid Res 58:809–819
15. Hines KM, Waalkes A, Penewit K, Holmes EA, Salipante SJ, Werth BJ, Xu L (2017) Characterization of the mechanisms of daptomycin resistance among gram-positive bacterial pathogens by multidimensional lipidomics. mSphere 2:e00492–e00417
16. Kliman M, May JC, McLean JA (2011) Lipid analysis and lipidomics by structurally selective ion mobility-mass spectrometry. Biochim Biophys Acta Mol Cell Biol Lipids 1811:935–945
17. Fenn LS, McLean JA (2008) Biomolecular structural separations by ion mobility-mass spectrometry. Anal Bioanal Chem 391:905–909
18. McLean JA, Ruotolo BT, Gillig KJ, Russell DH (2005) Ion mobility-mass spectrometry: a new paradigm for proteomics. Int J Mass Spectrom 240:301–315
19. Mason EA, Schamp HW Jr (1958) Mobility of gaseous ions in weak electric fields. Ann Phys 4:233–270
20. Mason EA, McDaniel EW (1988) Transport properties of ions in gases. Wiley, New York, NY
21. Hines KM, May JC, McLean JA, Xu L (2016) Evaluation of collision cross section calibrants for structural analysis of lipids by traveling wave ion mobility-mass spectrometry. Anal Chem 88:7329–7336
22. Ruotolo BT, Benesch JL, Sandercock AM, Hyung SJ, Robinson CV (2008) Ion mobility-mass spectrometry analysis of large protein complexes. Nat Protoc 3:1139–1152
23. Forsythe JG, Petrov AS, Walker CA, Allen SJ, Pellissier JS, Bush MF, Hud NV, Fernandez FM (2015) Collision cross section calibrants for negative ion mode traveling wave ion mobility-mass spectrometry. Analyst 140:6853–6861
24. Fenn LS, Kliman M, Mahsut A, Zhao SR, McLean JA (2009) Characterizing ion mobility-mass spectrometry conformation space for the analysis of complex biological samples. Anal Bioanal Chem 394:235–244
25. May JC, Goodwin CR, Lareau NM, Leaptrot KL, Morris CB, Kurulugama RT, Mordehai A, Klein C, Barry W, Darland E, Overney G, Imatani K, Stafford GC, Fjeldsted JC, McLean JA (2014) Conformational ordering of biomolecules in the gas phase: nitrogen collision cross sections measured on a prototype high resolution drift tube ion mobility-mass spectrometer. Anal Chem 86:2107–2116

26. Paglia G, Angel P, Williams JP, Richardson K, Olivos HJ, Thompson JW, Menikarachchi L, Lai S, Walsh C, Moseley A, Plumb RS, Grant DF, Palsson BO, Langridge J, Geromanos S, Astarita G (2015) Ion mobility-derived collision cross section as an additional measure for lipid fingerprinting and identification. Anal Chem 87:1137–1144
27. Kyle JE, Zhang X, Weitz KK, Monroe ME, Ibrahim YM, Moore RJ, Cha J, Sun X, Lovelace ES, Wagoner J, Polyak SJ, Metz TO, Dey SK, Smith RD, Burnum-Johnson KE, Baker ES (2016) Uncovering biologically significant lipid isomers with liquid chromatography, ion mobility spectrometry and mass spectrometry. Analyst 141:1649–1659
28. Zhou Z, Tu J, Xiong X, Shen X, Zhu ZJ (2017) LipidCCS: prediction of collision cross-section values for lipids with high precision to support ion mobility-mass spectrometry-based lipidomics. Anal Chem 89:9559–9566
29. Zhou Z, Shen X, Chen X, Tu J, Xiong X, Zhu ZJ (2019) LipidIMMS analyzer: integrating multi-dimensional information to support lipid identification in ion mobility- mass spectrometry based lipidomics. Bioinformatics 35:698–700. https://doi.org/10.1093/bioinformatics/bty1661

Chapter 8

A Workflow for the Identification of Mycotoxin Metabolites Using Liquid Chromatography–Ion Mobility-Mass Spectrometry

Laura Righetti and Chiara Dall'Asta

Abstract

The structural identification of phase-I and phase-II metabolites of mycotoxins is a difficult task, mostly due to the lack of standards and because of the large number of isomeric forms. Here, we describe the use of ion mobility-mass spectrometry to analyze cereal extracts and how structural information on newly discovered mycotoxins metabolites could be obtained.

Key words Mycotoxins, CCS, Metabolite identification, Ion mobility-mass spectrometry, TWIMS

1 Introduction

Mycotoxins are one of the major food contaminants, causing important losses on the agricultural market, as well as being a great concern for food safety due to their toxicity [1].

These compounds may undergo modification in plants and animals, leading to the formation of a large number of possible modified forms, whose toxicological relevance and occurrence in food and feed are still largely unexplored. The analysis of modified mycotoxins by liquid chromatography–mass spectrometry is a challenging task because of their low occurrence levels, the complexity of the matrices, their chemical diversity, the large number of isomeric forms, and the lack of analytical standards.

The detection and identification of these unknown modified forms are permitted by taking advantage of the accurate mass, the isotopic pattern distribution provided by HRMS, and the study of the fragmentation pattern to identify positional isomers. However, in the case of phase-II conjugated metabolites, this approach has been shown to be poorly effective, as the conjugated moiety is the first to be lost in the fragmentation [2].

Giuseppe Paglia and Giuseppe Astarita (eds.), *Ion Mobility-Mass Spectrometry: Methods and Protocols*,
Methods in Molecular Biology, vol. 2084, https://doi.org/10.1007/978-1-0716-0030-6_8,

In this frame, ion mobility-mass spectrometry (IM-MS) is emerging as a promising technique to overcome the lack of analytical standards and the limitation of HRMS. In particular, IM-MS expands the analyte coverage of existing workflows by providing an additional separation dimension as well as a physiochemical property for characterization and identification of molecules, the collision cross section (CCS) [3]. Recently, the utility of this approach in supporting emerging research fields such as metabolomics [3–5] has been demonstrated providing a rapid and effective filtering method in the untargeted workflow. Research groups have used IM-MS to build CCS libraries in which the measured values serve as additional molecular descriptors for assigning identities to unknown target analytes [6].

Such approach is highly promising but depends on data from analysis of appropriate pure mycotoxin compounds. This represents a significant limitation because authentic chemical standards are not available for the majority of mycotoxins and their metabolites. These obstacles hinder the inclusion of CCS values in the mycotoxin identification workflow.

Very recently, a CCS database containing more than 100 TWIMS-derived values have been proposed [7], but it is far away from covering the wide range of new mycotoxins (more than 400) and their metabolites that can be present in food samples.

Therefore, further studies to extend and populate existing database are encouraged.

The most practical approach for increasing the size of reference libraries is through in silico calculation of theoretical CCS. They can be obtained by applying computational modeling or machine learning-based prediction strategies [8, 9]. An alternative strategy is the use of an in silico chemical library (ISiCLE) workflow which utilizes a combination of molecular dynamics and ion mobility calculations to predict CCS values based on theoretical structure information [10].

As a result, unknown mycotoxin metabolites may be discovered and characterized by matching theoretical and experimental rotationally averaged cross-sectional areas, despite the lack of analytical standards.

In this protocol, we describe how IMS might be integrated into traditional LC-HRMS mycotoxin workflows (Fig. 1). The main aim is to guide the reader through the identification of unknown metabolites. In particular, three steps are presented: (1) the protocol for extracting mycotoxins from cereal samples. (2) Then the appropriate condition for the mycotoxins detection and CCS value measurement using two different TWIMS mass spectrometers will be described. (3) Lastly, the prediction of the theoretical CCS by using different molecular descriptor will be introduced. In this way, it will be demonstrated that the comparison of the theoretical and experimentally derived collision cross sections could represent a

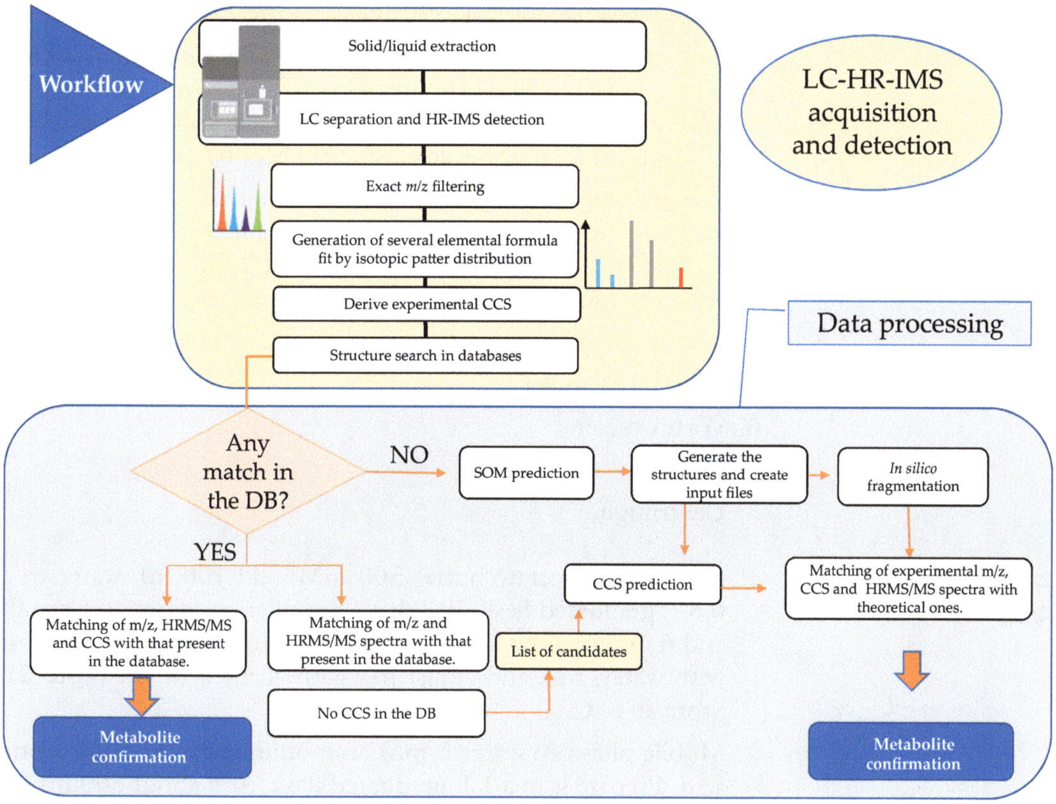

Fig. 1 Scheme of the suggested workflow for the identification of new mycotoxin metabolites

promising method to gain confidence in the metabolite identification and implement the traditional LC-HRMS workflow (Fig. 1).

2 Materials

1. Prepare all solutions using ultrapure water (prepared by purifying deionized water, to obtain a sensitivity of 18 MΩ cm at 25 °C) and HPLC grades (or higher purity) solvents only.
2. Prepare and store all reagents at room temperature unless indicated otherwise.
3. To avoid potential contact with toxic substances, perform all work with appropriate personal protection equipment, including gloves and glasses.
4. Vendor commercial software are indicated for acquisition and data visualization (Subheading 2.2) while free-online software packages are suggested for the steps needed for metabolite identification (i.e., in silico fragmentation, SOM prediction, drawing structures). The reader will choose among the large number of open-source software specifically designed.

2.1 Wheat Extraction

1. Store wheat samples at −20 °C (*see* **Note 1**).
2. Extraction mixture: prepare 500 mL of acetonitrile/water mixture (80/20; v:v) 0.1% acetic acid (*see* **Note 2**). Transfer 400 mL of acetonitrile in a graduated cylinder and add 100 mL of ultrapure water. Add 500 μL of acetic acid.
3. Prepare analytical standard of mycotoxin in acetonitrile. Make a dilution from the stock standard solution (approximately 100 μg/mL). Leave one aliquot at 4 °C for current use and store the remaining aliquots at −20 °C (*see* **Note 3**).
4. Sodium hypochlorite.
5. Laboratory miller.
6. Vortex mixer.
7. Shaker.
8. Centrifuge.

2.2 UHPLC-HRMS Analysis

1. Ammonium acetate buffer 500 mM: add 100 mL water to a 0.5 L graduated flask. Weigh 1.92 g of ammonium acetate salt and transfer to the flask. Add water up to a volume of 500 mL with water. Mix and adjust pH with acetic acid (*see* **Note 2**). Store at 4 °C.
2. Mobile phase A: water 5 mM ammonium acetate, 0.1% acetic acid. Prepare it in a 1 L graduated flask. Add about 500 mL of ultrapure water, 10 mL of 500 mM ammonium acetate, and 500 μL of acetic acid. Make up to 1 L water (*see* **Note 4**).
3. Mobile phase B: methanol 5 mM ammonium acetate, 0.1% acetic acid. Prepare a 1 L solution as in previous step (*see* **Note 5**).
4. LC column: ACQUITY UPLC BEH C18 column, 100 × 2.1 mm (inner diameter), particle size 1.7 μm.
5. 1.5 mL screw cap vial.
6. Poly-DL-alanine mix.
7. CCS Major Mix calibration solution.
8. Leukine enkephalin.
9. Use the following equipment.
   - LC system, e.g., ACQUITY UPLC I-Class System.
   - MS system, e.g., SYNAPT G2 HDMS and VION IMS QToF.
   - MassLynx software, used to acquire data for the SYNAPT HDMS system (Waters).
   - UNIFI software, used to acquire and processing data for the VION IMS QToF system (Waters).

- DriftScope software, used to extract regions of interest for different molecules on the basis of selective drift time extraction of mass spectra (Waters).

2.3 CCS Prediction

1. ChemAxon software [11], used to draw the chemical structure and to calculate 12 molecular descriptors.
2. ALOGPS, used to calculate log*P* and log*S* from the SMILES (simplified molecular-input line-entry system) string.
3. MetCCS, freely available webserver used to provide the predicted CCS values derived from nitrogen buffer gas.

3 Methods

Diligently follow all waste disposal regulations when disposing waste materials. Never dispose any mycotoxins by pouring their aqueous solutions down the drain or their organic solutions into a waste container. Prior to their disposal, the solutions should be treated with at least one-tenth their volume of 5% sodium hypochlorite (household bleach) for 12 h according to the laboratory's chemical hygiene plan.

3.1 Mycotoxin Extraction

1. Ground the entire lot sample to a fine particle size (*see* **Note 6**) using a laboratory miller.
2. Weight 1 g of ground wheat in 15 mL falcon tubes and add 4 mL of extraction mixture acetonitrile/water mixture (80:20, v/v) acidified with 0.1% acetic acid. Suspension is horizontally shaken for 90 min at 200 strokes/min.
3. Centrifuge the extract samples at 9.5 × *g* for 10 min to bring down the condensate. Then transfer 1 mL of supernatant in a screw cap vial and evaporate to dryness under a gentle stream of nitrogen.
4. Redissolve the residue in 1 mL of water/methanol (80:20, v/v) prior to LC-MS injection. The compound must be dissolved in a solvent HPLC grade (or higher purity) where it is fully soluble. Solution must be free of precipitate and floating particles.
5. Reconstructed extracts have to be injected the same day or stored in deep freezers immediately after extraction until measurement (*see* **Note 7**).

3.2 Setup of the LC for Analyzing Mycotoxins

1. Prepare mobile phases A and B.
2. Mobile phase A: water with 5 mM ammonium acetate and 0.1% (vol/vol) acetic acid.

   Mobile phase B: methanol with 5 mM ammonium acetate and 0.1% (vol/vol) acetic acid.

Table 1
Gradient elution suggested for the separation of mycotoxins and their metabolites

Time (min)	Mobile phase (%B)
0	10
3	40
15	90
17	90
17.1	10
20	10

3. Insert the column into the column compartment. We suggest using the ACQUITY UPLC BEH C18 column, 100 × 2.1 mm (inner diameter), particle size 1.7 μm.
4. Set the column temperature at 40 °C, the flow rate at 0.4 mL/min, the injection volume at 5 μL, and the autosampler temperature at 10 °C.
5. Set gradient conditions. Use the elution gradient reported in Table 1.

3.3 TWIMS-MS Acquisition

1. Optimize TWIMS-MS settings by directly injecting a mycotoxin mixture (1 mg/L in acetonitrile) both in positive and negative ionization modes. Optimize mycotoxins detection and IMS separation starting from the information given in Table 2.
2. To perform calibration for Synapt HDMS or Vion IMS QToF, read the following paragraphs.

3.3.1 Calibration of Synapt G2 HDMS

1. Acquire data using MassLynx software.
2. Perform calibration of the TOF analyzer by infusion (10 μL/min) of the corresponding calibration solution according to the manufacturer's guidelines.
3. Perform CCS calibration by infusion (10 μL/min) of the calibration solution (*see* **Note 8**). Ensure that you clearly see the complete range of *m/z* and drift times corresponding to the calibrants you are using (i.e., poly-DL-alanine).
4. Inspect the drift time of each calibrant both in positive and negative ionization modes. Experimental effective drift times have to be manually converted into estimated collision cross sections fitting the power trendline obtained plotting the corrected drift values (dt′) versus the normalized cross section (Ω′) of the calibrants (*see* **Note 9**).

Table 2
TWIMS-MS optimized settings for mycotoxins detection

HDMS^E acquisition parameters	
Ionization	ESI+ and ESI−
Acquisition mode	HDMS^E
Analyzer mode	Sensitivity
Mass-to-charge range (m/z)	50–1000
Scan time (s)	0.15
Low energy (V)	6
Collision energy ramp (V)	28–52
Wave velocity in the TWIMS cell (m/s)	650
Wave height in the TWIMS cell (V)	40
IMS gas	Nitrogen

3.3.2 Calibration of Vion IMS QT of Mass Spectrometer

1. Acquire data using UNIFI software.
2. Inject lock mass solution (Leucine Enkephalin), CCS calibration solution (Major Mix), and run the automated setup (*see* **Note 10**). It will calibrate both the Tof analyzer and the IMS system to allow direct measurement of the CCS values. In the event of check failure, the UNIFI software will alert the user.

3.3.3 Sample Analysis

1. Run the sample extract using UHPLC-TWIMS-MS. Set the same settings used for the calibration (*see* **Note 11**) both in positive and negative ionization modes. Experiments should be carried out in triplicate, and the CCS of each analyte should be monitored to check the reproducibility of the measurement.
   - Convert experimentally acquired drift time to CCS value using the calibration curve.
   - Read the CCS values generated from the automatic calibration performed by UNIFI software.

3.4 HR-IMS Spectra Interpretation

1. Import the LC-TWIMS-MS data into UNIFI software.
2. Search in the scan spectrum the *m/z* corresponding to the suspect metabolite. Generate the putative elemental formula by isotopic patter distribution fitting.
3. Compare retention time of parent mycotoxin with that of the putative metabolite of interest to check weather the retention time is anticipated due the increased polarity.
4. Inspect the HRMS/MS and make hypothesis on its chemical structure on the basis of its fragmentation spectrum.

5. Visualize the data in the DriftScope/UNIFI software via the 2D plot (*m/z* value versus drift time) to check the presence and the separation of isomers and/or isobars along the driftogram (*see* **Note 12**).
6. Search the structure against online database containing *m/z*, HRMS fragments and CCS values (*see* **Note 13**). Input both *m/z* value and CCS of the unknown metabolite and define a proper tolerance together with polarity information (*see* **Note 14**).
7. If there are no metabolite candidates, draw the putative chemical structure using software suitable for chemical structure drawing.
8. Perform SOM analysis to make hypothesis on its chemical structure (*see* **Note 15**).
9. Perform *in silico* fragmentation using MetFrag (http://c-ruttkies.github.io/MetFrag/) to find a match between measured and predicted HRMS/MS spectra (*see* **Note 16**).
10. Use MetCCS Predictor for predicting theoretical CCS for tentatively identify metabolites not present in the database.

3.5 CCS Predictions

1. Generate the SMILES string from the sketch of the chemical compound (*see* **Note 17**) using MarvinSketch.
2. Load the SMILES string in ChemAxon to get the molecular descriptors (i.e., polar surface area, rotatable bond count, etc.) required for CCS prediction.
3. Use ALOGPS to calculate logP and logS from the SMILES string.
4. Enter the 12 molecular descriptors to predict the CCS value by the machine learning tool "MetCCS Predictor" (http://www.metabolomics-shanghai.org/MetCCS/) (*see* **Note 18**).

4 Notes

1. An important aspect of the whole sampling procedure is the accurate storage of samples, especially if they have been collected in the field and need to be transported to a laboratory for analysis. The underlying principle is to handle samples in such a way as to prevent spoilage before analysis and thus ensure that analyses reflect the mycotoxin levels in the commodity at the time of sampling.

   If samples are already dry, it is advisable to store them in polyethylene or glass bottles, so that no moisture enters the sample. If possible, the samples should be kept in cold storage at <5 °C until ready for analysis. In the case of visibly moist samples, it is preferable to place these at −20 °C as soon as possible after collection and to maintain the samples at this temperature until ready for analysis.

2. Acetic acid is corrosive and highly flammable; it should be handled in a fume hood.
3. Weighing and transferring mycotoxins in dry form should be avoided; they should be dissolved in a solvent. The electrostatic nature of a number of the mycotoxins in dry form results in a tendency for them to be easily dispersed in the working area, and to be attracted to exposed skin and clothes. Mycotoxins present several toxic effects depending on the classed. Trichothecenes are potent dermal irritants and orally toxic. Ocratoxin A has been classified as a possible human carcinogen (group 2B) by the International Agency for Research on Cancer (IARC), while aflatoxins are classification as Group-1 because of collected evidence on its carcinogenic effect in humans.

   Therefore, protective clothing, gloves and safety glasses should always be worn, and all standard solutions and sample preparation stages should be carried out in a fume hood. Swab accidental spills of mycotoxins with 1% NaOCl bleach, leave for 10 min, and then add 5% aqueous acetone. Rinse all glassware exposed to mycotoxins with methanol, add 1% NaOCl solution, and after 2 h, add acetone to 5% of total volume. Let it react for 30 min and then wash thoroughly.
4. Prepare the aqueous solution fresh before the analysis and use amber flask to protect against the light. Buffer solution is added to promote the adduct formation. Indeed, several mycotoxins are poorly ionized if no modifiers are added and some of them are monitored as $[M + NH_4]^+$, $[M + Na]^+$ $[M + HCOO]^-$ $[M + CH_3COO]^-$.
5. Methanol is harmful and highly flammable; it should be handled in a fume hood.
6. The procedures for grinding and homogenization should ensure the smallest particle size and greatest homogenization possible, to effectively control mycotoxin extraction. The laboratory, taking in account its facilities, must validate the grinding and homogenization procedures.

   A finer sample particle size could actually facilitate the solvent access to the matrix, thus disrupting the matrix-toxin complexes involving noncovalent binding (*see* ref. Damiani et al. [12]).
7. We recommend storing sample extracts in deep freezers immediately after extraction until measurement to avoid the formation and detection of artifacts. Sauerschnig and coworkers [13] demonstrated that sample extracts stored at 10 °C for several days, as can typically be the case during longer measurement

sequences, led to an increase in both the number and abundance of methylated artifacts. In contrast, frozen sample extracts were relatively stable during a storage period of one week (*see* ref. Sauerschnig et al. [13]).

8. CCS derivation by TWIMS is achieved by calibration against a series of molecules whose CCS are known and ideally are of similar 'class' to the analyte being studied. Poly-DL-alanine is often used as IMS calibrant for deriving CCS. Since peptides have unique physical properties and gas phase conformations, it might not be ideal to calculate the accurate CCS values for all metabolites. For instance, alternative calibrants have been proposed for pesticides, which better reflect their chemical structure (Regueiro et al. [14]). We recommend using Major Mix Calibration instead of poly-DL-alanine for mycotoxins CCS derivation.
9. When using Synapt HDMS, the CCSs are derived after calibration with compounds of known CCS values. For a much more detailed description of the procedure, the reader is referred to the comprehensive Nature Protocol previously published (Paglia et al. [15]).
10. In the Vion IMS QTof, compared to Synapt HDMS, the procedure used for deriving CCS is automated, to facilitate its use for routine applications.
11. CCS measurements are influenced by few parameters that must be standardized along the same measurements to get comparable results. The buffer gas (N_2 or He or CO_2) is the main parameter used to influence separation, in addition to T-Wave velocity and height. Also the adduct type influences the ion structure and thus the time taken to pass through the drift tube. Therefore, we recommend comparing results acquired using the same mobile phases or at least to monitor the same adduct species to compare CCS values.
12. Phase-II mycotoxin metabolites are fragile molecules and might originate in source fragmentation. On the other hands, T-wave cell is also reported to produce significant ion heating that might break the covalent bonds. Therefore, when checking the driftogram, the analyst should pay attention to the formation of in source fragments to avoid misinterpretation of the data.

    The novel geometry of the Vion system allows to elucidate fragmentations that might occur for labile mycotoxin metabolites during different stages of the ESI-TWIMS QTOF analysis. Indeed, it is possible to state if dissociation occurs before (fragments would present shorter drift times than precursor ion) or after the TWIMS cell (drift time of the fragment ions would match that of the precursor ion). This is useful to understand which parameter has to be properly set to avoid/promote analyte fragmentation.

13. Few searchable databases containing *m/z* and CCS values for small molecules are publicly available, including MetCCS (http://www.metabolomics-shanghai.org/MetCCS/) and the one proposed by the Bush Lab (https://depts.washington.edu/bushlab/ccsdatabase/). Users are required to input both of the *m/z* and CCS values of an unknown metabolite and find a match with the values present in the database. Recently, MetCCS has been implemented including also some residues such as pesticides.

    On the other hand, database containing mycotoxin *m/z* values, fragments, and rt are in-house built but they are not publicly available for the scientific community. The most common mycotoxins can be found on Pub Chem database (https://pubchem.ncbi.nlm.nih.gov/) together with some human and plant metabolites such as DON and ZEN glucoside and glucuronide or ZEN sulfate. Recently, we developed an online database containing 3D structures, chemical formula, molecular weight, SMILES code, and LogP values of 20 ZEN biotransformation products. The database is freely available for download at https://sites.google.com/site/zearalenonesdf/home, with new molecules being added monthly.

14. Search metabolites by selecting the tolerance of *m/z* and CCS values as 5 ppm and 3%, respectively.

15. Several free web servers are available to predict the sites of metabolism (SOM). Based on the structural formula and by setting phase-I and phase-II transformation (i.e., dehydration, desaturation, hydration, oxidation, reduction, glucoside conjugation, sulfation, GSH conjugation), a list of theoretical metabolites will be generated, and they can be compared with the putative metabolites found.

16. HRMS/MS spectra for small molecules can be predicted with reasonable accuracy given appropriate training data, enabling the generation of short lists of molecules whose MS/MS spectra might match to experimental spectra. There is a large number of open-source software specifically designed. MetFrag (Ruttkies et al. [16]) is a freely available web tool (http://c-ruttkies.github.io/MetFrag/) for the annotation of high-precision tandem mass spectra. Start by uploading candidate structures as SDF (structure data file). Candidates are then fragmented in silico, and the fragments are compared with the products ion in the measured mass spectrum. A score calculated using the fragment peak matches gives hints to the quality of the candidate spectrum assignment.

17. The SMILES structures of new mycotoxin metabolites have to be created by drawing the chemical structure. However, for metabolites and lipids, they can be obtained from the LIPID

MAPS Structure Database (LMSD) or other similar databases. While for other residue compounds, such as steroids, all the molecular descriptors required for CCS prediction can be directly obtained from DrugBank database.

18. MetCCS Predictor provides the predicted CCS values derived from nitrogen buffer gas, denoted as ΩN_2. Currently, the prediction of CCS values in helium gas (ΩHe) or in other buffer gases is not supported. MetCCS Predictor is applicable primarily for metabolites (<1000 Da) and other small molecules like synthetic drugs, pesticides, nature products.

References

1. Berthiller F, Crews C, Dall'Asta C, Saeger SD, Haesaert G, Karlovsky P, Oswald IP, Seefelder W, Speijers G, Stroka J (2013) Masked mycotoxins: a review. Mol Nutr Food Res 57:165–186
2. Righetti L, Paglia G, Galaverna G, Dall'Asta C (2016) Recent advances and future challenges in modified mycotoxin analysis: why HRMS has become a key instrument in food contaminant research. Toxins 8:361
3. Picache JA, Rose BS, Balinski A, Leaptrot KL, Sherrod SD, May JC, McLean JA (2019) Collision cross section compendium to annotate and predict multi-omic compound identities. Chem Sci 10:983–993. https://doi.org/10.1039/C8SC04396E
4. Schrimpe-Rutledge AC, Sherrod SD, McLean JA (2028) Improving the discovery of secondary metabolite natural products using ion mobility–mass spectrometry. Curr Opin Chem Biol 42:160–166
5. Paglia G, Williams JP, Menikarachchi LC, Thompson JW, Tyldesley- Worster R, Halldorsson S, Rolfsson O, Moseley MA, Grant DF, Langridge J, Palsson BØ, Astarita G (2014) Ion mobility derived collision cross sections to support metabolomics applications. Anal Chem 86:3985e3993
6. Chalet C, Hollebrands B, Janssen HG, Augustijns P, Duchateau G (2018) Identification of phase-II metabolites of flavonoids by liquid chromatography–ion-mobility spectrometry-–mass spectrometry. Anal Bioanal Chem 410:471–482
7. Righetti L, Bergmann A, Galaverna G, Rolfsson O, Paglia G, Dall'Asta C (2018) Ion mobility-derived collision cross section database: application to mycotoxin analysis. Anal Chim Acta 1014:50–57
8. Zhou Z, Shen X, Zhu ZJ (2016) Advancing the large-scale CCS database for metabolomics and lipidomics at the machine learning era. Anal Chem 88:11084–11091
9. Zhou Z, Xiong X, Zhu ZJ (2017) MetCCS predictor: a web server for predicting collision cross section values of metabolites in ion mobility-mass spectrometry based metabolomics. Bioinformatics 33:2235–2237
10. Colby SM, Thomas DG, Nunez JR, Baxter DJ, Glaesemann KR, Brown JM, Pirrung MA, Govind N, Teeguarden JG, Metz TO, Renslow RS (2018) ISiCLE: a molecular collision cross section calculation pipeline for establishing large in silico reference libraries for compound identification. Quant Biol arXiv:1809.08378 [qbio.BM]
11. ChemAxon (2017) cxcalc, 16.11
12. Damiani T, Righetti L, Galaverna G, Suman M, Dall'Asta C (2019) Analytical issue related to fumonisins: a matter of sample comminution? Food Control 95:1–5
13. Sauerschnig C, Doppler M, Bueschl C, Schuhmacher R (2018) Methanol generates numerous artifacts during sample extraction and storage of extracts in metabolomics research. Meta 8:1
14. Regueiro J, Negreira N, Berntssen MHG (2016) Ion-mobility-derived collision cross section as an additional identification point for multiresidue screening of pesticides in fish feed. Anal Chem 88:11169e11177
15. Paglia G, Astarita G (2017) Metabolomics and lipidomics using traveling-wave ion mobility mass spectrometry. Nature 12:797e813
16. Ruttkies C, Schymanski EL, Wolf S, Hollender J, Neumann S (2016) MetFrag relaunched: incorporating strategies beyond in silico fragmentation. J Cheminform 8:3

Chapter 9

Quantitation of Cyclosporin A in Cell Culture Media by Differential Mobility Mass Spectrometry (DMS-MS/MS)

Amol Kafle, James Glick, Stephen L. Coy, and Paul Vouros

Abstract

Cell permeability is an important factor in determining the bioavailability of therapeutics that is usually measured by cell culture testing. The concentration of pharmaceutical in a medium such as Hank's Balanced Salt Solution with HEPES organic buffer (HBSS-HEPES) is measured at a series of time points, making simplicity and high throughput of the analytical method important characteristics. We report an electrospray differential mobility spectrometry mass spectrometry method (nanoESI-DMS-MS) for the rapid determination of cyclosporin A (CsA, cyclosporine) concentration in such a buffer. DMS technology provides gas phase atmospheric pressure ion filtration for small-molecule bioanalytical methods that suppresses interfering ions and reduces chemical noise, without the use of chromatography. This allows simplified sample preparation, fast calibration curve development, and shortened analysis times. It has also been noted that the DMS prefilter can reduce contamination of the mass spectrometer by salts, thereby extending mass spectrometer system uptime.

In the application described here, DMS-MS/MS is applied to cyclosporine A (CsA) in cell medium. Sample preparation is limited to dilution with an ammonium acetate-methanol-water mobile phase and the addition of CsA-d4 internal standard. The isotope ratio data are obtained in DMS-MS MRM mode observing NH_3 loss from the ammonium adduct of the two species. A calibration curve with high linearity ($R^2 = 0.998$) is rapidly obtained with nearly zero intercept, while it was found that a liquid chromatography LC-MS method required a preliminary SPE step to obtain a linear calibration curve. The time for data acquisition in the DMS-MS MRM method with flow injection (FIA) or infusion introduction at ESI flow of 400 nL/min is typically 30 s leading to a cycle time of less than 1 min.

Key words Cyclosporine, Cyclosporine A, Differential mobility spectrometry, Mass spectrometry, DMS, DMS-MS, Bioavailability, Cell permeability

1 Introduction

Several key applications in the field of mass spectrometry involve the analysis of drugs, metabolites, as well as endogenous/exogenous compounds in matrices of varying complexity. Advances in liquid chromatography and integration with online or offline cleanup methods have greatly facilitated these analyses, frequently bringing down the analysis time to a few minutes. However, it

Giuseppe Paglia and Giuseppe Astarita (eds.), *Ion Mobility-Mass Spectrometry: Methods and Protocols*, Methods in Molecular Biology, vol. 2084, https://doi.org/10.1007/978-1-0716-0030-6_9,

remains that method development for analytes in complex matrices is challenging and can require sample cleanup steps prior to a long chromatography step. Hence, an analytical method such as differential mobility mass spectrometry (DMS-MS) capable of removing interfering electrospray ions and suppressing chemical noise, and remaining effective with reduced sample preparation can be of great utility. There are a number of recent applications of DMS-MS in the literature, and Sciex now offers the technology as SelexION on their time of flight and triple-quad systems. The most recent publications from our group, the Vouros group at Northeastern University, apply DMS-MS to radiation exposure biodosimetry [1, 2], to quantitation of cancer-linked DNA adducts [3], and to improve the understanding the DMS method [4]. Reviewing the extensive differential ion mobility literature must be omitted here, but an overview of the technology can be found in broader publications on differential mobility methods [5, 6] or in some of the many other journal articles [7–9].

Measuring the cell membrane permeability of different types of molecules is an important consideration in optimizing the drug delivery and evaluating the uptake of molecules by various cells. Caco-2 cells (human colon epithelial cancer cell line) are widely used as a model system for studying intestinal absorption of drugs and other compounds [10], and it has also been demonstrated that Caco-2 cells in transport media under culture conditions can be used to predict in vivo human absorption [11]. Hence, in vitro studies of drug candidates in Caco-2 cells have become standard in early phases of drug development to help streamline the selection of drug candidates [12].

Analysis of permeability studies for drug candidates comes with significant analytical challenges. The compound usually has to be detected at a low concentration in a matrix like Hanks' balanced salt solution (HBSS) with additional HEPES buffer [11]. Furthermore, permeability is usually measured as a function of time which makes the throughput of the analytical technique important. Liquid chromatography with ultra-violet detector (LC-UV) has been widely used in permeability studies; however, the method largely depends on the absorbance of the analyte and matrix in the UV region. To overcome some of the detection issues using LC-UV, radioactive conjugates of the candidate molecules have been used in the permeability studies to enable detection [13], but radioactive labeling adds additional costs and necessitates special safety procedures during the synthesis, study, and disposal. While LC-MS has established itself as the preferred detection method for the free molecule in the complex buffer-based system, we found that obtaining a linear calibration cure required and additional pretreatment step like SPE, further increasing the time requirement. A rapid method eliminating sample cleanup steps prior to analysis could prove highly beneficial.

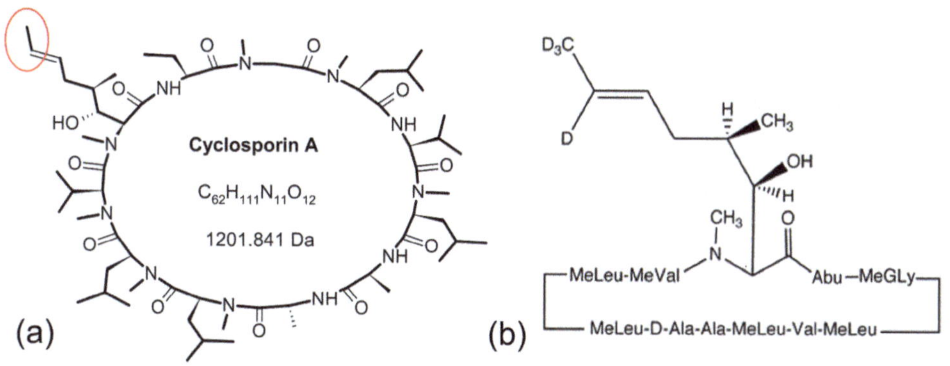

Fig. 1 Structure of cyclosporin A and cyclosporin A-d4. This cyclic peptide is important in avoiding rejection of transplants. The d4 deuterium-labeled compound is the only one commonly available. It provided an optimal approach to MS/MS analysis based on NH_3 loss from the ammonium ion adduct induced by a high level of ammonium acetate in the mobile phase

In the current application, we describe a DMS-MS/MS stable isotope dilution (SID) method for the determination of cyclosporin A (CsA) in the HBSS-HEPES cell culture medium that contains Na, K, Ca, and Mg metal cations, chloride, sulfate, phosphate, and bicarbonate anions, glucose, and the organic pH buffer HEPES. Cyclosporin A ($C_{62}H_{111}N_{11}O_{12}$, 1202.62 Da, Fig. 1) is a cyclic peptide that has revolutionized organ transplantation because of its effectiveness in the prevention of graft-versus-host disease (GVHD) and following graft rejection. It has additional applications for arthritis and in autoimmune diseases.

The current analysis method for a DMS-equipped triple-quad mass spectrometer is briefly summarized as follows. The HBSS/HEPES cell medium sample to be analyzed for cyclosporine A (CsA) level is combined with CsA-d4 internal standard in 40 mM ammonium acetate in 80:20 methanol/water. The NH_3-loss MS/MS transitions in both target (1219.88 → 1202.85) and internal standard (1223.90 → 1206.87) can be used without interference from metal ion adducts. The buffering action of ammonium acetate ensures that the mobile phase is of similar pH to the cell culture medium. The CsA-d4 stock solution (IS) is made up in the ammonium acetate mobile phase at a concentration of approximately 1000 pg/μL (0.83 μM), or any value in the 0.5–1.0 μM range. The sample and IS (sample:IS:mobile phase::1:1:18) are infused or flow injected to the ion source for 30 s. A low flow rate of 0.4 μL/min is used to ensure the desolvation required for DMS ion filtration, and to minimize ion suppression, and solvent consumption. Although the CsA mass range is somewhat above the range where chemical noise is especially troublesome [14], the DMS interface is able to reduce the baseline noise and enhance the ion signal, as shown in Fig. 2. The overall improvement in S/N estimated from Fig. 2 is about 4.5×. A comparative analysis of the

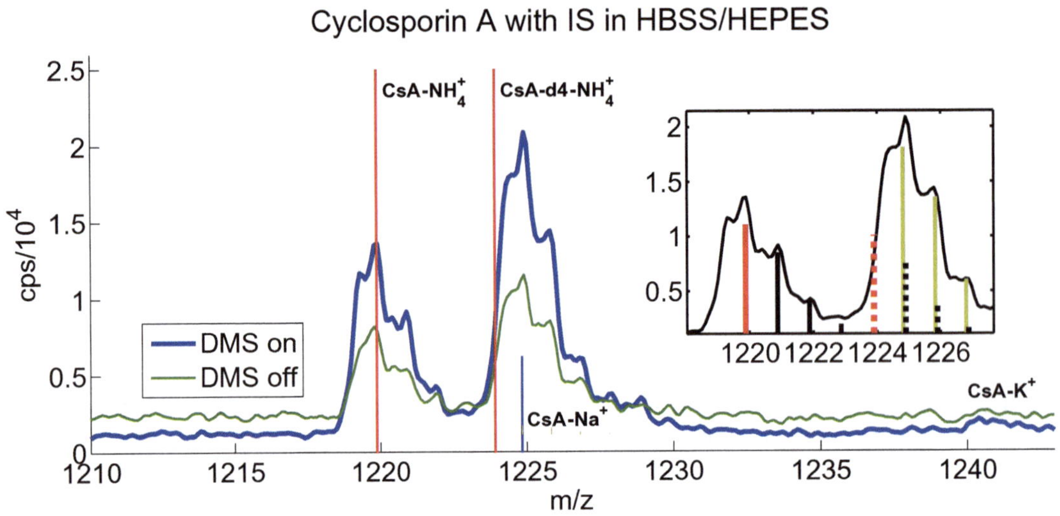

Fig. 2 DMS improvement in signal/noise ratio. There is a reduction in chemical noise in the *m/z* range by about a factor of two, and an increase in signal intensity due to ion motion in the gas flow of a similar factor, with an overall improvement of about 4.5×. A more detailed analysis based on the MS/MS data calibration results in an improvement factor of 3.6×

calibration curve data of Fig. 3, to the same samples analyzed without DMS filtration, gives a more accurate overall improvement in LOD and LOQ due to the DMS filtration of 3.6×.

2 Materials

1. Cyclosporin A (Fig. 1). Store solid cyclosporin at 2–8°°C and protect it from light (the compound is stable for 2 years).
2. Stock solutions of cyclosporin A should be stored at −20°°C, protected from light.
3. Cyclosporin A—d4. The CsA-d4-deuterated internal standard identifying substituted locations is illustrated in Fig. 1.
4. Hanks' Balanced Salt Solution (HBSS), sterile-filtered, endotoxin-free, suitable for cell culture, can be obtained preconstituted in liquid. The formulation includes 1 gm/L D-glucose, bicarbonate buffer, and Ca, Mg, K, and Na salts (chloride, sulfate, phosphate, and bicarbonate anions).
5. Organic pH buffer, HEPES, in a form rated as suitable for cell culture, such as BioXtra-labeled 1 M solutions.
6. Ultra-pure dry nitrogen, oil-free, filtered (ESI desolvation and nebulization, curtain gas, DMS transport gas).
7. Isopropanol 99.5% purity, and free from metal ion and organic contaminants, is used as gas phase modifier.

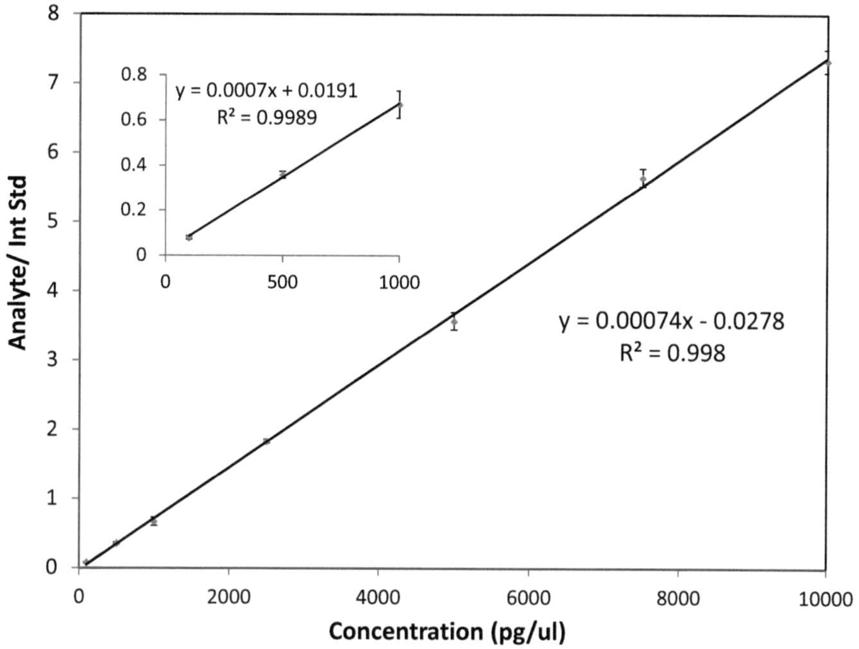

Fig. 3 Calibration curve prepared for DMS-MS/MS analysis of cyclosporine A in cell culture buffer. The DMS-MS/MS combination results in a highly linear curve with near-zero intercept. With the current instrumentation, the LOD is estimated to be 33 pg/μL, LOQ 100 pg/μL. Due to the absence of SPE and LC steps, the calibration curve can be generated in about an hour. Individual samples observed in MS/MS require about 30 s, making sample cycle times less than 1 min possible.

8. Methanol and water UHPLC grade.
9. Ammonium acetate prefiltered 5 M solution.
10. Mobile phase of 40 mM ammonium acetate in 80:20 methanol:water is used for dilution of the sample and internal standard and the ESI mobile phase.

3 Methods

3.1 Mass Spectrometer and DMS Instrumentation and Parameters

A triple-quad mass spectrometer equipped with positive-ion-mode low-flow electrospray ionization and DMS capability is used. Commercial and custom systems have been reported in the literature. The Sciex SelexION system is the most common commercial system, while Heartland Mass Spectrometry (https://heartlandms.wordpress.com) produces custom systems. We have also used simple home-built systems based on a Sionex design in a number of publications. This method was developed for a Sciex preproduction prototype SelexION interface hardware and software integrated on a Sciex API-3000 triple-quad system but would be expected to work on current SelexION systems with adjustments described in Subheadings 3 and 4 (*see* **Notes 1** and **2**).

1. A nebulized microflow ion source using the New Objective Metal Taper Tip stainless steel emitter was used for this method development, with flow at 400 nL/min. The Sciex Opti-flow Turbo V which supports flows from 1 to 200 μL/min, when operating at 1 μL/min, would be an appropriate source for use with a commercial SelexION instrument. Other microflow or nano-flow ESI sources would be expected to work as well, as long as ESI conditions such as flow rate, spray quality, spray geometry, temperature, and curtain gas or nebulizer gas flow remain stable.

3.2 Prepare Mobile Phase

1. Prepare the mobile phase of ammonium acetate-methanol-water. One liter of the mobile phase (80:20 (v/v) methanol water with 40 mM ammonium acetate) may be prepared at about 20 °C by combining 200 mL water, 8 mL 5 M ammonium acetate, and 800 mL methanol.
2. Select a clean HDPE 1 L bottle with closure. Using a graduated cylinder, add 200 mL water. Introduce by pipette 8 mL of 5 M ammonium acetate solution. Swirl briefly. Using a graduated cylinder measure and add 800 mL methanol. At 20 °C, the result is 978 mL of 80:20 v/v methanol water with ammonium acetate at 40.9 mM. Close and store at room temperature.

3.3 Instrument Calibration Check

1. Test the *m/z* tuning accuracy for the mass spectrometer Q1 and Q3 by infusing CsA 1000 pg/μL in mobile phase in a Q1 scan and in a Q3 scan from 1200 to 1250 *m/z*, or by using a tuning mix in the same mass range. Adjust as necessary for 0.1 Da accuracy.

3.4 Set Up Modifier Introduction

Isopropanol is the modifier used for this analysis. A gas phase modifier is introduced in the heated curtain gas or in a mixing chamber before DMS filtration, depending on the instrument configuration. The amount of liquid required is small, since the irate of introduction of the liquid into the heated curtain gas is approximately 20 μL/min, or an amount sufficient to provide a concentration of about 1% (v/v) (*see* **Note 3**). The interface depends on the instrumentation, with examples of different configurations available in the literature [6, 7]. Current commercial Sciex SelexION systems allow introduction at 1.5% and at 3% (v/v), while the system used here uses a syringe pump to introduce liquid isopropanol to a heated curtain gas line.

1. Determine the rate of flow of isopropanol required to produce 1–1.5% v/v in the curtain gas using the formula in Subheading 4 (*see* **Note 3**). The gas flow used in **Note 3** is the total modifier-added gas flow rate. This includes the gas flow to vacuum of the mass spectrometer as well as any desolvation outflow from the region preceding the DMS cell. Current

SelexION commercial instruments have a low setting which provides 1.5% modifier by volume.

2. Other configurations will require use of a separate liquid pump or syringe pump and syringe with fused silica line into the curtain gas line. The introduction area of a custom system must be warmed to ensure that the modifier enters into the gas stream.

3.5 Set Up Multiple Reaction Monitoring (MRM)

1. MRM transitions for NH_3 loss from CsA and CsA-d4 ammonium adducts are listed here (*see* **Notes 4** and **5**). Each transition should be recorded for 100 ms (including 5 ms dwell time), with the sequence repeated for a total of 30 s. The MRM m/z values are the following.
   (a) $CsA{\cdot}H^+ \leftarrow CsA{\cdot}NH_4^+$ is 1202.849 ← 1219.875
   (b) $CsA\text{-}d4{\cdot}H^+ \leftarrow CsA\text{-}d4{\cdot}NH_4^+$ is 1206.873 ← 1223.900
2. Prepare internal standard solution (IS) The internal standard solution is made by dissolving 1 mg of CsA-d4 in 1 L of mobile phase. This solution is at a concentration of 1000 pg/μL, which is 0.83 μM.
3. Reference sample. A reference sample is prepared that is used to develop or to verify correct configuration and correct tuning of the system. The reference sample is a 1:1:18 mixture with 1 part of 1000 pg/μL of CsA in HBSS/HEPES, 1 part of 1000 pg/μL CsA-d4 in mobile phase, and 18 parts of mobile phase. The procedure follows the description in the next section which describes calibration curve development.
4. Verify system using the reference sample. Scanning m/z from 1210 to 1250 m/z, step 0.1 m/z, in DMS-MS single-ion monitoring (SIM) with isopropanol modifier flow to add 1–1.5% v/v (*see* **Note 6**), and DMS parameters SV = 4500 V, COV = −1.0 V. For high-altitude locations, modify the SV for local atmospheric pressure as described in Subheading 4 (*see* **Note 7**).
5. Compare results to Fig. 2. The main part of Fig. 2 shows the DMS on/off improvement in S/N, and the inset shows the isotope patterns for CsA adducts, $CsA{\cdot}NH_4^+$, $CsA\text{-}d4{\cdot}NH_4^+$, and $CsA{\cdot}Na^+$. A similar pattern should be observed.
6. Select the best COV. Set the MRM transitions as in **step 1** above. Set the isopropanol modifier flow to 1% to 1.5% ("Low" on SelexION) (*see* **Notes 4** and **5**). Set SV to 4500 V, or as modified by the discussion in Subheading 4 (*see* **Notes 4** and **5**).
7. Scan COV from −5 to +5 V, to identify the best peak COV value.

8. The settings for SV, COV, and modifier flow are now completed. The COV value (typically about −1.0 V) is used in creating the final MRM method used for developing the calibration curve and for recording unknown concentrations.

3.6 Prepare the Cyclosporin A Calibration Curve

1. Internal standard stock solution. The CsA-d4 1000 pg/μL internal standard solution in mobile phase has already been prepared and will be used for the calibration. Cyclosporine concentrations may drop due to adsorption to the container wall and so may need to be tested.
2. Normal species cyclosporine A (CsA) stock solution. Prepare a stock CsA solution in HBSS/HEPES cell buffer, by dissolving CsA 1 mg/100 mL (10,000 pg/μL) in HBSS/HEPES (*see* **Note 8**).
3. Analytical samples: The calibration range 100–10,000 pg/μL is covered using a 1:2:5 sequence from 100 to 10,000 pg/μL. An additional sample of 0 pg/μL is also measured. Select eight 1.5 mL tubes.
4. In each of the eight tubes, add 50 μL internal standard (IS) solution.
5. To four tubes, add CsA stock solution, 50, 25, 10, 5 μL, respectively, for the 10,000, 5000, 2000, 1000 pg/μL labeled samples. Add makeup volumes of HBSS-HEPES buffer of 0, 25, 40, and 45 μL.
6. Dilute 200 μL of CsA stock solution with 800 μL of HBSS/HEPES cell buffer. Vortex to mix, add 25, 10, 5 μL to three tubes for 500, 200, 100 pg/μL. Add 25, 40, 45 μL of makeup volume of HBSS-HEPES. These tubes are labeled as 500, 200, 100 pg/μL.
7. In the final tube, labeled 0 pg/μL, add 50 μL HEBES-HEPES solution.
8. Add 900 μL mobile phase to each of the eight tubes and vortex.
9. Calibration data: Record the MRM calibration data for the samples in random order, in triplicate. Conditions are SV 4500 V, COV −0.9 V.
10. Select the same time range for each MRM intensity for averaging. A full FIA curve may show a peak or a tailing off at the edges of the injected flow. Only the central time range with approximately constant signal should be used. The correct time range can be identified from one of the more concentrated calibration samples. The averaged MRM intensity of CsA and CsA-d4 are recorded, and the CsA/CsA-d4 ratio computed.
11. The calibration curve is generated from *x* values of CsA concentration [CsA], and *y* values of the observed ratio of MRM intensities, *Y* = I(CsA)/I(CsA-d4). Create a linear calibration

curve with vendor software such as MultiQuant or using a spreadsheet program.

12. Compare this calibration for linearity and fit quality with that shown in Fig. 3.
13. Combine 50 μL unknown sample, 50 μL IS solution, and 900 μL of mobile phase (Volumes can be reduced proportionately; *see* **Note 9**).

3.7 Analyze Samples

1. All MRM conditions are the same as those of Subheading 3.5, **step 1** used in preparing the calibration curve. The MRM transitions are repeated here.
   (a) $CsA{\cdot}H^+ \leftarrow CsA{\cdot}NH_4^+$ is 1202.849 ← 1219.875
   (b) $CsA\text{-}d4{\cdot}H^+ \leftarrow CsA\text{-}d4{\cdot}NH_4^+$ is 1206.873 ← 1223.900
2. Sample introduction is by flow injection analysis (FIA) or by infusion.
3. MRM transition intensities for CsA and CsA-d4 are averaged for the full period of about 30 s omitting any transient edge effects in FIA at the beginning or end of the injection.

3.8 Data Analysis

1. Chromatographic peak integration is not necessary in DMS since particular ions are totally transmitted at the COV peak; the averaged CsA and CsA-d4 intensities are used directly as the ratio.
2. The observed ratio of averaged intensities Υ = (CsA)/(CsA-d4) is used with the calibration curve parameters to determine the CsA concentration [CsA]. If the calibration curve is Υ = (slope)·[CsA] + (intercept), then [CsA] = (Υ-intercept)/slope.
3. Concentration uncertainties can be determined by propagation of errors if the uncertainties of the slope and intercept are known. Those uncertainties are provided by LINEST in Excel or by other statistical software.

4 Notes

1. DMS filtration. DMS is a rapid gas phase ion mobility ion filtration technique that separates ions at near-atmospheric pressure based on the ion's mobility through an asymmetrical electric field which differs from the uniform electric field employed in conventional ion mobility techniques. This filtration occurs prior to mass analysis. An alternating RF voltage, known as the separation voltage (SV), is applied between two planar parallel electrodes through which ions pass on the weight to the MS inlet. In the asymmetric waveform, ions

experience a short period of high field followed by a longer period of low field causing a net ion displacement. In addition to the SV, a DC voltage also called the compensation voltage (COV) is adjusted to transmit selected ions of interest through the electric field, leaving most unwanted ions to be neutralized by encountering the electrodes. This continuous, non-pulsed ion transmission offers efficient coupling with continuous-sampling mass spectrometers such as a triple-quad or a high sampling rate TOF. Ion signals are recorded at the best COV chosen for maximum ion transmission at a SV that provides good separation of the analyte from other ion signals. Continuously recorded signals can be treated in the same way as those from standard flow injection analysis (FIA).

2. DMS intensity increase (Fig. 2). The increase of peak intensity when the DMS field is applied is a flow-dynamic effect resulting from the transverse oscillating electric field at the entrance to the DMS cell. Ions are entrained in the higher gas flow near the axis of the DMS ion filter entrance, so that the effective acceptance angle for ions is increased. This effect is useful in bringing more ions into the mass spectrometry and is now a feature of the Sciex SelexION+ system, by the addition of an additional oscillating field.
3. Commercial Sciex SelexION system has two settings for modifier delivery. The low setting corresponds to 1.5% v/v, while the high setting is 3.0% v/v. It is also possible to provide an alternative delivery system. We used a syringe pump introducing a flow of liquid modifier into the curtain gas. The required liquid flow rate can be calculated from the following expression: Flow(μL/min) =

$$120 \cdot F_{\mathrm{LPM}} M_{\mathrm{Da}} P_{\mathrm{bar}} / d_{\mathrm{g/cc}} T_{\mathrm{K}}$$

Notation:

(a) F_{LPM} (DMS transport gas flow in L/min).

(b) M_{Da} (molecular weight of modifier).

(c) P_{bar} (pressure in the DMS region in bar).

(d) $d_{\mathrm{g/cc}}$ (liquid density of modifier in g/cm^3).

(e) T_{K} (gas temperature in the DMS region).

In the case described here (1.0% isopropanol), the liquid flow was 20 μL/min with a total curtain gas flow of 750 mL/min. This flow consists of a vacuum drag of 600 mL/min and a curtain gas outflow of 150 mL/min.

4. The following adducts are observed in MS or MS/MS ($\cdot H^+$, $\cdot NH4^+$, $\cdot Na^+$, $\cdot K^+$, $-d4\cdot NH4^+$). The MS/MS transitions that were most effective are the same as those used by Keevil et al.

[15]. They correspond to the loss of NH_3 from the ammonium adduct of both the internal standard and the target compound.

5. The sodium and potassium adducts are also observed in DMS-MS (Fig. 2). In Fig. 2, the first peak group following *m/z* 1220 is the ammonium adduct peak group. As shown in detail in the inset in Fig. 2, the second peak group includes the CsA-d4 internal standard and the Na adduct of the normal species. The lowest CsA-d4 (*m/z* 1223.900) is 1 Da below the first Na adduct peak (*m/z* 1224.831), so the internal standard is separated from Na adducts in MS/MS in two ways: by *m/z* of the precursor (1024, 1025) and by *m/z* of the fragment (*m/z* 1207, *m/z* 1203). There is no interference on the CsA detection from the Na adduct by Q1 filtering which selects 1219.875 (5 Da below $CsA{\cdot}Na^+$), and no interference from the Na adduct on the CsA-d4 transition by Q3 filtering which selects 1206.873 *m/z* (4 Da above $CsA{\cdot}H^+$). Zero intercepts of the calibration curves of Fig. 3 also indicate no interference.
6. Electrical breakdown is more likely to occur at decreased pressure. For ambient pressure much different than 1 bar, reduce the SV value by a factor of (1 bar)/*P*, where *P* is the mean local pressure in bar. The two-harmonic waveform [4, 16] is now the most common. For the two-harmonic waveform with COV small compared to SV, the peak voltage is two-thirds of the setting. For SV = 4500 V, in a system with a 1 mm gap, and for a transport gas temperature of 100 °C, and 1 bar pressure, the field is 154.6 townsend (Td) units. Conditions with equal Td values have the same electric field contribution to the ion-neutral collision energy at different temperature and pressure conditions. The bulk gas temperature makes a separate contribution which must also be considered, so comparisons must be made for equal bulk temperature and Td [4, 17].
7. The COV value for this system is low and should not require readjustment day-to-day, as has been noted in publications with large COV shifts [1].
8. The buffer used in this sample (HBSS + HEPES) is among those typically used in cell culture assays. HBSS: The essential function of Hanks' Balanced Salt Solution (HBSS) is to maintain osmotic balance, also providing water and essential inorganic ions to cells. The classic formulation includes glucose and other salts. The composition is 0.14 M NaCl; 5 mM KCl; 1 mM $CaCl_2$; 0.4 mM $MgSO_4{\cdot}7H_2O$; 0.5 mM $MgCl_2{\cdot}6H_2O$; 0.3 mM Na_2HPO_4; 0.4 mM KH_2PO_4; 6 mM glucose; and 4 mM $NaHCO_3$. The ratio of Na/K in HBSS is 0.14/0.54 $\approx$ 26. This is consistent with an estimate of the peak height ratio CsA·Na/CsA·K seen in the Q3-MS DMS on

recording of Fig. 2. HEPES is an organic pH-buffering agent that is used in cell cultures because it maintains physiological pH more effectively with changes in cell respiration and temperature than alternatives the bicarbonate buffer present in HBSS.

9. Unknown samples are prepared with the same 1:1:18 dilution used to create the calibration curve. The same internal standard solution (IS) used to create the calibration curve must be used. If a new IS solution is prepared, a new calibration curve should be generated. Depending on the equipment, a smaller initial volume can be used for the sample than the 50 μL used in preparing the calibration samples. If 10 μL is used for the sample, then 10, 10, 180 μL volumes are combined and vortexed, and analyzed immediately or stored cold.

Acknowledgements

Development of this method was supported by NIH: RO1 CA 069390-16 (Paul Vouros, PI).

References

1. Vera NB, Chen Z, Pannkuk E, Laiakis EC, Fornace AJ Jr, Erion DM, Coy SL, Pfefferkorn JA, Vouros P (2018) Differential mobility spectrometry (DMS) reveals the elevation of urinary acetylcarnitine in non-human primates (NHPs) exposed to radiation. J Mass Spectrom 53:548–559. https://doi.org/10.1002/jms.4085
2. Chen Z, Coy SL, Pannkuk EL, Laiakis EC, Fornace AJ Jr, Vouros P (2018) Differential mobility spectrometry-mass spectrometry (DMS-MS) in radiation biodosimetry: rapid and high-throughput quantitation of multiple radiation biomarkers in nonhuman primate urine. J Am Soc Mass Spectrom 29(8):1650–1664. https://doi.org/10.1007/s13361-018-1977-z
3. Kafle A, Klaene J, Hall AB, Glick J, Coy SL, Vouros P (2013) A differential mobility spectrometry/mass spectrometry platform for the rapid detection and quantitation of DNA adduct dG-ABP. Rapid Commun Mass Spectrom 27(13):1473–1480. https://doi.org/10.1002/rcm.6591
4. Kafle A, Coy SL, Wong BM, Fornace AJ Jr, Glick JJ, Vouros P (2014) Understanding gas phase modifier interactions in rapid analysis by differential mobility-tandem mass spectrometry. J Am Soc Mass Spectrom 25(7):1098–1113. https://doi.org/10.1007/s13361-013-0808-5
5. Shvartsburg AA (2008) Differential ion mobility spectrometry: nonlinear ion transport and fundamentals of FAIMS. CRC Press, Boca Raton, FL. https://doi.org/10.1201/9781420051070
6. Schneider BB, Nazarov EG, Londry F, Vouros P, Covey TR (2016) Differential mobility spectrometry/mass spectrometry history, theory, design optimization, simulations, and applications. Mass Spectrom Rev 35(6):687–737. https://doi.org/10.1002/mas.21453
7. Coy SL, Krylov EV, Schneider BB, Covey TR, Brenner DJ, Tyburski JB, Patterson AD, Krausz KW, Fornace AJ Jr, Nazarov EG (2010) Detection of radiation-exposure biomarkers by differential mobility prefiltered mass spectrometry (DMS-MS). Int J Mass Spectrom 291(3):108–117. https://doi.org/10.1016/j.ijms.2010.01.013
8. Coy SL, Cheema AK, Tyburski JB, Laiakis EC, Collins SP, Fornace AJ (2011) Radiation metabolomics and its potential in biodosimetry. Int J Radiat Biol 87(8):802–823. https://doi.org/10.3109/09553002.2011.556177
9. Campbell JL, Yang AMC, Melo LR, Hopkins WS (2016) Studying gas-phase interconversion

of tautomers using differential mobility spectrometry. J Am Soc Mass Spectrom 27 (7):1277–1284. https://doi.org/10.1007/s13361-016-1392-2

10. van Breemen RB, Li Y (2005) Caco-2 cell permeability assays to measure drug absorption. Expert Opin Drug Metab Toxicol 1 (2):175–185. https://doi.org/10.1517/17425255.1.2.175
11. Yee S (1997) In vitro permeability across Caco-2 cells (colonic) can predict in vivo (small intestinal) absorption in man—fact or myth. Pharm Res 14(6):763–766. https://doi.org/10.1023/a:1012102522787
12. Hidalgo IJ, Raub TJ, Borchardt RT (1989) Characterization of the human colon carcinoma cell line (Caco-2) as a model system for intestinal epithelial permeability. Gastroenterology 96(3):736–749
13. Grès M-C, Julian B, Bourrié M, Meunier V, Roques C, Berger M, Boulenc X, Berger Y, Fabre G (1998) Correlation between oral drug absorption in humans, and apparent drug permeability in TC-7 cells, a human epithelial intestinal cell line: comparison with the parental Caco-2 cell line. Pharm Res 15 (5):726–733. https://doi.org/10.1023/a:1011919003030
14. Covey TR, Thomson BA, Schneider BB (2009) Atmospheric pressure ion sources. Mass Spectrom Rev 28(6):870–897. https://doi.org/10.1002/mas.20246
15. Keevil BG, Tierney DP, Cooper DP, Morris MR (2002) Rapid liquid chromatography-tandem mass spectrometry method for routine analysis of cyclosporin A over an extended concentration range. Clin Chem 48(1):69–76
16. Krylov EV, Coy SL, Vandermey J, Schneider BB, Covey T, Nazarov E (2010) Selection and generation of waveforms for differential mobility spectrometry. Rev Sci Instrum 81 (2):024101. https://doi.org/10.1063/1.3284507
17. Mason EA, McDaniel EW (1988) Transport properties of ions in gases. Wiley-Interscience, New York, NY. https://doi.org/10.1002/3527602852

Part II

Proteomics and Glycomics

Chapter 10

An Analytical Perspective on Protein Analysis and Discovery Proteomics by Ion Mobility-Mass Spectrometry

Johannes P. C. Vissers and Michael McCullagh

Abstract

Ion mobility combined with mass spectrometry (IM-MS) is a powerful technique for the analysis of biomolecules and complex mixtures. This chapter reviews the current state-of-the-art in ion mobility technology and its application to biology, protein analysis, and quantitative discovery proteomics in particular, from an analytical perspective. IM-MS can be used as a technique to separate mixtures, to determine structural information (rotationally averaged cross-sectional area) and to enhance MS duty cycle and sensitivity. Moreover, IM-MS is ideally suited for hyphenating with liquid chromatography, or other front-end separation techniques such as, GC, microcolumn LC, capillary electrophoresis, and direct analysis, including MALDI and DESI, providing an semiorthogonal layer of separation, which affords the more unambiguous and confident detection of a wide range of analytes. To illustrate these enhancements, as well as recent developments, the principle of in-line IM separation and hyphenation to orthogonal acceleration time-of-flight mass spectrometers are discussed, in addition to the enhancement of biophysical MS-based analysis using typical proteomics and related application examples.

Key words Ion mobility-mass spectrometry, Proteomics, Protein analysis, Cyclic TWIMS

1 Introduction

The use of ion mobility as an analytical technique to detect and separate biomolecules dates back to the break of the century with the application of the method for proteomics [1–3], glycomics, [4–6] and metabolomics [7]. It is a technique that separates gas phase ions based upon their mobility in a buffer gas. The separation is related to ion size, shape, as well as *m/z* and charge. The basis for separation by traditional drift-tube ion mobility at a low electric limit can be derived from the Mason–Champ equation,

$$K = \frac{3}{16}\left(\frac{2\pi}{\mu k_{\mathrm{B}} T}\right)^{1/2} \frac{ze}{N\Omega}$$

Giuseppe Paglia and Giuseppe Astarita (eds.), *Ion Mobility-Mass Spectrometry: Methods and Protocols*, Methods in Molecular Biology, vol. 2084, https://doi.org/10.1007/978-1-0716-0030-6_10,

where K = drift velocity v_d/electric field strength E, μ = is the reduced mass of the ion (neutral given by $(m_{neutral}m_{ion})/(m_{neutral} + m_{ion})$), k_B is the Boltzmann constant, T is the temperature, z is the charge state of the analyte ion, e is the charge on an electron, N is the number density of the drift gas, and Ω = average collisional cross section. The hyphenation of ion mobility spectrometry (IMS) with MS is often referred to as ion mobility-mass spectrometry (IM-MS). The most common mass analyzer coupled with IMS comprises a time-of-flight (TOF) instrument due to the inherent high sampling rate, although other mass detection systems have been described [8]. Four different methods of ion mobility separation are currently used in combination with MS, including drift-time ion mobility spectrometry (DTIMS), aspiration ion mobility spectrometry (AIMS), differential-mobility spectrometry (DMS), also called field-asymmetric waveform ion mobility spectrometry (FAIMS), trapped ion mobility spectrometry (TIMS), structures for lossless ion manipulations (SLIM) IM-MS, and traveling wave ion mobility spectrometry (TWIMS). A description of these methods is beyond the scope of this chapter, particularly since the majority of them have been reviewed [8, 9] and classified (in terms of nomenclature) [10] in great detail elsewhere.

The innovative demonstration of protein conformer separation by means of IMS by Clemmer et al. [11] has prompted instrumental IM-MS designs and the broader application of IMS as an analytical tool. The designs by Pringle et al. [12], Baker et al. [13], and Fernandez-Lima et al. [14], all orthogonal acceleration TOF (oa-TOF)-based IM-MS platforms, but utilizing different IMS geometries, have been commercialized. Application of IMS continues to expand and evolve. This utility has been applied in numerous areas of research including drug metabolism/metabolites [15, 16], lipids [17], trace impurities [18], carbohydrates [19–21], macromolecular protein species and viruses [22, 23], discovery proteomics [24, 25], protein ligand-binding studies [26], metal-based anticancer drugs [27], PEGylated conjugates [28], and MS imaging proteomics [29, 30]. In this chapter, the application of IMS for the identification, quantification, and characterization of proteins and proteomes are illustrated by application examples that demonstrate the benefits of integrating IMS into the analytical schema in terms of increased resolution and sensitivity, as well as those obtained from collision cross-sectional measurements.

2 Traveling Wave Ion Mobility-Mass Spectrometry

The principle of TWIMS is briefly discussed as it provides a basis and a reference for subsequent sections and discussions. A schematic of the device is shown in Fig. 1a. The design and principle details of TWIMS are described by Pringle et al. [12] and Giles et al.

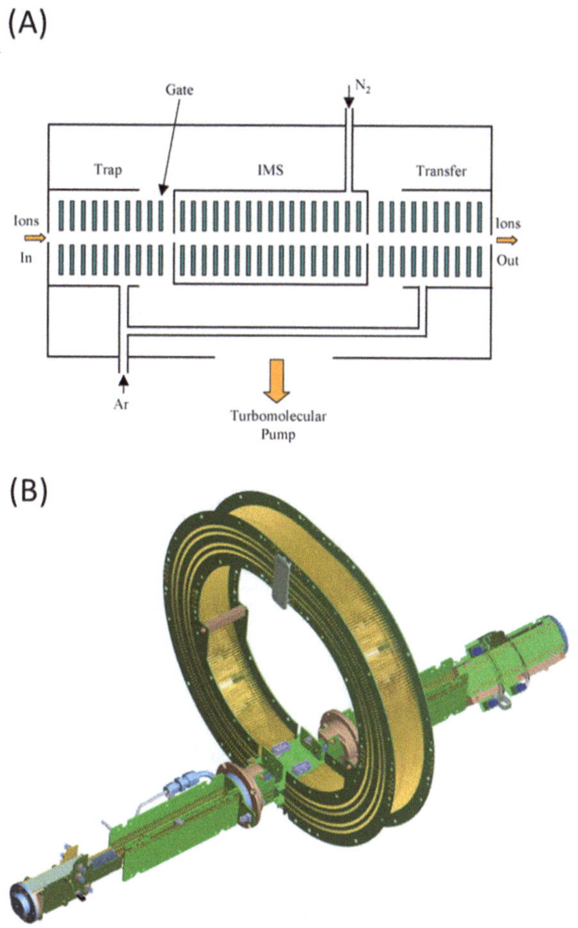

Fig. 1 (**a**) Original schematic diagram of TWIMS section incorporated in an oa-TOF geometry, comprising three traveling wave ion guides labeled, trap, IMS, and transfer. TriWave ion mobility optics consisting of a trap, helium cell, ion mobility separator and transfer, and (**b**) a CAD drawing of a cyclic TWIMS, comprising the same basic elements as the original TWIMS design, with the exception of a helium gate and a multifunction region for ion entry, separation, and exit. (Reprinted (**a**) with permission of Elsevier [12])

[31]. Ions are formed by electrospray ionization (ESI) in the source and subsequently pass through a quadrupole for mass selection before injection into the ion mobility cell. Unlike other designs, which use a uniform electric field across the cell for ion mobility experiments, so-called drift-tube IMS (DTIMS), segmented ion funnels, or a trapped ion mobility spectrometer (TIMS) analyzer, where an electric field hold ions stationary against a moving gas so that the drift force is compensated by the electric field, this device uses traveling wave (T-wave) technology.

The T-wave cell consists of a stacked-ring radio frequency (RF) ion guide which incorporates a repeating sequence of transient voltages applied to the ring electrodes. These voltage pulses result in a traveling electric field that propels ions through the background gas present in the mobility cell. The total time taken for an ion to drift through the cell depends on its mobility, the wave period and height, as well as the gas pressure. Ions with high mobility are better able to keep up with traveling waves and are pushed more quickly through the cell. Ions with low-mobility crest over the waves more often and have to wait for subsequent waves to push them forward, resulting in longer drift times. To measure an arrival time distribution (ATD), ions are gated into the mobility cell using an up-front stacked-ring RF device that traps ions before releasing them into the IMS cell. The oa-TOF pulses in an asynchronous manner, sending ions that have exited the mobility cell into the TOF mass analyzer and finally to the detector. Resolution enhancements to the device are recently described [32].

More recently, a multipass cyclic design was developed based on the use of TWIMS [33, 34], avoiding the high voltages needed in DTIMS, but still affording increased path length and consequently increased IM resolution. The resolution (resolving power) of a cyclic TWIMS device can be approximated by, [35, 36]

$$R^{\sim}\sqrt{\frac{zEL}{T}}$$

where R = resolving power, z = charge state of the analyte ion, E = electric field strength, L = path length, and T = temperature.

A schematic of the ion optics of the device is shown in Fig. 1b. A single pass provides an IM path length of 100 cm. Reverse five amino acid peptide separations up to 50 passes have been demonstrated from which a resolution ($\Omega/\Delta\Omega$) of ~550 is indicated [33]. A key feature within the design is the multifunction ion entry/separation/exit region consisting of an electrode array [37]. It can switch function to facilitate to the injection and ejection of ions into and out of the device, respectively, following mobility separation. Species can be reinjected under both nonactivating and activating, generated by collisional activation on re-entry into the array, conditions. An example of the increased IM resolving power that can be achieved with a cyclic TWIMS design is shown in Fig. 2, demonstrating the separation of two coeluting tryptic peptides with overlapping isotopic distributions using a fast chromatographic method. Shown inset are the arrival time distributions of the individual peptides and their corresponding, IM-separated MS spectra. TWIGs have also been integrated on printed circuit boards aiming at increasing path length on a small foot print as well [38, 39]. Proof-of concept results of TW SLIM devices have been demonstrated for the separation of peptides from an albumin tryptic digest and isomeric sugars.

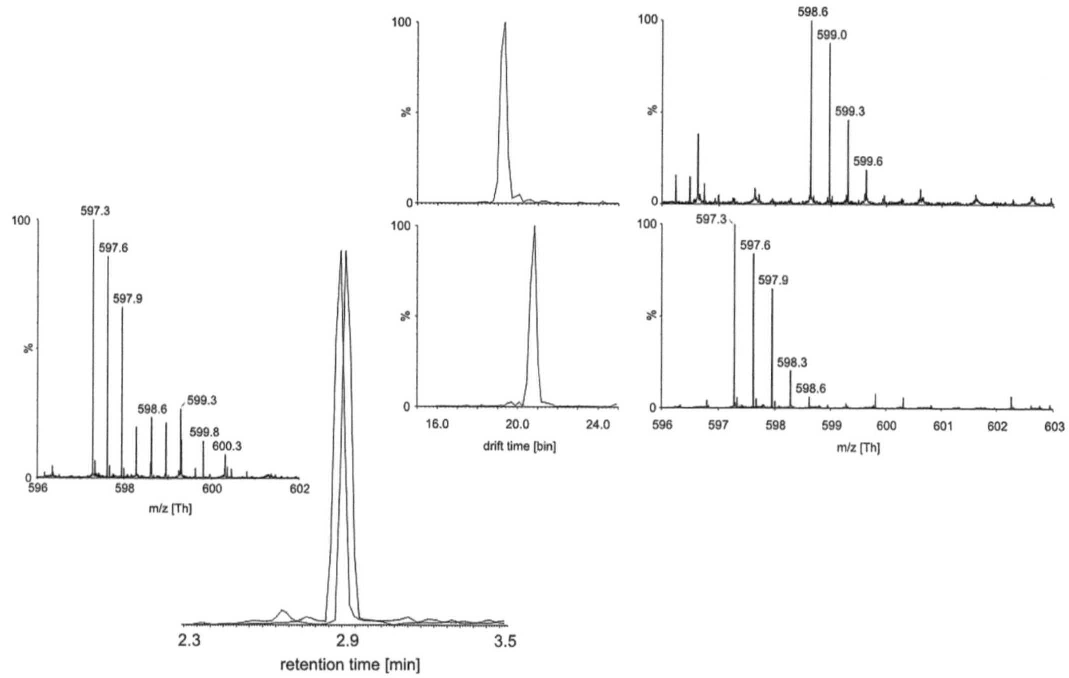

Fig. 2 Reversed phase chromatographic separation of a tryptic digest of enolase showing two near-isobaric coeluting tryptic peptides (middle), with overlapping isotopic distributions (left), their corresponding arrival time distributions (ATD) (left), and the resulting IM-separated MS spectra using cyclic TWIMS

3 IM-MS and LC-IM-MS Analysis of Simple and Complex Mixtures

3.1 Cross Section and Structure

By measuring the mobility of an ion, information about the rotationally averaged collision cross section, which is given by shape and size, can be determined. The relationship between the mobility of an ion and its collision cross section has been derived in detail using kinetic theory [40]. When all experimental IM parameter values are held constant, a dependence of the ion mobility constant results only from the average cross section with $K \sim 1/\Omega$ [41–43], where K = drift velocity v_d/electric field strength E and Ω = average collisional cross section. A detailed description of kinetic theory is however beyond the scope of this discussion. Ruotolo et al. [22] were among the first describing the use of IM-MS to decipherer protein complex structure. The analysis of the Trp RNA-binding attenuation protein provided compelling evidence that many features of protein assemblies, including quaternary structure, can be preserved in the absence of solvent molecules. TWIMS was coupled to a modified oa-TOF mass spectrometer to measure the CCS of four charge states of an eleven-mer complex, demonstrating that the lowest charge state exhibited the largest CCS, with a value in close agreement to that estimated for a ring structure determined by X-ray crystallography. IM-MS has since proved to be extremely

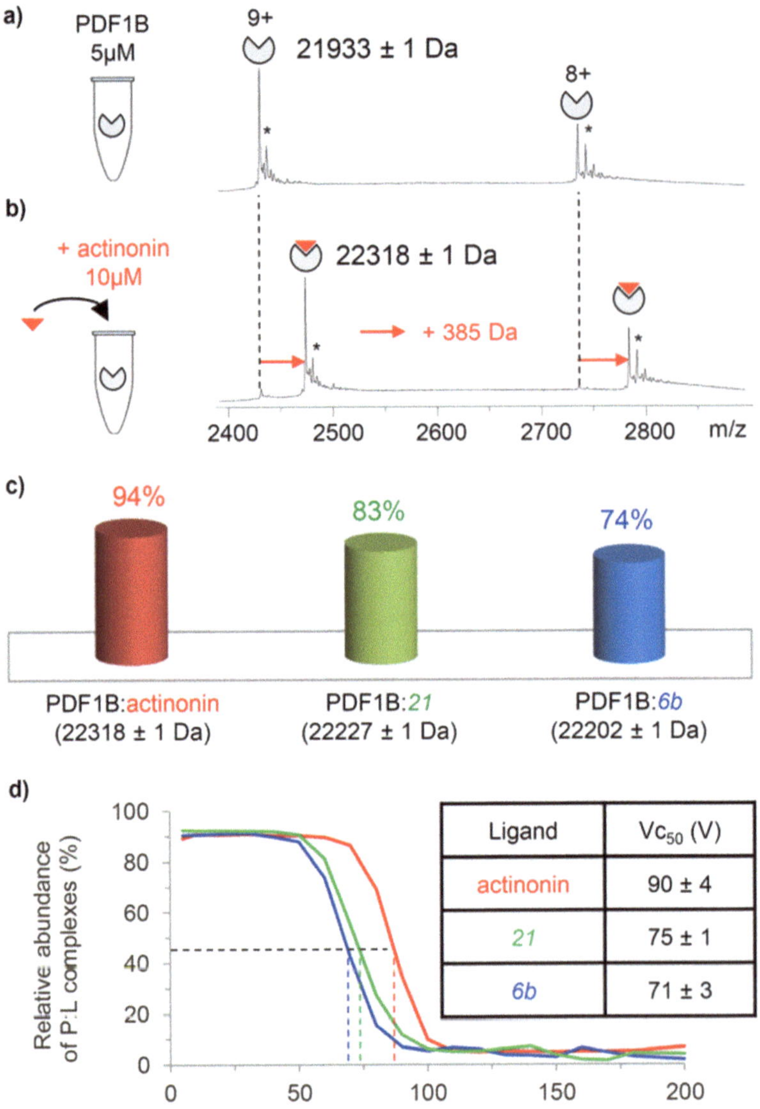

Fig. 3 Principle of native MS characterization of protein:ligand interaction. Native mass spectra of peptide deformylase (PDF1B) (**a**) alone or (**b**) after 10 min incubation with a ligand/drug (actinonin; $* = Ni^{2+}$ adducts). (**c**) Relative proportions of the PDF1B:ligand complexes, as determined by native MS from the intensities of the 8+ and 9+ charge states after incubation of PDF1B with ligand. (**d**) Protein:ligand ratios for three complexes as a function of the cone voltage in the ion source. The dashed lines indicate the corresponding V_c50 (cone voltage that indicates achieve 50% dissociation of the complex) values. (Reproduced with permission of Royal Society of Chemistry [47])

useful for the structural analysis of proteins and protein assemblies as illustrated in a number of recent reviews [9, 44–46]. A recent example from a related application area, is the use of IM-MS for protein–ligand (protein–drug) binding studies determining rates constants/kinetics. Shown in Fig. 3 is the principle of a protein–

ligand binding/interaction study using IM-MS [47], providing binding constant information that cannot be readily or easily obtained with other analytical techniques. In short, the change in conformation is monitored as a function of dosing concentration and a rate or binding/dissociation constant derived. A number of papers have been recently published that overview and described the current state of the art of IM-MS for studying protein–ligand [48–50], and protein–protein interactions, protein (complex) unfolding [51–53].

Collisional cross section measurements and structure IM-MS experiments are not restricted to the analysis of large molecules but have been applied to other molecule classes and applications as well. For example, Valentine et al. [54] used IMS to measure collision cross sections for 660 peptide ions generated by tryptic digestion proteins. Measured cross sections were compiled into a database that contains peptide molecular weight and sequence information and can be used to generate average intrinsic contributions to cross section for different amino acid residues. This was achieved by relating unknown contributions of individual residues to the sequences and cross sections of database peptides. Size parameters were combined with information about amino acid composition to calculate cross sections for database peptides. Figures 4a and 4b summarize the work showing cross sections as a function of molecular weight for the singly and doubly charged database peptides,

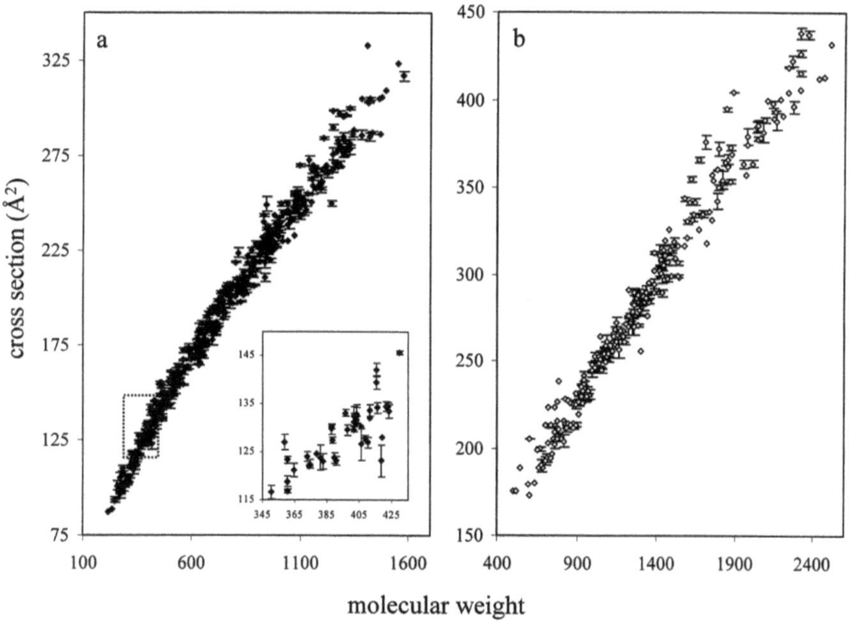

Fig. 4 (**a**) Cross sections for 420 $[M + H]^+$ peptides (solid diamonds) as a function of molecular weight. Uncertainties correspond to one standard deviation or to a range. The inset shows variations in cross sections for $[M + H]^+$ peptides over a smaller molecular weight range (defined by the dashed-line box). (**b**) Cross section measurements for 240 $[M + 2H]^{2+}$ peptides (open diamonds) as a function of molecular weight. (Reprinted with permission of Springer Nature [54])

respectively [54]. A strong correlation of increasing cross section with increasing molecular weight was observed, suggesting that (predicted) cross section can be used as an additional search parameter for peptide identification. A follow-up study proposed that a method that employs intrinsic amino acid size parameters to obtain ion mobility predictions can be used to rank candidate peptide ion assignments. The intrinsic amino acid size parameters were determined for doubly charged peptide ions from the complete annotated yeast proteome. The subsequent use of the predictive enhancement, as a means to aid peptide ion identification, was discussed and a simple peptide ion scoring scheme presented.

3.2 Separation

The work of Clemmer and coworkers [1, 55, 56] demonstrates the use of IMS for the separation and profiling of plasma proteins. The integration of IMS into an LC-MS schema is described as an example to increase the separation power of a platform. The set-up comprises off-line strong cation exchange (SCX) and in-line LC-IM-MS separation of trypsin-digested plasma proteins. The SCX-LC-IM-MS set-up is described in great detail as well as how the additional IMS separation dimension increases the available experimental peak capacity. The experimental two-dimensional LC-IM peak capacity was estimated to be ~6000–9000 obtained from a partial $t_r(t_d)$ base-peak plot derived from a single LC-IM-MS analysis, which greatly exceeds that of a single LC or IMS experiment. Also discussed was the use and creation of a relational table or database that comprises physicochemical analyte information such as SCX-retention time $t_{r,SCX}$, reversed phase (RP) retention time $t_{r,RP}$, drift time t_d, and m/z. This information can be stored in a multidimensional space as shown in Fig. 5. Knowledge of the positions of peaks will further corroborate assignments of other data sets. In addition, the accumulation of data provides valuable information for future work that would aim to predict SCX-retention times, LC-retention times, as well as mobility values based on sequences and charge states. The contribution of IM for the identification peptides as an additional search and identification parameter has been discussed [54, 56]. These concepts have been applied by Thalassinos et al. [57] for peptide and proteins identification and quantitation across two similar mammalian species and Paglia et al. [58] for the identification of the key metabolites potentially involved in cancer. A recent paper of Shah et al. [59] discusses the application of a machine-learning model that uses the physicochemical properties of peptides to accurately predict drift time directly from an amino acid sequence. The authors mentioned that the high-throughput nature of IM benefits from accurate calculation of cross sections, mobility values, and the associated drift times of peptides, thereby enhancing downstream data analysis, as a search metric, for the elucidation of secondary structure information, the identification of post-translational modifications,

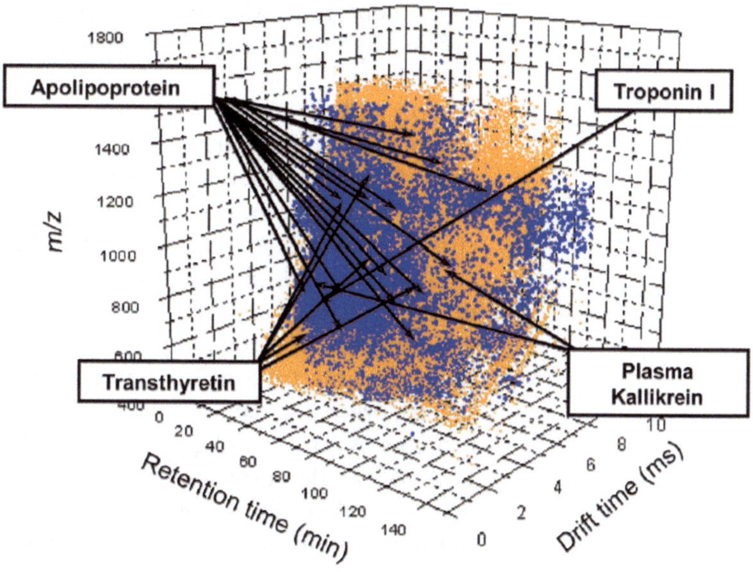

Fig. 5 3D dot plot representation of the positions of peaks (in the retention time, drift time, and *m/z* dimensions) that are obtained from the 1×10^5 most intense features (orange) observed during the triplicate LC-IM-MS analyses of all SCX fractions associated with Sample 1. Superimposed on the plot are the positions for >10,000 features that have been assigned to peptides (blue). The arrows indicate some of the precursor ion positions of peptides identified for the four proteins labeled. This representation is intended to provide the reader with the impression that the possible existence of abundant protein in plasma could be tested at many positions in the map, and therefore upon comparison, there should be little ambiguity regarding its detection, whereas a low-abundance protein may be represented at only a single position, leading to uncertainty about its detection. (Reprinted with permission of Springer Nature [55])

or to develop/apply peptide CCS databases. The increase in system peak capacity, experimentally derived, for a multidimensional LC-IM-MS system has been described and demonstrated by Rodríguez-Suárez et al. [60]. An increase in the number detected precursor and product ions of 130% was noted when including IM in the schema at fixed oa-ToF resolution. This particular figure of merit was independent of the LC mode, i.e., 1D, three fraction 2D or five fraction 2D nanoscale LC. The average concurrent increase in deconvoluted charge state groups (charge state groups combined), peptide, and proteins was 125%, 142%, and 142%, respectively.

Ion mobility-assisted data-independent analysis (DIA), LC-MS [61, 62] can be seen as an extension to the work of Clemmer and coworkers. Here, however, IMS is additionally used to align precursor and product ions to increase the specificity of a DIA workflow using TWIMS. In other words, it not only increases system peak capacity but also enhances the selectivity of DIA. In this experiment, to maximize duty cycle, peptide precursor ions are not isolated by the quadrupole mass analyzer positioned in front of the TWIMS cell. The ions undergo separation first in the mobility section and are either not fragmented or collision-induced dissociated (CID) in the transfer region. This process is repeated at a fixed frequency thereby generating so-called low and elevated energy precursor and product ion spectra, respectively. Thus, precursor and product ions share identical t_d which can be used to entangle the multiplexed product ion spectra. Briefly, precursor and product ion mass extracted chromatograms are created in the t_r and t_d domains. Precursor and product ion that share the same drift and retention time are correlated, which simplifies the multiplexed CID spectra prior to a database search for identification of peptides and proteins. As an example, Yang et al. [63] applied label-free LC-IM-DIA-MS to demonstrate that RSL3 binds to and inhibits GPX4 which regulates ferroptotic cancer cell death. Figure 6 contains a 3D representation of the isotopic clusters of peptide from GPX4 analyzed by LC-IM-DIA-MS. Detection and identification was conducted by dedicated software. The results shown in Fig. 6 illustrate the presence of GPX4 with RSL3 active probe treatment and its absence when the probe was inactive or a competitor was present. It was concluded that RSL3 inhibits GPX4, a protein essential for cancer cell viability. Numerous applications describe the use of LC-IM-DIA-MS for the label-free quantification, as described in a recent review describing DIA and its application [64]. Other IMS geometries, DTIMS in particular, generating

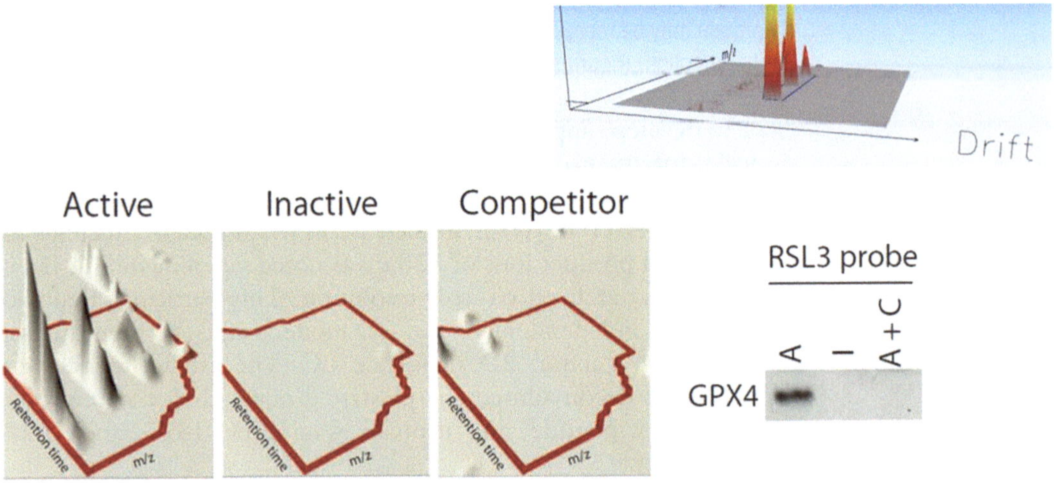

Fig. 6 Confirmation of GPX4 binding to an active affinity probe. The right panel represent cell lysates prepared from cells treated with active probe (A), inactive probe (I), or active probe in the presence of competitor (A + C) that were affinity purified and probed by western blot using a GPX4-specific antibody. Left panel is a 3D visualization of isotopic clusters of peptide ILAFPCNQFGK from GPX4 as analyzed by LC-IM-DIA-MS. Shown inset top right is the ability of the analysis software to discriminate between overlapping isotopic distributions by means of IMS, thereby improving quantitative precision. (Adapted with permission from Elsevier [63])

multiplexed data sets as well, have also been applied for quantitative discovery experiments [65]. The development and application of software to analyze multiple dimensional (retention time, ion mobility, m/z, and intensity) DTIMS data have been recently demonstrated [66], which is a critical component for making the analysis of IM-enabled proteomics, not only DTIMS but also any IMS data set, including TWIMS and TIMS, more mainstream. IM-assisted DIA is not exclusively applied in proteomics discovery experiments. Iacob et al. [67] determined the utility of adding ion mobility spectrometry to studies probing the solution-phase hydrogen/deuterium exchange of proteins and concluded that additional dimension of separation allowed for the deconvolution of overlapping isotopic patterns for coeluting peptides and extraction of valuable deuterium incorporation data for those peptides.

3.3 Sensitivity

A more recent application of IM-MS is described by Helm et al. [24] who used the technique to increase MS/MS sensitivity in untargeted data-dependent analysis (DDA) and targeted parallel reaction monitoring (PRM) like proteomic LC-MS experiments on a commercial hybrid quadrupole–ion mobility–time of flight mass spectrometer. This technique, as will be demonstrated, enhances the duty cycle of the oa-ToF analyzer and thus sensitivity. Briefly, as shown previously, TWIMS separation is strongly dependent upon ion charge z. Moreover, ions are nested for a given charge state by mass and drift time. This charge state separation and nesting can be used to discriminate against single charge background and to exclusively select multiply charged peptides for tandem MS. Subsequently, precursor ions are sequentially selected by the quadrupole mass analyzer and fragmented by CID in the first stacked-ring ion guide of the triwave device and prior to reaching the ion mobility cell. Products ions are trapped within this first traveling wave region of the triwave device and gated into the high-pressure ion mobility cell where they are separated according to their gas phase mobility within the cell. As a result, as illustrated in Fig. 7, fragment ions of the same mobility exit the cell as a series of compact packets. Hence, by synchronizing the pusher pulse that accelerates the fragment ions into the oa-TOF mass analyzer with the arrival of product ions from the TWIMS cell into the pusher region, fragment ions are sequentially injected into the TOF analyzer with greatly enhanced duty cycle (~100%) across the mass scale. This synchronization leads to a concomitant increase in sensitivity where the percent identified DDA spectra versus amount protein digest on column is contrasted. On average, a tenfold increase in peptide MS/MS sensitivity can be observed [24]. Since the ion mobility time frame is in the order of milliseconds, it nests well between the second timeframe of liquid chromatography and that of the oa-TOF mass spectrometer that operates in the microsecond timeframe. This particular DDA

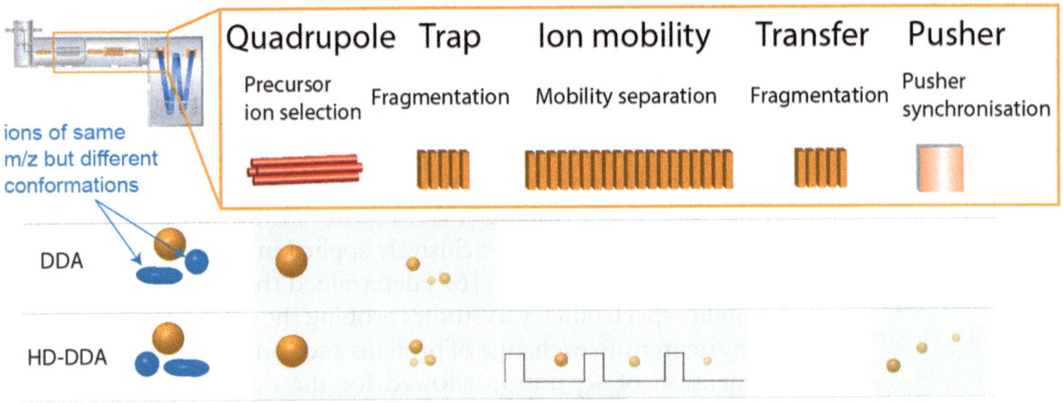

Fig. 7 Implementation of DDA (top) and IM-enhanced DDA (bottom) methods as implemented on an oa-ToF mass spectrometer. (Adapted with permission from Konstantinos Thalassinos (Institute of Structural and Molecular Biology, Department of Biological Sciences, Birkbeck, University College London))

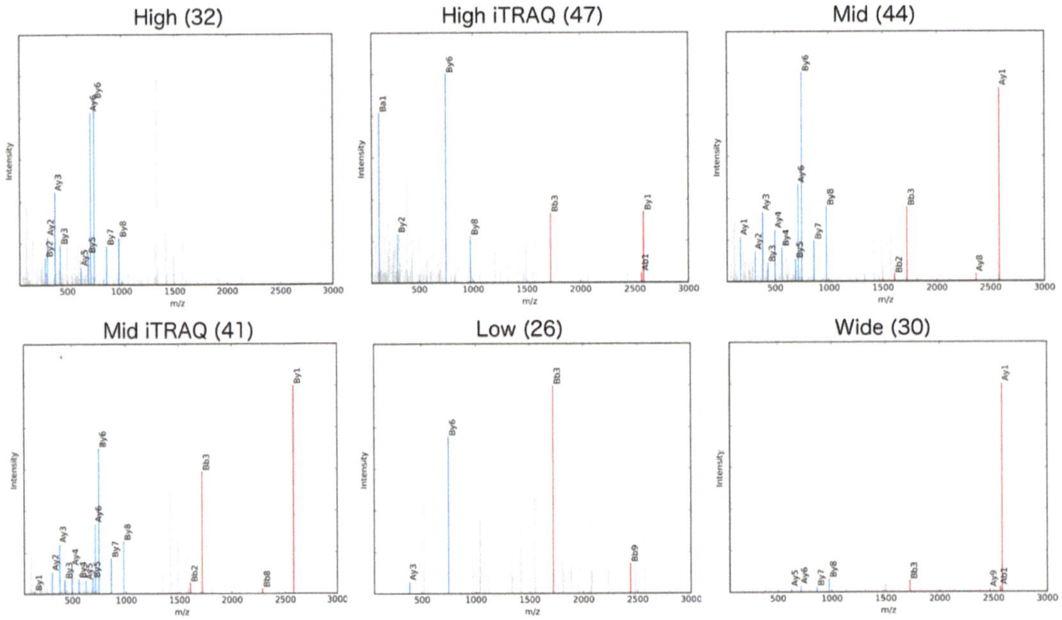

Fig. 8 Example spectra for cross-link identification of DTHKSEIAHR-FKDLGEEHFK (a4, b2) from tryptic digested bovine serum albumin. Isotopically labeled bis(sulfosuccinimdyl)suberate cross-lined fragment ions are shown in red and linear fragment ions in blue. The xQuest identification scores are shown in brackets. (Reprinted with permission of ACS publications [68])

method has been applied by James et al. [68] for the analysis of chemically cross-linked peptides for structural analysis of proteins and protein complexes. A dual CID collision energy ramp method was applied to obtain the highest cross-lined and linear fragment ion coverage possible. Shown in Fig. 8 are the results of the CID optimization using single and dual ramp collision energy experiments. The accumulation, or enrichment of ions, peptide precursor

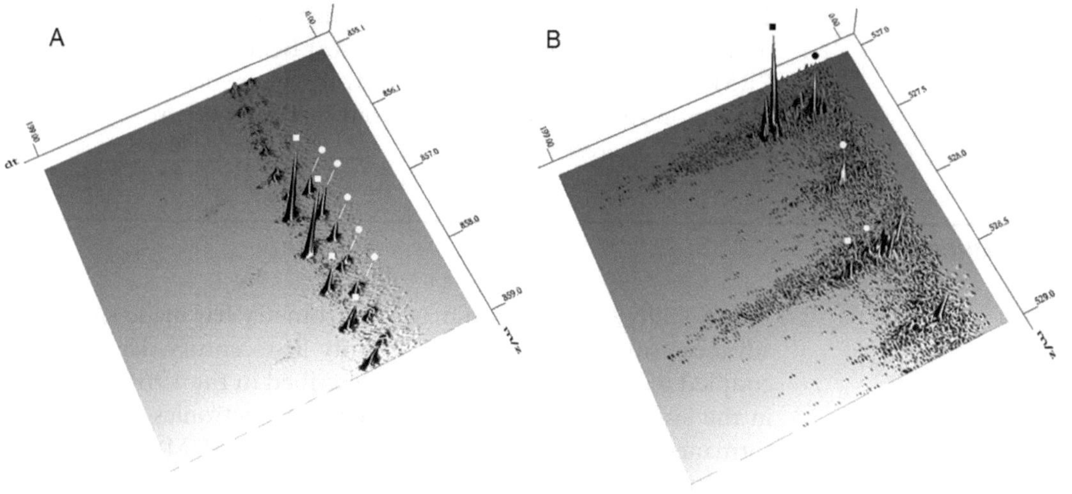

Fig. 9 (**a**) Precursor isotope IM separation of two peptides, using an IM enabled oa-ToF MRM (PRM) method, with A0 m/z 856.8 (spheres) and A0 m/z 857.0 (squares) coeluting and coisolated, respectively, followed by (**b**) IM separation of an associated product ion to both precursors with m/z 527.3. A third product ion with m/z 528.3, originating from precursor m/z 856.8 was detected (not annotated) that partially m/z overlapped with the isotopes of the 527.3 product ion but was separated in the IM domain. (Reprinted with permission of John Wiley and Sons [71])

ions in this particular instance, can also be utilized in DDA schema of TIMS analyzers in order to improve the duty cycle and sensitivity of MS/MS experiments [25, 69]. Here, ion mobility-separated ions are released sequentially from the IM analyzer. Next, the TIMS analyzer and the quadrupole are synchronized to resolve precursor ions of different *m/z* that are subsequently CID fragmented.

An example of an IM-enabled targeted high-resolution multiple reaction monitoring (HR-MRM) experiment is shown in Fig. 9 where both peptide precursor and product ions are mobility separated. In high-resolution MRM, also known as parallel reaction monitoring (PRM), the last quadrupole of a tandem quadrupole instrument is substituted with a high-resolution mass analyzer to allow parallel detection of all product ions in a single, high-resolution, accurate mass experiment. Here, unlike the previously described experiment, peptide precursor masses, including internal standards, are predefined, along with their retention time and CID collision energy profile. The principle of product ion enrichment to increase duty cycle and MS/MS sensitivity is identical to the IM-enhanced DDA case described in the previous paragraph. In this particular example, a number of putative cardiovascular disease plasma proteins were quantified [70]. It was demonstrated that for Apolipoprotein A-I, one of the candidate markers of interest, a serum concentration of 1.403 mg/ml can be calculated. The latter compared well with an average reported literature value of

1.400 mg/ml. The reported linear dynamic range was at least three orders of magnitude; however, the great benefit of high-resolution PRM is its high selectivity in the mass-to-charge domain since high-resolution MS can often separate out interferences from the product ion of interest [71].

4 Outlook

Ion mobility coupled with mass spectrometry has made significant advances since the turn of the twenty-first century. Despite IMS coupled with TOF MS first being described in the 1960s, it is only in the last 20 years that improvements in electronics and the performance of TOF mass analyzers has allowed IM-MS to become a mainstream platform. As electronic improvements still incline to track Moore's law [72], this should allow concomitant improvements in the underlying performance of both IMS and oa-TOF technologies, resulting in higher resolving power, as well as faster sampling rates. Improvements in IMS resolution have already been described, and the unique hybrid combination of IMS with TOF should allow the comprehensive profiling of complex heterogeneous samples.

The outlook for the future is promising. The use of rotationally averaged collisional cross sections as a means for identification and confirmation of compound identity is an intriguing prospect, providing a physicochemical supplement to retention time and tandem MS information. The limiting factor is currently the lack of CCS measurements populated into compound libraries, and the lack of computational tools to rapidly generate theoretical, either calculated or predicted, CCS values from compound structure. Although there has been development across multiple areas of research [73–75], including pesticides, veterinary drugs, mycotoxins, metabolism, steroids, and steviol glycosides, undoubtedly further progress will make the routine use of IMS information for identification purposes a powerful technology. Methods have been proposed and evaluated to calculate or predict theoretical CCS using molecular modeling and machine-learning technologies, respectively. In summary, the next few years should see significant improvements in the technology, workflows, prediction algorithms, and the required analysis (bio)informatics to apply and use the information generated for both qualitative and quantitative analyses.

Acknowledgements

The Scientific Operations, MS Research, and Advanced Mass Spectrometry Technology groups of Waters Corporation are kindly acknowledged for their support and help in obtaining experimental data and information. Martin Palmer is acknowledged for providing example cyclic IM-MS data.

References

1. Valentine SJ et al (2006) Toward plasma proteome profiling with ion mobility-mass spectrometry. J Proteome Res 5:2977–2984
2. McLean JA, Ruotolo BT, Gillig KJ, Russell DH (2005) Ion mobility–mass spectrometry: a new paradigm for proteomics. Int J Mass Spectrom 240:301–315
3. Gabryelski W, Froese KL (2003) Rapid and sensitive differentiation of anomers, linkage, and position isomers of disaccharides using high-field asymmetric waveform ion mobility spectrometry (FAIMS). J Am Soc Mass Spectrom 14:265–277
4. Bruno VM et al (2010) Comprehensive annotation of the transcriptome of the human fungal pathogen Candida albicans using RNA-seq. Genome Res 20:1451–1458
5. Jin L, Barran PE, Deakin JA, Lyon M, Uhrín D (2005) Conformation of glycosaminoglycans by ion mobility mass spectrometry and molecular modelling. Phys Chem Chem Phys 7:3464–3471
6. Hoaglund CS, Valentine SJ, Clemmer DE (1997) An ion trap interface for ESI–ion mobility experiments. Anal Chem 69:4156–4161
7. Dwivedi P et al (2008) Metabolic profiling by ion mobility mass spectrometry (IMMS). Metabolomics 4:63–80
8. Kanu AB, Dwivedi P, Tam M, Matz L, Hill HH (2008) Ion mobility–mass spectrometry. J Mass Spectrom 43:1–22
9. Lanucara F, Holman SW, Gray CJ, Eyers CE (2014) The power of ion mobility-mass spectrometry for structural characterization and the study of conformational dynamics. Nat Chem 6:281
10. Gabelica V et al (2019) Recommendations for reporting ion mobility mass spectrometry measurements. Mass Spectrom Rev 38:291–230
11. Clemmer DE, Hudgins RR, Jarrold MF (1995) Naked protein conformations: cytochrome C in the gas phase. J Am Chem Soc 117:10141–10142
12. Pringle SD et al (2007) An investigation of the mobility separation of some peptide and protein ions using a new hybrid quadrupole/travelling wave IMS/oa-ToF instrument. Int J Mass Spectrom 261:1–12
13. Baker ES et al (2007) Ion mobility spectrometry–mass spectrometry performance using electrodynamic ion funnels and elevated drift gas pressures. J Am Soc Mass Spectrom 18:1176–1187
14. Fernandez-Lima F, Kaplan DA, Suetering J, Park MA (2011) Gas-phase separation using a trapped ion mobility spectrometer. Int J Ion Mobil Spectrom 14. https://doi.org/10.1007/s12127-011-0067-8
15. Dear GJ et al (2010) Sites of metabolic substitution: investigating metabolite structures utilising ion mobility and molecular modelling. Rapid Commun Mass Spectrom 24:3157–3162
16. Chalet C, Hollebrands B, Janssen HG, Augustijns P, Duchateau G (2018) Identification of phase-II metabolites of flavonoids by liquid chromatography–ion-mobility spectrometry–-mass spectrometry. Anal Bioanal Chem 410 (2):471–482. https://doi.org/10.1007/s00216-017-0737-4
17. Kliman M, May JC, McLean JA (2011) Lipid analysis and lipidomics by structurally selective ion mobility-mass spectrometry. Biochim Biophys Acta Mol Cell Biol Lipids 1811:935–945
18. Eckers C, Laures AM-F, Giles K, Major H, Pringle S (2007) Evaluating the utility of ion mobility separation in combination with high-pressure liquid chromatography/mass spectrometry to facilitate detection of trace impurities in formulated drug products. Rapid Commun Mass Spectrom 21:1255–1263
19. Vakhrushev SY, Langridge J, Campuzano I, Hughes C, Peter-Katalinić J (2008) Ion mobility mass spectrometry analysis of human glycourinome. Anal Chem 80:2506–2513
20. Schenauer MR, Meissen JK, Seo Y, Ames JB, Leary JA (2009) Heparan sulfate separation, sequencing, and isomeric differentiation: ion mobility spectrometry reveals specific iduronic and glucuronic acid-containing hexasaccharides. Anal Chem 81:10179–10185
21. Fasciotti M et al (2012) Separation of isomeric disaccharides by traveling wave ion mobility mass spectrometry using CO2 as drift gas. J Mass Spectrom 47(12):1643–1647. https://doi.org/10.1002/jms.3089

22. Ruotolo BT et al (2005) Evidence for macromolecular protein rings in the absence of bulk water. Science (80) 310:1658–1661
23. Bereszczak JZ et al (2014) Sizing up large protein complexes by electrospray ionisation-based electrophoretic mobility and native mass spectrometry: morphology selective binding of Fabs to hepatitis B virus capsids. Anal Bioanal Chem 406:1437–1446
24. Helm D et al (2014) Ion mobility tandem mass spectrometry enhances performance of bottom-up proteomics. Mol Cell Proteomics 13:3709–3715
25. Meier F et al (2015) Parallel accumulation–serial fragmentation (PASEF): multiplying sequencing speed and sensitivity by synchronized scans in a trapped ion mobility device. J Proteome Res 14:5378–5387
26. Göth M, Pagel K (2017) Ion mobility–mass spectrometry as a tool to investigate protein--ligand interactions. Anal Bioanal Chem 409:4305–4310
27. Williams JP et al (2009) Isomer separation and gas-phase configurations of organoruthenium anticancer complexes: ion mobility mass spectrometry and modeling. J Am Soc Mass Spectrom 20:1119–1122
28. Bagal D, Zhang H, Schnier PD (2008) Gas-phase proton-transfer chemistry coupled with TOF mass spectrometry and ion mobility-MS for the facile analysis of poly(ethylene glycols) and PEGylated polypeptide conjugates. Anal Chem 80:2408–2418
29. Towers MW, Karancsi T, Jones EA, Pringle SD, Claude E (2018) Optimised desorption electrospray ionisation mass spectrometry imaging (DESI-MSI) for the analysis of proteins/peptides directly from tissue sections on a travelling wave ion mobility Q-ToF. J Am Soc Mass Spectrom 29:2456–2466
30. Sans M, Feider CL, Eberlin LS (2018) Advances in mass spectrometry imaging coupled to ion mobility spectrometry for enhanced imaging of biological tissues. Curr Opin Chem Biol 42:138–146
31. Giles K et al (2004) Applications of a travelling wave-based radio-frequency-only stacked ring ion guide. Rapid Commun Mass Spectrom 18:2401–2414
32. Giles K, Williams JP, Campuzano I (2011) Enhancements in travelling wave ion mobility resolution. Rapid Commun Mass Spectrom 25:1559–1566
33. Giles K et al (2017) Design and performance of a second-generation cyclic ion mobility enabled Q-ToF. In 65th ASMS conference on mass spectrometry and allied topics
34. Ujma J et al (2019) Cyclic ion mobility mass spectrometry distinguishes anomers and open-ring forms of pentasaccharides. J Am Soc Mass Spectrom 30(6):1028–1037. https://doi.org/10.1007/s13361-019-02168-9
35. Shvartsburg AA, Smith RD (2008) Fundamentals of traveling wave ion mobility spectrometry. Anal Chem 80:9689–9699
36. Richardson K, Langridge D, Giles K (2018) Fundamentals of travelling wave ion mobility revisited: I. Smoothly moving waves. Int J Mass Spectrom 428:71–80
37. Eldrid C et al (2018) Gas phase stability of protein ions in a cyclic ion mobility spectrometry travelling wave device. Anal Chem 91 (12):7554–7561. https://doi.org/10.26434/chemrxiv.7388723.v1
38. Deng L et al (2016) Ultra-high resolution ion mobility separations utilizing traveling waves in a 13 m serpentine path length structures for lossless ion manipulations module. Anal Chem 88(18):8957–8964. https://doi.org/10.1021/acs.analchem.6b01915
39. Deng L et al (2017) Serpentine ultralong path with extended routing (SUPER) high resolution traveling wave ion mobility-MS using structures for lossless ion manipulations. Anal Chem 89(8):4628–4634. https://doi.org/10.1021/acs.analchem.7b00185
40. Graves DB (2018) Transport properties of ions in gases by Edward A. Mason and Earl W. McDaniel, John Wiley and Sons, New York, NY, 1988, 560 + xvi pp. AIChE J 35:701
41. Bowers MT, Marshall AG, McLafferty FW (1996) Mass spectrometry: recent advances and future directions. J Phys Chem 100:12897–12910
42. Henderson SC, Valentine SJ, Counterman AE, Clemmer DE (1999) ESI/ion trap/ion mobility/time-of-flight mass spectrometry for rapid and sensitive analysis of biomolecular mixtures. Anal Chem 71:291–301
43. Verbeck GF, Ruotolo BT, Sawyer HA, Gillig KJ, Russell DH (2002) A fundamental introduction to ion mobility mass spectrometry applied to the analysis of biomolecules. J Biomol Tech 13:56–61
44. Zhong Y, Hyung S-J, Ruotolo BT (2012) Ion mobility-mass spectrometry for structural proteomics. Expert Rev Proteomics 9:47–58
45. Uetrecht C, Rose RJ, van Duijn E, Lorenzen K, Heck AJR (2010) Ion mobility mass spectrometry of proteins and protein assemblies. Chem Soc Rev 39:1633–1655
46. Snijder J, Heck AJR (2014) Analytical approaches for size and mass analysis of large

protein assemblies. Annu Rev Anal Chem 7:43–64

47. Stojko J et al (2015) Ion mobility coupled to native mass spectrometry as a relevant tool to investigate extremely small ligand-induced conformational changes. Analyst 140 (21):7234–7245. https://doi.org/10.1039/c5an01311a
48. Hoffmann W, von Helden G, Pagel K (2017) Ion mobility-mass spectrometry and orthogonal gas-phase techniques to study amyloid formation and inhibition. Curr Opin Struct Biol 46:7–15
49. Eschweiler JD, Kerr R, Rabuck-Gibbons J, Ruotolo BT (2017) Sizing up protein–ligand complexes: the rise of structural mass spectrometry approaches in the pharmaceutical sciences. Annu Rev Anal Chem 10:25–44
50. Eyers CE, Vonderach M, Ferries S, Jeacock K, Eyers PA (2018) Understanding protein–drug interactions using ion mobility–mass spectrometry. Curr Opin Chem Biol 42:167–176. https://doi.org/10.1016/j.cbpa.2017.12.013
51. Hopper JTS, Oldham NJ (2009) Collision induced unfolding of protein ions in the gas phase studied by ion mobility-mass spectrometry: The effect of ligand binding on conformational stability. J Am Soc Mass Spectrom 20:1851–1858
52. Migas LG, France A, Bellina B, Barran P (2017) ORIGAMI: a software suite for activated ion mobility mass spectrometry (aIM-MS) applied to multimeric protein assemblies. bioRxiv 427:20–28
53. Beveridge R et al (2016) Mass spectrometry locates local and allosteric conformational changes that occur on cofactor binding. Nat Commun 7:12163
54. Valentine SJ, Counterman AE, Clemmer DE (1999) A database of 660 peptide ion cross sections: use of intrinsic size parameters for bona fide predictions of cross sections. J Am Soc Mass Spectrom 10:1188–1211
55. Liu X et al (2007) Mapping the human plasma proteome by SCX-LC-IMS-MS. J Am Soc Mass Spectrom 18:1249–1264
56. Valentine SJ, Kulchania M, Barnes CAS, Clemmer DE (2001) Multidimensional separations of complex peptide mixtures: a combined high-performance liquid chromatography/ion mobility/time-of-flight mass spectrometry approach. Int J Mass Spectrom 212:97–109
57. Thalassinos K et al (2012) Design and application of a data-independent precursor and product ion repository. J Am Soc Mass Spectrom 23:1808–1820
58. Paglia G et al (2014) Ion mobility derived collision cross sections to support metabolomics applications. Anal Chem 86:3985–3993
59. Shah AR et al (2010) Machine learning based prediction for peptide drift times in ion mobility spectrometry. Bioinformatics 26:1601–1607
60. Rodriguez-Suarez E et al (2013) An ion mobility assisted data independent LC-MS strategy for the analysis of complex biological samples. Curr Anal Chem 9:199–211
61. Geromanos SJ, Hughes C, Ciavarini S, Vissers JPC, Langridge JI (2012) Using ion purity scores for enhancing quantitative accuracy and precision in complex proteomics samples. Anal Bioanal Chem 404:1127–1139
62. Distler U et al (2014) Drift time-specific collision energies enable deep-coverage data-independent acquisition proteomics. Nat Methods 11:167–170
63. Yang WS et al (2014) Regulation of ferroptotic cancer cell death by GPX4. Cell 156:317–331
64. Distler U, Kuharev J, Tenzer S (2014) Biomedical applications of ion mobility-enhanced data-independent acquisition-based label-free quantitative proteomics. Expert Rev Proteomics 11:675–684
65. Kedia K et al (2018) Application of multiplexed ion mobility spectrometry towards the identification of host protein signatures of treatment effect in pulmonary tuberculosis. Tuberculosis 112:52–61. https://doi.org/10.1016/j.tube.2018.07.005
66. MacLean BX et al (2018) Using skyline to analyze data-containing liquid chromatography, ion mobility spectrometry, and mass spectrometry dimensions. J Am Soc Mass Spectrom 29:2182–2188
67. Iacob RE, Murphy JP, Engen JR (2008) Ion mobility adds an additional dimension to mass spectrometric analysis of solution-phase hydrogen/deuterium exchange. Rapid Commun Mass Spectrom 22:2898–2904
68. James JMB, Cryar A, Thalassinos K (2019) An optimization workflow for the analysis of cross-linked peptides using a quadrupole time-of-flight mass spectrometer. Anal Chem 91 (3):1808–1814. https://doi.org/10.1021/acs.analchem.8b02319
69. Meier F et al (2018) Online parallel accumulation – serial fragmentation (PASEF) with a novel trapped ion mobility mass spectrometer. bioRxiv 17(12):2534–2545. https://doi.org/10.1101/336743

70. Domanski D et al (2012) MRM-based multiplexed quantitation of 67 putative cardiovascular disease biomarkers in human plasma. Proteomics 12:1222–1243
71. Mbasu RJ et al (2016) Advances in quadrupole and time-of-flight mass spectrometry for peptide MRM based translational research analysis. Proteomics 16:2206–2220
72. Waldrop MM (2016) The chips are down for Moore's law. Nature 530(7589):144–147. https://doi.org/10.1038/530144a
73. McCullagh M, Douce D, Van Hoeck E, Goscinny S (2018) Exploring the complexity of steviol glycosides analysis using ion mobility mass spectrometry. Anal Chem 90(7):4585–4595. https://doi.org/10.1021/acs.analchem.7b05002
74. Righetti L et al (2018) High resolution-ion mobility mass spectrometry as an additional powerful tool for structural characterization of mycotoxin metabolites. Food Chem 245:768–774
75. Hernández-Mesa M, Le Bizec B, Monteau F, García-Campaña AM, Dervilly-Pinel G (2018) Collision cross section (CCS) database: an additional measure to characterize steroids. Anal Chem 90(7):4616–4625. https://doi.org/10.1021/acs.analchem.7b05117

Chapter 11

Ion Mobility-Mass Spectrometry to Evaluate the Effects of Protein Modification or Small Molecule Binding on Protein Dynamics

Lauren J. Tomlinson and Claire E. Eyers

Abstract

Ion mobility-mass spectrometry (IM-MS) of intact protein complexes under native conditions is a powerful tool for the analysis of protein complexes and protein–ligand interactions, permitting insight into ligand-induced changes in protein conformation. Here we describe a procedure for analyzing the effects of phosphorylation and/or inhibitor binding on protein kinase conformational flexibility using Protein Kinase A (PKA) as a model system. By calculating the protein collision cross section (CCS) before and after inhibitor binding, and additionally by performing collision-induced unfolding (CIU), we can establish the effects of protein modification or small molecule binding on protein dynamics.

Key words Ion mobility-mass spectrometry, Native mass spectrometry, Collision cross section, Collision-induced unfolding

1 Introduction

Ion mobility-mass spectrometry (IM-MS) of intact protein complexes under native electrospray ionization (ESI) conditions is advancing the structural analysis of proteins, protein complexes, and protein–ligand interactions. The benefits of rapidly obtaining information on protein dynamics and conformational flexibility from small amounts of (partially) purified material means that IM-MS is confidently being implemented alongside more traditional structural techniques such as X-ray crystallography and nuclear magnetic resonance (NMR) to gain insight into the structural alterations associated with protein modification, protein–ligand binding, and protein complex formation [1–5].

Many elegant IM-MS studies are now being published that evaluate the effect of different types of ligands on protein complex formation and protein dynamics, including small molecules [6–13], RNA and DNA oligomers [14, 15], and (proteo)glycans [16, 17].

Giuseppe Paglia and Giuseppe Astarita (eds.), *Ion Mobility-Mass Spectrometry: Methods and Protocols*, Methods in Molecular Biology, vol. 2084, https://doi.org/10.1007/978-1-0716-0030-6_11,

There are four primary forms of ion mobility that are coupled with MS: drift-tube ion mobility spectrometry (DTIMS), traveling wave ion mobility spectrometry (TWIMS), trapped ion mobility (TIMS), and differential-mobility spectrometry (DMS), a version of which is field-asymmetric ion mobility spectrometry (FAIMS) [18, 19]. DTIMS and TWIMS (and more recently TIMS) are preferentially used in the investigation of protein–ligand interactions, as they can be used to define analyte rotationally average collision cross section (CCS) values permitting direct comparison of conformers, e.g., before and after ligand binding. If required, CCS values can also be correlated with data obtained using software programs, such as MOBCAL [7].

Protein kinases are important drug targets, in large part because of their rate-limiting roles in numerous diseases [20]. Consequently, understanding the structural implications upon binding of small (and large) molecule drugs to these proteins, and the effect of regulatory post-translational modifications, such as phosphorylation, is important. Investigating protein conformational dynamics and the effects on stability is particularly relevant as the field moves toward development of, arguably more specific, "type II" small molecule allosteric inhibitors that bind regions of the protein outside of the relatively conserved ATP-binding site [21, 22].

Here we present a method of utilizing IM-MS to assess the structural effects of phosphorylation and ligand binding on cAMP-dependent protein kinase (PKA) [7, 23]. The described method is readily transferable to the investigation of phosphorylation and ligand-induced changes in other proteins, including other members of the protein kinase superfamily.

2 Materials

Prepare all solutions using HPLC-grade water. All protein solutions should be kept at 4 °C unless otherwise stated.

2.1 Buffer Exchange

1. Ammonium Acetate: 50 mM. Weigh out 0.39 g ammonium acetate and dissolve in 100 mL HPLC-grade water to make 50 mM ammonium acetate buffer. Prechill to 4 °C.
2. HPLC-MS water.
3. Amicon spin filter columns.
4. Bench top centrifuge.
5. Gel loading tips.

2.2 Inhibitor Assay

1. Protein(s) of interest: exemplified here are different forms of the catalytic subunit of PKA (PKAc)—wild-type (WT; hyperphosphorylated), λ protein phosphatase (λPP)-treated PKAc,

catalytically inactive K72H PKAc, and the non-PKI-binding R133A PKAc variant.

2. Inhibitors: Prepare all stocks of inhibitors (staurosporine (STS), H89 and AT13148) by diluting to 10 mM final concentration in DMSO.

2.3 Ion Mobility-Mass Spectrometry

1. Ion mobility-mass spectrometer: Waters G2-Si Synapt.
2. API Calibration (NaCsl) Solution Kit (Waters).
3. TW120–4 Thin-Wall Capillary (4 × 1.2 mm), 10 G.
4. Capillary puller: Sutter P-1000 Puller (*see* **Note 1**).
5. Ceramic capillary cutter.
6. Platinum (Pt) wire (0.125 mm × 5 m), one per capillary, cut to 4 cm lengths.
7. Protein calibrants for CCS determination: β-lactoglobulin A, avidin, transthyretin, concanavalin A, and serum albumin (Sigma-Aldrich) prepared as 5–10 μM in 200 mM ammonium acetate (*see* **Note 2**).

3 Methods

3.1 IM-MS Setup

1. Prepare the capillaries for nano electrospray ionization (nESI) using the capillary puller and the program as shown in Table 1 (*see* **Note 1**).
2. To calibrate the ToF, infuse the Lockspray Flow Control with NaCsl. Start infusing at 20 μL/min, decreasing to 10 μL/min once a strong signal is visible.
3. Calibrate using NaCsl over a m/z range of 500–8000 in resolution mode (*see* **Note 3**).

3.2 Protein Buffer Exchange

1. Prepare the protein solution for IM-MS analysis by buffer exchanging into 50 mM ammonium acetate (*see* **Note 4**) using amicon 0.5 mL spin filter columns (*see* **Note 5**). Dilute the required amount of protein solution (*see* **Note 6**) in ammonium acetate to the capacity of the amicon spin filter (*see*

Table 1
Sutter P-1000 Puller settings

Line	Heat	Pull	Vel.	Delay	Pressure	Ramp
1 × 1	489	0	20	120	235	491
2 × 1	489	0	18	120	Delay mode	X
3 × 1	489	0	18	120	Safe heat	X
4 × 1					Jaw temp. 19 °C	

Note 7). Centrifuge at 14,000 × *g* for 10 min, in a precooled bench top centrifuge at 4 °C. Discard the flow through. Repeat the dilution and centrifugation step twice (*see* **Note 8**).

2. Following the final buffer exchange spin, invert the filter in to a new low-bind centrifuge tube and spin for 3 min at 14,000 × *g*, 4 °C.
3. Determine the concentration of the final protein solution in ammonium acetate using a NanoDrop spectrophotometer (*see* **Note 9**).

3.3 IM-MS Analysis

1. Cut off ~2 cm from the back end of the capillary using a ceramic capillary cutter (*see* **Note 10**).
2. Place the capillary in the secure clasp holder. Using gel-loading tips, add between 1–3 μL (2–5 μM) of protein sample (either protein CCS calibrant or protein of interest) directly in to the nESI capillary.
3. Spin down the clasp holder containing the capillary for a few seconds in a microfuge to ensure that all of the protein solution has reached the end of the capillary tip.
4. Remove the capillary from the clasp and place the piece of platinum wire inside the capillary (*see* **Note 11**). Affix the metal clasp and the protective rubber insert to the outside of capillary and mount onto the source stage of the Synapt G2-Si instrument.
5. Use the camera function to position the capillary, so that it is in the optimal position (*see* **Note 12**).
6. Increase the nano flow gas to 0.05 bar and the sampling cone to 20 V.
7. Set the instrument to TOF/MS mode and increase the capillary voltage to ~1.1–1.6 kV until a stable signal is observed (*see* **Note 13**).
8. Determine the optimal TriWave settings for the protein of interest by adjusting the Wave Velocity (m/s) and Wave Height (V) (*see* **Note 13**).
9. Once a good signal is established, and TriWave settings optimized, ensure the 'Add Drift Time' function button is ticked and acquire data over the required *m/z* range (*see* **Note 14**). *See* Fig. 1 for native MS and IM-MS data for the exemplar hyperphosphorylated PKAc, before and after treatment with λ protein phosphatase (PKA λPP), as well as two protein variants: K72H PKAc which is catalytically inactive and therefore not phosphorylated, and R133A PKAc which is unable to bind PKI (the inhibitor of PKA).

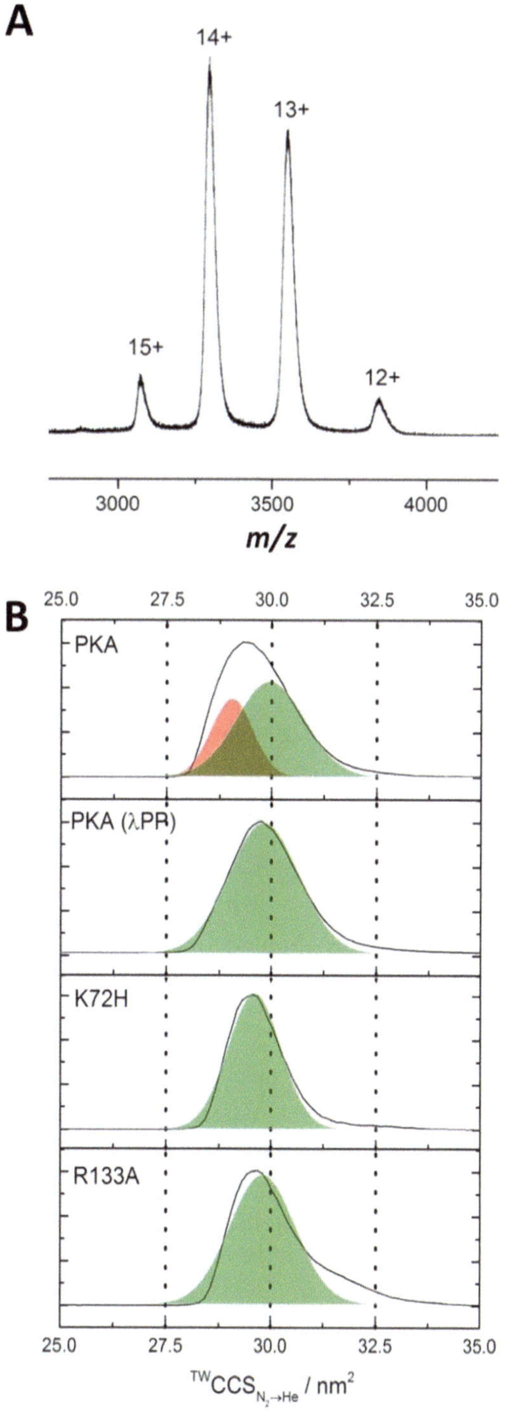

Fig. 1 Structural analysis of variants of the PKA catalytic subunit (PKAc) by IM-MS. (**a**) ESI mass spectrum of PKAc obtained under native conditions; charge states are indicated. (**b**) $^{TW}CCS_{N_2\rightarrow He}$ for the $[M+13H]^{13+}$ form of untreated WT PKAc (PKA), PKAc following treatment with Mn^{2+}–λ protein phosphatase (PKA

3.4 Calibration of the TWIMS for CCS Determination

1. To determine CCS values for the protein of interest, the drift time through the instrument must first be calibrated under the optimal MS acquisition parameters used for the protein. Calibrate the drift time using the CCS Calibration setting on the Intellistart program. Acquire IM-MS data as described in Subheading 3.3.
2. Open the data file in the MassLynx software using the 2: TOFMS (500:8000) ES+ function.
3. Extract the peak of interest from Total Ion Chromatogram (TIC) to open spectrum.
4. Right click across the peak in the spectrum view to open another chromatogram window.
5. In the chromatogram window, click edit and copy the chromatogram list. Paste these values (drift times and the corresponding intensities) in to Microsoft Excel or a similar program of choice.
6. Refer to the Collision Cross Section (CCS) database (https://depts.washington.edu/bushlab/ccsdatabase/) provided by the Bush Lab [24] to determine the native-like ion value for CCS (He)/nm^2 of the chosen CCS calibrants.
7. Calculate the reduced cross section Ω' of the protein calibrants using Eq. 1:

$$\Omega' = \Omega \frac{\sqrt{u}}{q} \quad (1)$$

where μ is the reduced mass and q is the charge state.

8. Plot (Ω') versus (tD') and fit a straight line using Eq. 2 to convert the drift time scale to a CCS (He) scale:

$$\Omega' = \Omega \frac{q}{\sqrt{u}}^{\circ} tDr^{A^\circ} exp(C) \quad (2)$$

where A is the determined slope of the fit and C is the intercept.

9. Plot the converted nonlinear scale against the corresponding intensity values.

Fig. 1 (continued) PP), and the K72H and R133A variants. The hyperphosphorylated PKAc had a $^{TW}CCS_{N_2 \rightarrow He}$ value of 29.4 nm^2. $^{TW}CCS_{N_2 \rightarrow He}$ value increased by 1.5% following treatment with λPP. Half-height width of the CCS distribution reduced from 2.2 nm^2 (28.4–30.6 nm^2) to 1.8 nm^2 (28.8–30.6 nm^2) with addition of λPP. Two overlapping conformations of PKA are indicated in red and green. (Reproduced from Byrne et al. 2016 with permission from the Biochemical Journal)

10. Using OriginPro9 (or similar), plot the CCS values calculated in **step 9** (*x*-axis) against the intensity values for each conformer (*y*-axis).
11. Normalize the inputted data by clicking on Analysis–Mathematics–Normalize Columns–Normalize to [0,1] and plot the best fit line. This can then be used to determine the CCS value for all similar analytes according to their measured drift times under the same conditions (Fig. 1).

3.5 Evaluating the Effect of Inhibitor Addition on Protein Conformation

1. Mix the protein solution (in ammonium acetate) with the required concentration of inhibitor, to achieve a molar ratio ~1:10 protein:inhibitor. Leave the mixture at room temperature for ~10 min to facilitate binding prior to IM-MS analysis.
2. Prepare a capillary and collect IM-MS data as described in Subheading 3.4.
3. Calculate the CCS value of the different charge states of the ligand-bound protein (Fig. 2) according to the procedure outlined in Subheading 3.5 (*see* **Note 15**).

3.6 Collision-Induced Unfolding

1. Determine the lowest charge state of suitable intensity for CIU analysis following IM-MS data acquisition of the protein as described in Subheading 3.4.
2. Isolate this single charge state in the quadrupole for further investigation.
3. Increase the CID trap energy gradually from ~20 to 41 V, acquiring IM-MS data (as detailed in Subheading 3.4) in two-volt intervals.
4. Repeat **steps 2–4** for each observed charge state.

3.7 CIU Data Analysis

1. Open the CIU data files in the MassLynx software using the 2: TOFMS (500:8000) ES+ function.
2. Extract the peaks of interest from Total Ion Chromatogram (TIC) to open the spectra.
3. Right click across the peak in the spectrum view to open another chromatogram window.
4. In the chromatogram window, click edit and copy the chromatogram list. Paste these values (drift times and the corresponding intensities) in to Microsoft Excel or similar program of choice.
5. Copy data for all the acquired voltage settings in to OriginPro9.
6. Normalize the inputted data by clicking on Analysis–Mathematics–Normalize Columns–Normalize to [0,1].
7. Copy all and add to matrix.
8. Create a CIU plot by going to Plot-Contour-Color (Fig. 3).

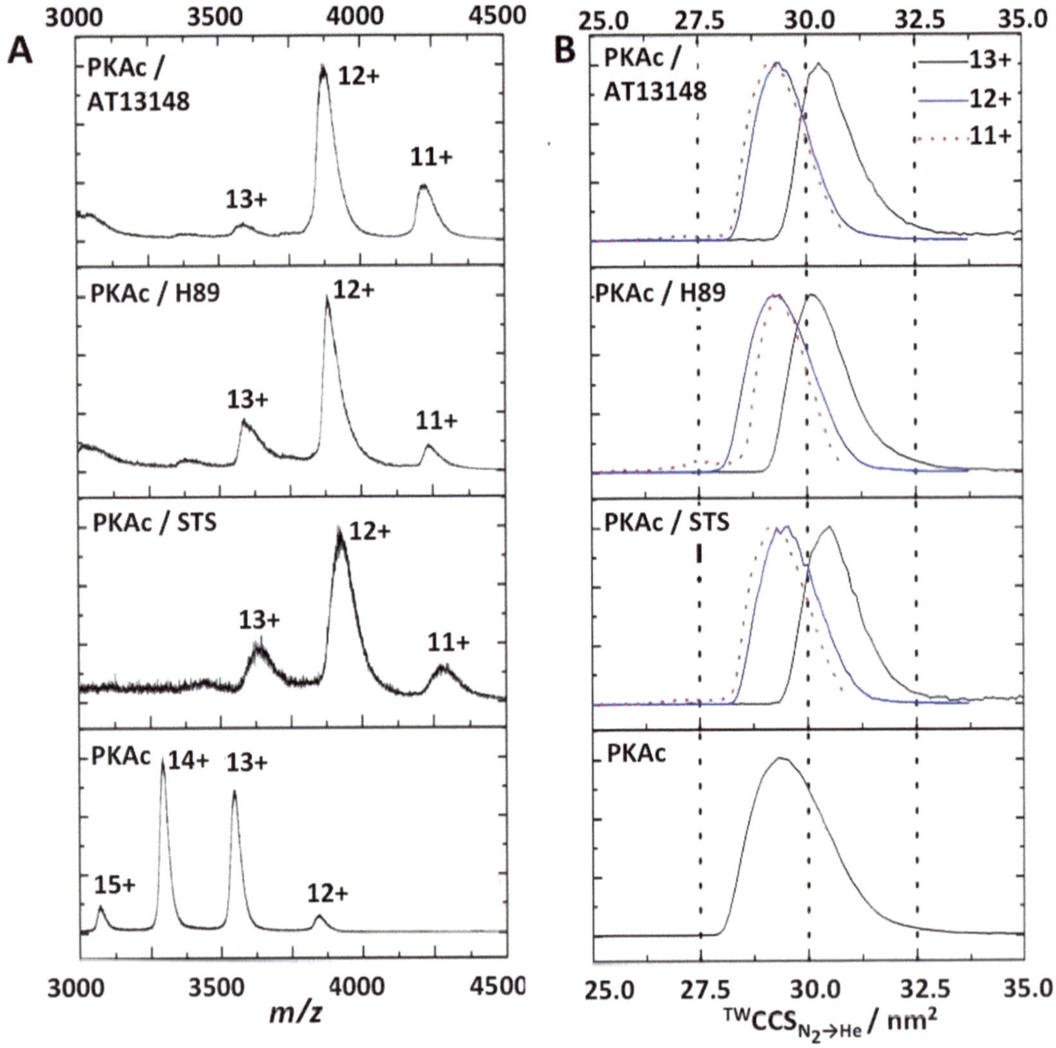

Fig. 2 Small-molecule inhibitor binding to PKAc. (**a**) Native ESI mass spectra. (**b**) $^{TW}CCS_{N_2 \rightarrow He}$ values in the presence of DMSO vehicle or with tenfold molar excess of staurosporine (STS), H89, or AT13148. CCS distributions are presented for $[M+11H]^{11+}$ (red dotted line), $[M+12H]^{12+}$ (blue line), and $[M+13H]^{13+}$ (black line). (Reproduced from Byrne et al. 2016 with permission from the Biochemical Journal)

4 Notes

1. The program in Table 1 should be used as a starting guide and adapted to achieve the optimal capillary shape. *See* manufacturer's instructions for the use of other puller models. The fine end of the capillary tip should be ~0.4 cm in length.
2. In order to determine CCS values, it is an absolute requirement that the drift time through the instrument be calibrated under the same conditions as used for the analyte protein (and ligand

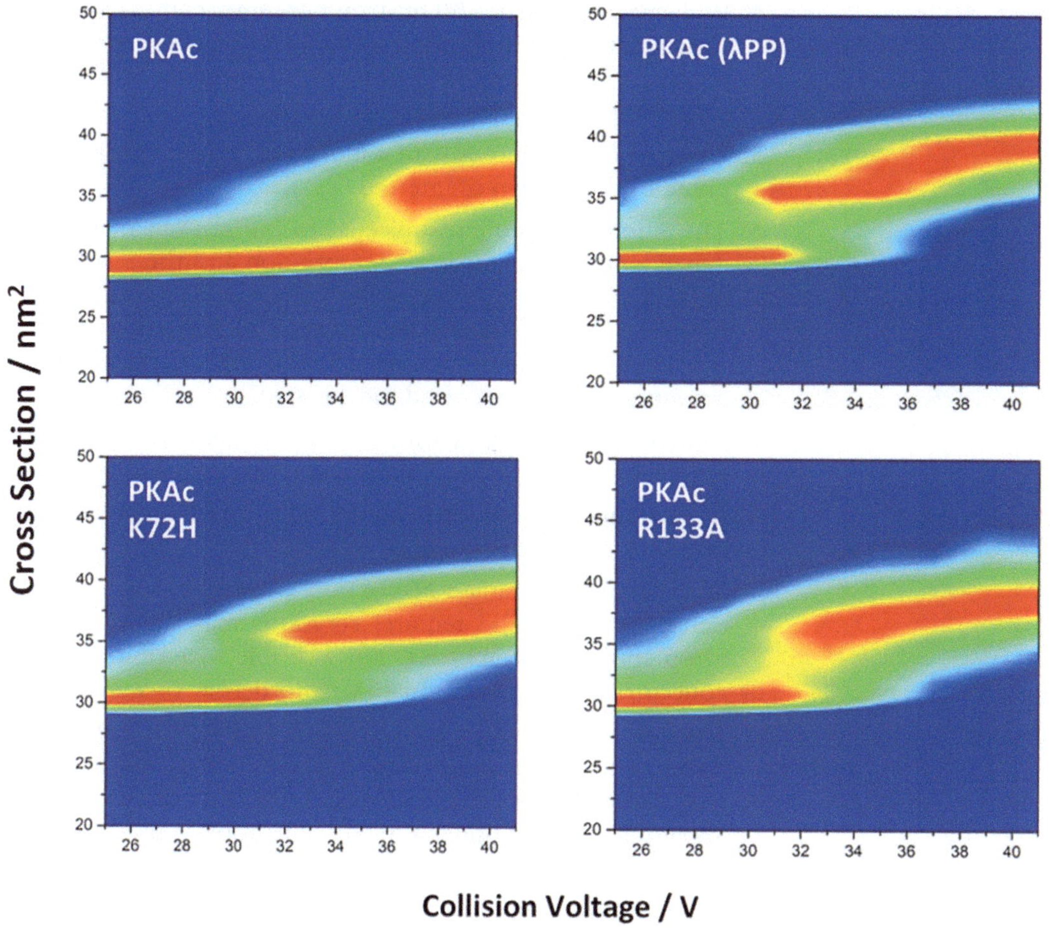

Fig. 3 Collision-induced unfolding of PKAc variants. CIU profiles of PKAc wild-type (WT), λPP-treated, K72H and R133A. PKAc is more stable and requires higher collision energy to begin unfolding at ~36 V. λPP-treated PKA starts to unfold at ~31 V. Both PKAc variants are less stable than PKAc WT and begin unfolding at ~32 V. (Reproduced from Byrne et al. 2016 with permission from the Biochemical Journal)

complex). It is ideal to use protein calibrants that are of a similar mass to the one under investigation. We typically use β-lactoglobulin A, avidin, transthyretin, concanavalin A, and serum albumin. It should be noted that determination of CCS values are not an absolute requirement to evaluate ligand-induced protein conformational changes, but they are required to undertake cross-comparison studies using different techniques, or for analyses undertaken on different instruments, or under different conditions.

3. The ToF should be calibrated to achieve an acceptable tolerance of 1 ppm or below. If the mass accuracy is greater than 1 ppm, calibration should be repeated.

4. While we recommend 50 mM ammonium acetate as a starting concentration, this should be evaluated for your protein of interest, with ranges of 25–250 mM ammonium acetate typically being employed.
5. Use spin filter columns with a molecular weight cutoff appropriate to the protein of interest.
6. The amount of protein subjected to buffer exchange will be dependent on the starting concentration and the relative stability of the protein during buffer exchange. The final concentration of protein required for analysis is ~1 mg/mL or greater; this should be taken into consideration when deciding on the amount of protein required for this step.
7. Amicon spin filters should be pre-equilibrated with the required concentration of ammonium acetate prior to use.
8. Ammonium acetate buffer should be kept cold on ice throughout the buffer exchange step.
9. We recommend using a NanoDrop for protein concentration determination as this requires very little material. If a NanoDrop spectrophotometer is not available, determine protein concentration using a standard Bradford assay or similar.
10. Extreme care should be taken when cutting the capillary to ensure a clean cut, and that no glass enters the capillary which may cause the capillary to block. Both ends of the capillary should be checked under a microscope prior to use.
11. Ensure the platinum wire goes to the tip of the capillary without pushing through the tip.
12. The optimal position of the nESI capillary with respect to the instrument orifice is instrument dependent. Using the camera as a guide, the position of the capillary should be adjusted to ~1 cm away from the orifice such that signal is optimized.
13. To determine optimal spectra, adjust the wave height and velocity settings until the peak observed in the drift time spectra is at the center of the *x*-axis.
14. We typically acquire over a TOF MS range of 500–8000 *m/z* for ~10 min, using a scan time of 5 s.
15. Evaluation of the effects of ligand binding on protein conformation requires comparison of the CCS values for the same protein charge states ± ligand.

References

1. Göth M, Pagel K (2017) Ion mobility–mass spectrometry as a tool to investigate protein--ligand interactions. Anal Bioanal Chem 409:4305. https://doi.org/10.1007/s00216-017-0384-9
2. Eyers CE, Vonderach M, Ferries S, Jeacock K, Eyers PA (2018) Understanding protein–drug interactions using ion mobility–mass spectrometry. Curr Opin Chem Biol 42:167–176.

https://doi.org/10.1016/j.cbpa.2017.12.013

3. Eschweiler J, Kerr R, Rabuck-Gibbons J, Ruotolo BT (2017) Sizing up protein–ligand complexes: the rise of structural mass spectrometry approaches in the pharmaceutical sciences. Annu Rev Anal Chem 10:25–44. https://doi.org/10.1146/annurev-anchem-061516-045414
4. Ben-Nissan G, Sharon M (2017) The application of ion-mobility mass spectrometry for structure/function investigation of protein complexes. Curr Opin Chem Biol 42:25–33. https://doi.org/10.1016/j.cbpa.2017.10.026
5. Shuai N, Rabuck JN, Ruotolo BT (2013) Ion mobility-mass spectrometry of intact protein-ligand complexes for pharmaceutical drug discovery and development. Curr Opin Chem Biol 17(5):809–817. https://doi.org/10.1016/j.cbpa.2013.06.019
6. Nshanian M, Lantz C, Wongkongkathep P et al (2019) Native top-down mass spectrometry and ion mobility spectrometry of the interaction of tau protein with a molecular tweezer. J Am Soc Mass Spectrom 30(1):16–23. https://doi.org/10.1007/s13361-018-2027-6
7. Byrne DP, Vonderach M, Ferries S, Brownridge PJ, Eyers CE, Eyers PA (2016) cAMP-dependent protein kinase (PKA) complexes probed by complementary differential scanning Fluorimetry and ion-mobility mass spectrometry. Biochem J 473(19):3159–3175
8. Wongkongkathep P, Han JY, Choi TS, Yin S, Kim HI, Loo JA (2018) Native top-down mass spectrometry and ion mobility MS for characterizing the cobalt and manganese metal binding of α-synuclein protein. J Am Soc Mass Spectrom 29(9):1870–1880. https://doi.org/10.1007/s13361-018-2002-2
9. Yuwei T, Lippens J, Netirojjanakul C, Campuzano IDG, Ruotolo B (2018) Quantitative collision-induced unfolding differentiates model antibody–drug conjugates. Protein Sci 28(3):598–608. https://doi.org/10.1002/pro.3560
10. Rabuck-Gibbons JN, Lodge JM, Mapp AK, Ruotolo BT (2019) Collision-induced unfolding reveals unique fingerprints for remote protein interaction sites in the KIX regulation domain. J Am Soc Mass Spectrom 30(1):94–102. https://doi.org/10.1007/s13361-018-2043-6
11. Rabuck JN, Suk-Joon H, Ko KS, Fox CC, Soellner MB, Ruotolo BT (2013) Activation state-selective kinase inhibitor assay based on ion mobility-mass spectrometry. Anal Chem 85(15):6995–7002. https://doi.org/10.1021/ac4012655
12. Saunders C, Young LM, Mahood RA, Jackson MP, Revill CH, Foster RJ, Smith AD, Ashcroft AE, Brockwell DJ, Radford SE (2016) An in vivo platform for identifying inhibitors of protein aggregation. Nat Chem Biol 12:94–101. https://doi.org/10.1038/nchembio.1988
13. Hozefa A, Bate C, Williams A, Virdee J, Jeggo R, Spanswick D, Scopes DIC, Treherne JM, Mazzitelli S, Chawner R, Eyers CE, Doig AJ (2012) The N-methylated peptide SEN304 powerfully inhibits Aβ(1–42) toxicity by perturbing oligomer formation. Biochemistry 51(42):8338–8352. https://doi.org/10.1021/bi300415v
14. Vonderach M, Byrne DP, Barran PE, Eyers PA, Eyers CE (2018) DNA binding and phosphorylation regulate the core structure of the NF-κB p50 transcription factor. J Am Soc Mass Spectrom 30(1):128–138
15. Lixa C, Mujo A, de Magalhães MTQ, Almeida FCL, Lima LMTR, Pinheir AS (2018) Oligomeric transition and dynamics of RNA binding by the HuR RRM1 domain in solution. J Bimol NMR 72(3–4):179–192. https://doi.org/10.1007/s10858-018-0217-y
16. Wang Y, Park H, Lin H, Kitova EN, Klassen JS (2019) Multipronged ESI–MS approach for studying glycan-binding protein interactions with glycoproteins. Anal Chem 91(3):2140–2147. https://doi.org/10.1021/acs.analchem.8b04673
17. Zhao Y, Yang JY, Thieker DF, Xu Y, Zong C, Boons GJ, Liu J, Woods RJ, Moremen KW, Amster IJ (2018) A Traveling Wave Ion Mobility Spectrometry (TWIMS) study of the Robo1-heparan sulfate interaction. J Am Soc Mass Spectrom. https://doi.org/10.1007/s13361-018-1903-4
18. Lanucara F, Holman SW, Gray CJ, Eyers CE (2014) The power of ion mobility-mass spectrometry for structural characterization and the study of conformational dynamics. Nat Chem 6(4):281–294. https://doi.org/10.1038/nchem.1889
19. Gabelica V, Shvartsburg AA, Afonso C, Barran P, Benesch JLP, Bleiholder C, Bowers MT, Bilbao A, Bush MF, Larry Campbell J, Campuzano IDG, Causon T, Clowers BH, Creaser CS, De Pauw E, Far J, Fernandez-Lima F, Fjeldsted JC, Giles K, Groessl M, Hogan CJ, Hann S, Kim HI, Kurulugama RT, May JC, McLean JA, Pagel K, Richardson K, Ridgeway ME, Rosu F, Sobott F, Thalassinos K, Valentine SJ, Wyttenbach T (2017) Recommendations for reporting ion mobility Mass

Spectrometry measurements. Mass Spectrom Rev 38(3):291–320

20. Khajehali E, Malone DT, Glass M, Sexton PM, Christopoulos A, Leach K (2015) Biased agonism and biased allosteric modulation at the CB1 cannabinoid receptor. Mol Pharmacol 88 (2):368–379. https://doi.org/10.1124/mol

21. Verkhivker GM (2017) Leveraging structural diversity and allosteric regulatory mechanisms of protein kinases in the discovery of small molecule inhibitors. Curr Med Chem 24 (42):4838–4872. https://doi.org/10.2174/0929867323666161006113418

22. Wilson LJ, Linley A, Hammond DE, Hood FE, Coulson JM, MacEwan DJ, Ross SJ, Slupsky JR, Smith PD, Eyers PA, Prior IA (2017) New perspectives, opportunities, and challenges in exploring the human protein kinome. Cancer Res 78(1):1–16. https://doi.org/10.1158/0008-5472.CAN-17-2291

23. Masterson LR, Mascioni A, Traaseth NJ, Taylor SS, Veglia G (2008) Allosteric cooperativity in protein kinase A. Proc Natl Acad Sci U S A 105:506–511

24. Bush M (2019) Collision cross section database. Bush Lab. https://depts.washington.edu/bushlab/ccsdatabase/

Chapter 12

Liquid Extraction Surface Analysis (LESA) High-Field Asymmetric Waveform Ion Mobility Spectrometry (FAIMS) Mass Spectrometry for In Situ Analysis of Intact Proteins

Rian L. Griffiths, Klaudia I. Kocurek, and Helen J. Cooper

Abstract

Liquid extraction surface analysis (LESA) is an ambient mass spectrometry technique which enables direct analysis of analytes from solid substrates. LESA is particularly suitable for the analysis of intact proteins from biological substrates such as thin tissue sections and bacterial colonies growing on agar. Nevertheless, these substrates present a challenge for LESA protein analysis owing to their inherent complexity. It is therefore beneficial to hyphenate LESA with a gas phase separation technique. One such technique is high-field asymmetric waveform ion mobility (FAIMS), in which ions are separated according to their different mobilities in high and low electric fields. Here, we describe the protocols for LESA FAIMS mass spectrometry of intact proteins in thin tissue sections and bacterial colonies.

Key words LESA, FAIMS, Proteins, Bacterial colonies, Tissue

1 Introduction

Ambient ionization mass spectrometry offers the opportunity to analyze biomolecules directly from biological substrates. Of the many techniques available, liquid extraction surface analysis (LESA) mass spectrometry (MS) has emerged as a promising technique for the analysis of intact proteins directly from substrates such as thin tissue sections, dried blood spots, and bacterial colonies [1]. LESA involves the extraction of analytes of interest via the formation of a liquid junction between a robotically controlled conductive pipette tip and the surface of interest [2]. The liquid junction is maintained for a few seconds before the sample is reaspirated into the pipette tip and nanoelectrospray ionization (nESI) is performed (Fig. 1).

One of the challenges of in situ analysis is sample complexity; thus the integration of some form of separation is attractive. High-field asymmetric waveform ion mobility spectrometry (FAIMS) (*see*

Giuseppe Paglia and Giuseppe Astarita (eds.), *Ion Mobility-Mass Spectrometry: Methods and Protocols*, Methods in Molecular Biology, vol. 2084, https://doi.org/10.1007/978-1-0716-0030-6_12,

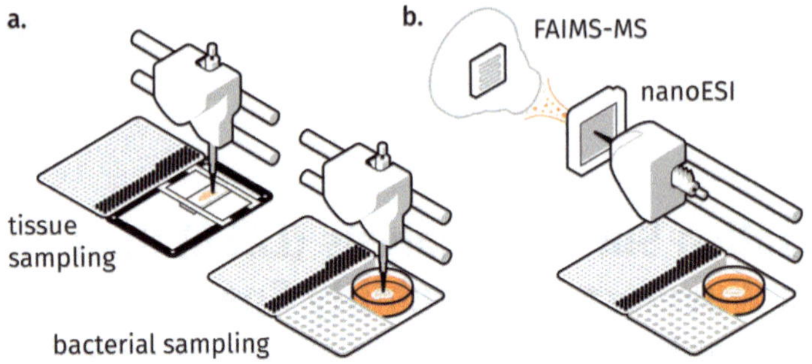

Fig. 1 (**a**) Schematic representation of liquid extraction surface analysis (LESA) from tissue sections and bacterial colonies. Solvent is aspirated into a robotically operated pipette tip, dispensed onto the sample, and then reaspirated. (**b**) Chip-based nanoelectrospray ionization is performed, and the ions enter the cone that holds the FAIMS device

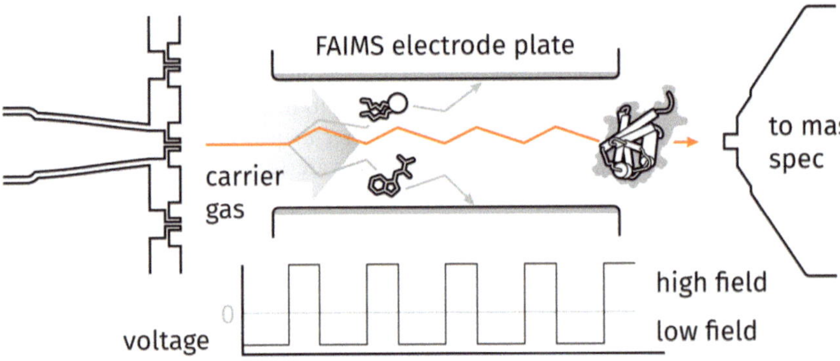

Fig. 2 Schematic representation of high-field asymmetric ion mobility spectrometry (FAIMS). Ions experience alternating high and low electric fields, only certain ions will be transmitted at particular settings

Fig. 2) is a gas phase separation technique in which ions are separated according to their different mobilities in electric fields [3, 4]. Ions traverse the device between two parallel electrodes. An oscillating electric field, consisting of a high field of short duration (known as the dispersion field, DF) and a low field of long duration, is applied between the electrodes. The differential mobility of ions in high and low electric fields results in the deflection of their trajectories through the device. A DC compensation voltage, giving rise to the compensation field (CF), is superimposed on the asymmetric waveform to correct the trajectories of a subset of ions and thus selectively transmit them through the device [4]. FAIMS coupled with LESA mass spectrometry has been shown to offer rapid molecular separation leading to improved signal-to-noise ratios and intact protein coverage from tissue sections and bacterial colonies [5–7].

2 Materials

2.1 Tissue Sample Preparation

1. Frozen tissue organ (*see* **Note 1**).
2. A Leica CM1810 Cryostat.
3. Glass-frosted microscope slides. We use VWR Super Premium Microscope slides (1.0–1.2 mm).
4. Washing solution, 80% ethanol/20% water. High purity (mass spectrometry grade) solvents should be used.

2.2 Bacterial Sample Preparation

1. A glycerol stock or donor plate containing bacteria of interest.
2. Liquid nutrient media (commonly lysogeny broth, LB; type dependent on bacterial strain), 5–10 ml per strain.
3. 60 mm diameter petri dishes containing agar media (commonly LB agar).
4. Sterile inoculation loops (1 μl).
5. A temperature-controlled incubator; a shaking incubator is required for liquid cultures.
6. A class II microbiological safety cabinet or a Bunsen burner.
7. A 96-well plate (Thermo Scientific), sealed and cut in half.
8. A 1:1 photocopy of the 1536 grid plate (supplied with the TriVersa NanoMate (*see* below), printed, and cut to fit inside the sampling tray).

2.3 LESA Solvents

1. 40% acetonitrile/60% water with the incorporation of 1% formic acid for analysis of tissue and Gram-negative bacteria. Only high-purity (mass spectrometry grade) solvents should be used.
2. 50% acetonitrile/45% water/5% formic acid (all mass spectrometry grade) for analysis of Gram-positive bacteria.

2.4 Analysis

1. Triversa Nanomate ion source (Advion) controlled by ChipSoft software with LESA capability (for tissue analysis). For analysis of bacterial colonies, the ChipSoft software needs to be AUI (Advanced User Interface) enabled.
2. UltraFAIMS device (Owlstone) controlled by UltraFAIMS Control software.
3. LTQ Orbitrap Elite™ mass spectrometer (Thermo Fisher Scientific) controlled by MS Tune software.

3 Methods

3.1 Tissue Sample Preparation

1. Tissue organs are stored at −80 °C. In order to perform tissue sectioning in a cryostat (Leica 1850 CM), place the tissue organ of interest into the cryostat.

2. Place a few drops of pure water (analytical grade) in the center of a room temperature chuck before putting this in the cryostat and mounting the tissue organ with tweezers. Once mounted (this is easier when the tissue is at −80 °C), leave to equilibrate to the cryostat temperature (~20 °C) (*see* **Note 2**).
3. Put the chuck into the holder and tighten the screw. Take care to avoid the blade below and orientate the tissue as desired.
4. Adjust the angle of the mounted chuck and the distance of the tissue away from the blade. Fine tune the distance between the tissue sample and the blade bringing the sample as close as possible without touching.
5. Set the cutting thickness. We typically use 10 μm.
6. Put the roll plate across the blade.
7. Rotate the handle to the right start in a smooth continuous motion away from yourself to start cutting.
8. Use the fine brushes to detach the tissue section from the blade and move to opposite side of the cool plate.
9. Use the large brush to remove debris away from the blade before cutting the next tissue section (*see* **Note 3**).
10. Thaw-mount section onto room temperature frosted glass slides by slowly bringing the glass slide close to the sections on the cold plate.
11. Tissue washing is not necessary for all tissue types, e.g., liver; however, in the case of brain tissue, it is beneficial to wash the slide in order to remove abundant lipids that may interfere with protein analysis. To wash the tissue, dip the slide in 80% ethanol for 10 s (*see* **Note 4**).

3.2 Bacterial Colonies Sample Preparation

3.2.1 Directly from Glycerol Stocks

1. Glycerol stocks are stored at −80 °C. Transfer stock vials to the microbiological safety cabinet or Bunsen burner using an ice box to delay thawing of the stock (repeated freeze-thawing of the stock will decrease bacterial viability).
2. If using a Bunsen burner, work within 20 cm of the burner, set to a blue flame.
3. Scrape the surface of the glycerol stock with a sterile inoculation loop and touch the surface of the agar plate where colony growth is desired. On a 60 mm agar plate, six or seven large colonies can be grown.
4. Place the agar plate upside down in the incubator. The growth temperature depends on the strain used; for *E. coli*, *P. aeruginosa*, *S. aureus*, and a number of other common human pathogens, 37 °C is optimal. Incubate for 24 h or until the colonies reach a size suitable for sampling (approx. 3–5 mm in diameter).

3.2.2 From Liquid Culture

1. Prepare a 5–10 ml aliquot of sterile nutrient broth in a 50 ml sterile tube (*see* **Notes 5** and **6**).
2. As above, retrieve glycerol stocks and scrape the surface of the stock with a sterile inoculation loop. Dip the loop in the nutrient broth.
3. Close the broth tube and swirl it gently. Leave the lid a quarter turn undone and secure with a strip of autoclave tape.
4. Place the tube in a shaking incubator at 200 rpm.
5. Incubate at optimum temperature overnight (up to 18 h). Ideally, the organisms should be in the log phase of growth; this can be checked if desired by measuring the optical density of the culture at 600 nm. An OD_{600} of <2.0 is usually fine.
6. Working under sterile conditions, dip an inoculation loop in the liquid culture and touch the agar plate to deposit the droplet where colony growth is desired. Leave to dry (5–10 min).
7. As above, place inverted agar plates in the incubator (stationary) at optimum growth temperature and incubate until ready.

3.3 Liquid Extraction Surface Analysis

3.3.1 Tissue Analysis

1. Mount the glass slide onto the LESA universal adapter plate.
2. Scan the slide using an Epson V300 Photo scanner. Load the scanned image into the LESA Points software.
3. Set up the following method in ChipSoft under the Method Manager tab: Set the solvent volume to 6 μl, depth to 0.2, dispense volume to 3 μl, repeat mix to 2, and reaspirate to 4 μl. The ESI voltage should be set to 1.7 kV, and the gas pressure should be set to 0.3 psi. Save this method (*see* **Note 7**).
4. Select locations for sampling in the LESA Points software, highlight all of the locations in the right hand panel (right click, Select All), and then select the above method (right click, add method, then browse to your saved method file). Save this as a sequence file.
5. Open the sequence file in the ChipSoft software in the sequence tab.
6. Optimize the height settings in the Method Manager tab in ChipSoft such that a liquid junction is formed without contacting the surface. The height can vary depending on the sample type and the calibration of the Triversa Nanomate. Trial and error is the best way to optimize this setting.

3.3.2 Bacterial Colonies

1. Navigate to the Interface Settings tab and set mode to Infusion.
2. Navigate to the Method Manager tab and change sample delivery duration to 30 min, ESI voltage to 1.75 kV, and gas pressure to 0.3 psi. Save this as a spray method file.

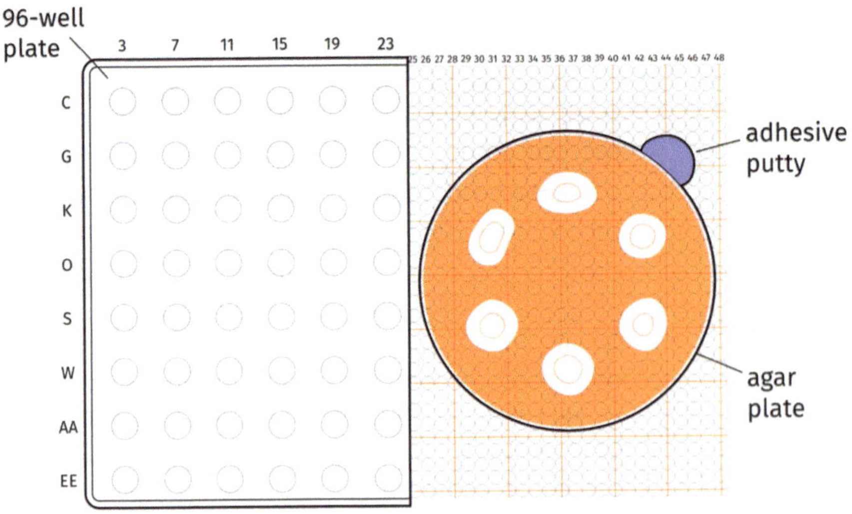

Fig. 3 A schematic representation of the set up used for the analysis of bacterial colonies. In the Advanced User Interface (AUI) mode, a solvent well must be defined within the normal plate holder area. Here we show half a 96-well plate on the left and the bacterial sample on the right. The grid locations used by the system are also indicated

3. Place the photocopy of the 1536 grid plate in the sampling tray of the TriVersa NanoMate.
4. Place half of a 96-well plate in the sampling tray of the TriVersa NanoMate. Deposit approximately 50 μl of LESA solvent in a well.
5. Place the petri dish containing bacterial colonies alongside the microtiter plate and secure with adhesive putty. *See* Fig. 3 for a schematic of the arrangement.
6. Navigate to the advanced user interface (Advanced UI) tab in ChipSoft.
7. Set up a new method containing three rows (steps).
8. Place an Aspirate step in the first row. Select the location corresponding to the solvent well; a translation of the universal adapter plate grid locations to well locations is provided in Fig. 3 (also *see* **Note 8**). Set the volume to 3 μl. Enable advanced options and set the aspiration height to approximately −13.4 mm.
9. Specify a Mix step in the second row. Select a location corresponding to a location on the colony to be sampled, using the 1536 grid plate copy as a guide. Set the dispense volume to 2 μl and the aspirate volume to 2.5 μl. Ensure the mix cycle is set to dispense first. Ensure the tip is not ejected by unticking the corresponding box. Set the aspirate delay time (the dwell time) to 3 s. Change all three depths (aspirate, dispense, touch-off) to the same value, determined by the

height of the agar plate and the colony. Approximately −10.0 mm is a good start point (the value is negative as it is measured from the top rim of the sampling tray) (*see* **Note 9**).

10. Specify a Spray step in the last row. Select the same location as specified in the previous row. Select the spray method created in **step 2**.
11. Validate and save the sequence.
12. Open the sequence in the Sequence tab and run it. Optimize sampling height (changing all three depth values in the Mix step simultaneously) until the pipette tip touches the colony surface but does not pierce through to the agar underneath (*see* **Notes 10** and **11**).

3.4 Configuration of High-Field Asymmetric Waveform Ion Mobility Spectrometry (FAIMS)

1. Place the FAIMS cone, housing the FAIMS chip, onto the mass spectrometer inlet and secure with grub screws. Attach the waveform generator to the chip housing by carefully pushing in the arm such that the connector pins contact the FAIMS chip housing and the plate appears to be flush. Secure with screws.
2. Attach the power cable to the waveform generator, connect the USB cable for the software control to the computer and to the waveform generator, and finally clip the earth crocodile clip to ground, e.g., the mass spectrometer housing. Refer to the user manual for further details.
3. Couple the TriVersa NanoMate to the FAIMS device using the brackets provided.
4. Open the Owlstone FAIMS software. Click Connect (top right). The system light in the software and on the FAIMS waveform generator will turn green once connected. Set the FAIMS chip region temperature to 100 °C.
5. Allow temperature to equilibrate for ~ 30 min.

3.5 FAIMS Method

The Owlstone FAIMS device can perform experiments in three different modes; 2D mode (varying dispersion and compensation voltages), 1D mode (a single dispersion field setting and varied compensation fields), or static mode (a selected dispersion and compensation field), as indicated by the tabs in the software.

3.5.1 Initial 2D Optimization Experiments

1. Set the Dispersion field to start at 130 Td and end at 270 Td, at 20 Td steps/intervals.
2. Set the Compensation field to scan from −1 Td to +4 Td.
3. Set the time to 180 s per dispersion field (for a 24-min experiment).
4. Press 'start scan' once the sample is electrospraying and data are acquiring.

5. Use the results to determine the optimal settings for transmission of your proteins of interest. There may be multiple combinations of settings that transmit species of interest. Data can be visualized as a map of transmission of all ions (Total Ion Transmission map) or single m/z (Single Ion Transmission map) using freely available software (http://www.biosciences-labs.bham.ac.uk/cooper/software.php) to aid determination of optimal settings.

3.5.2 1D Experiments

1. Set the Dispersion field to chosen value, e.g., 270 Td.
2. Set the Compensation field to scan from −1 Td to +4 Td.
3. Set the time for the experiment, e.g., 300 s.
4. Press 'start scan' once the sample is electrospraying and data are acquiring.

3.5.3 Static Experiments

1. Set the Dispersion field to the desired setting. (We typically use 270 Td).
2. Set the Compensation field to the desired setting. (We find that intact proteins are typically transmitted between 2–3.5 Td at a dispersion field of 270 Td.)
3. Set the time to the desired number of seconds. Typically, 60 s is sufficient. Click 'Add'. The settings should appear in the box on the right of the viewer. It is possible to add multiple steps by writing in the settings at the bottom and clicking 'Add'.
4. Click Start Scan when you are ready to acquire data.

3.5.4 Mass Spectrometry

1. Capillary temperature should be set to 350 °C for FAIMS experiments.
2. Define the mass spectrometry parameters in the MS Tune software in the 'Define Scan' window (bubbles icon) to the following (as a guideline):

Mass analyzer	FTMS
Mass range	Normal
Resolution	120,000
Scan type	Full
Microscans	1
Max injection time	1000 ms
Mass range	400–2000
AGC	On

3. In the 'Acquire Data' (camera icon) window, define which folder the data should be saved in, provide a File Name, and

acquire the data continuously from the moment you press 'Start'. Make sure start mode is 'Immediate'.

4. Mass spectrometry data can be visualized in Xcalibur software. This software opens automatically when 'View' is clicked in the 'Acquire Data' window.
5. It is possible to visualize the FAIMS data as a Total Ion Transmission (TIT) map (m/z vs. compensation field at a single dispersion field) or a Single Ion Transmission (SIT) map (compensation field vs. dispersion field for a single m/z) using freely available software (http://www.biosciences-labs.bham.ac.uk/cooper/software.php). Detailed instructions accompany the software download.

4 Notes

1. It is imperative to use freshly frozen and not fixed tissue samples for intact protein analysis.
2. When mounting tissue onto the chuck, it can be beneficial to hold the tissue gently in position with the tweezers to orientate the tissue as desired while the water freezes it into position. It is easiest to mount the tissue while it is still at −80 °C and then leave it to warm up to the cryostat temperature. Depending on the nature of the organ, the optimal cryostat temperature can vary between −25 to −18 °C. Refer to the manufacturers' manual for guidance.
3. It is easiest to cut a single section, move this across the cold plate, and then section the next section, repeating and lining up the serial sections ready for thaw mounting in the desired order and orientation. This approach enables a single thaw mount of multiple sections onto a single slide.
4. When marking up the slides, use pencil on the frosted part of the microscope slides. Use of ink will result in contamination if tissue washing is performed.
5. Preparation of colonies from liquid cultures requires an additional day of incubation but yields colonies of a uniform height and density, greatly facilitating LESA sampling. Using agar plates prepared with identical aliquots of media, and thus uniform in height, further reduces the need for extensive sampling height optimization.
6. Optimum growth conditions for different species may differ significantly and should be adjusted accordingly. For slow-growing species, larger volumes (e.g., 10 μl) of the starting liquid culture can be deposited onto agar plates prior to incubation.

7. The sampling protocol described for tissue sections should be considered a good starting point for general MS analysis. If MS/MS data are to be acquired, it may be necessary to use larger volumes (e.g., solvent 8 μl, dispense 5 μl, and reaspirate 6 μl). If imaging experiments are conducted, smaller volumes may be preferable (e.g., solvent 2 μl, dispense 1 μl, and reaspirate 1.5 μl) to minimize the size of the liquid junction.
8. The exact positions of solvent wells with regards to the 1536 grid plate template may vary with stage calibration and the dimensions/position of the grid template used and should therefore be verified and adjusted for the TriVersa NanoMate in use.
9. The sampling protocol described for bacterial colonies should be used as a starting point for optimization for the specific bacterial species under investigation. The dwell time can be extended to 15 or 30 s to increase protein extraction. If not immediately observed without FAIMS, the presence of protein signal can be verified with FAIMS at DF 270 Td–CF 3.0 Td. If no multiply charged peaks are observed within the first 5 min of data acquisition at high dwell times, increasing the proportion of acetonitrile up to 60% may help.
10. Use of a separate nanoelectrospray chip for bacterial analysis is recommended due to the tendency of aspirated colony matter to contaminate a large area of the chip around the nozzle used.
11. If the advanced user interface of the NanoMate is not available, bacterial colonies of interest can be excised from the agar plates by use of a sterile scalpel and a spatula and deposited onto glass slides. The glass slides can then be placed on the 1536 grid plate and analyzed by use of the LESA interface within ChipSoft. Use of LESA Points with a scanner is not recommended as the sample is likely to detach from the glass slide upon inversion and smear across the scanner surface.

Acknowledgements

H.J.C. and R.L.G. were funded by the EPSRC (EP/L023490/1). K.I.K. was in receipt of an EPSRC studentship in collaboration with the National Physical Laboratory.

References

1. Kocurek KI, Griffiths RL, Cooper HJ (2018) Ambient ionisation mass spectrometry for in situ analysis of intact proteins. J Mass Spectrom 53(7):565–578
2. Kertesz V, Van Berkel GJ (2010) Fully automated liquid extraction-based surface sampling and ionization using a chip-based robotic nanoelectrospray platform. J Mass Spectrom 45 (3):252–260

3. Buryakov IA, Krylov EV, Nazarov EG, Rasulev UK (1993) A new method of separation of multi-atomic ions by mobility at atmospheric-pressure using a high-frequency amplitude-asymmetric strong electric-field. Int J Mass Spectrom Ion Process 128(3):143–148
4. Guevremont R (2004) High-field asymmetric waveform ion mobility spectrometry: a new tool for mass spectrometry. J Chromatogr A 1058(1–2):3–19
5. Sarsby J, Griffiths RL, Race AM, Bunch J, Randall EC, Creese AJ et al (2015) Liquid extraction surface analysis mass spectrometry coupled with field asymmetric waveform ion mobility spectrometry for analysis of intact proteins from biological substrates. Anal Chem 87 (13):6794–6800
6. Griffiths RL, Creese AJ, Race AM, Bunch J, Cooper HJ (2016) LESA FAIMS mass spectrometry for the spatial profiling of proteins from tissue. Anal Chem 88:6758–6766
7. Griffiths RL, Simmonds AL, Swales JG, Goodwin RJA, Cooper HJ (2018) LESA MS imaging of heat-preserved and frozen tissue: benefits of multistep static FAIMS. Anal Chem 90 (22):13306–13314

Chapter 13

Ion Mobility-Mass Spectrometry of Glycoconjugates

Weston B. Struwe and David J. Harvey

Abstract

Glycoconjugates are diverse biomolecules that are dynamically assembled to regulate and fine-tune numerous cellular processes. Their biosynthesis is nontemplate-driven, achieved stepwise in discrete locations within the cell, giving rise to a range of complex branched structures that pose a significant challenge in structural biology. Mass spectrometry is the leading method for analysis of glycoconjugates, and the addition of ion mobility has proven valuable for improving structural assignments of individual glycans in complex biological mixtures. In this chapter, we briefly discuss recent applications of IM for glycomics and describe how to acquire, interpret, and analyze IM-MS data for the analysis of glycans.

Key words Glycosylation, Mass spectrometry, Ion mobility, Glycomics, Collision cross section

1 Introduction

Carbohydrates influence individual protein structure and function and play diverse roles in cell signaling and localization. Eukaryotic glycoconjugates are defined by type, comprising those attached to folded proteins (N-glycans, O-glycans, C-mannose, proteoglycans), lipids (glycosphingolipids, GPI-anchored proteins), as well as free oligosaccharides. Each class of glycoconjugate poses a separate analytical challenge, but in most cases, mass spectrometry (MS) is the principle technique used to examine their structures. Ion mobility (IM) is rapidly emerging as a valuable method that improves glycomics studies in two ways: (1) gas phase separation of isomeric structures and (2) enhanced sensitivity by exclusion of noncarbohydrate contaminant ions.

Ion mobility is based on the migration of ions through a neutral gas, typically nitrogen or helium, under the influence of a weak electric field, and separation is determined by their size, shape, and charge. The resulting drift time is directly related to the rotationally averaged collision cross section (CCS), a characteristic value for a given glycan ion, analogous to HPLC glucose units, and is a

Giuseppe Paglia and Giuseppe Astarita (eds.), *Ion Mobility-Mass Spectrometry: Methods and Protocols*, Methods in Molecular Biology, vol. 2084, https://doi.org/10.1007/978-1-0716-0030-6_13,

valuable metric for identifying glycan structures. And so, the development of curated glycan CCS databases derived from glycan standards or known structures can help structural characterization studies [1]. Ion mobility of glycans is largely influenced by instrument polarity (positive, negative), type of ion formed (e.g. protonated, deprotonated, or adducted), adduct characteristics (metal ion radius and valence), and the IM cell gas [2–4]. Therefore glycan separation, which is relevant for isomers of equal charge, depends on the degree to which gas phase conformations differ, particularly among glycan isomers.

Predicting the three-dimensional structures of glycan ions poses a significant challenge and is exemplified by limited reports of gas-phase modeling. One issue is the location of charge, specifically for deprotonated $[M-H]^-$ ions, as the exact position is ambiguous. A systematic ab initio molecular dynamics experiment of $[M-H]^-$ ions discovered that charge is dynamic and moves across monosaccharides on a given glycan ion, complicating in silico structural approaches [5]. Taken as a whole, density-functional theory methods are not only complicated and costly but do not yet cover the entire time scale of IM measurements (milliseconds) nor are realistic for exploring all known glycan structures in nature. Consequently, to understand gas-phase properties of glycan ions, researchers rely heavily on experimental data from standards and/or synthetic oligosaccharides. Substantial work has helped shed light on gas phase behavior of glycans and demonstrated the benefit of IM-MS for glycomics studies [6].

One of the strengths of IM lies in different gas phase properties of biomolecules, namely lipids, peptides, carbohydrates, and salts, resulting in class-specific localization [7] that enables improved sensitivity through extraction of glycan ions alone. Signal suppression by noncarbohydrate contaminants is one problem in glycomics that is solved by IM. A notable example is the removal of contaminant ions such as those from polyethylene glycol whose singly charged molecular ions fall on a different *m/z*:CCS trend line to those from N-glycans [8]. IM-MS also enabled characterization of HIV gp120 N-glycans directly from an incubation mixture containing endoglycosidase (PNGase F), deglycosylated gp120 protein, salt, and detergent [9]. For targeted fragmentation studies, the quadrupole-selected *m/z* value for a glycan analyte may also capture additional species with the same *m/z* value leading to a mixed and sometimes difficult to interpret spectrum. Again, IM can remove the contaminating ions leaving a clean spectrum. There are multiple approaches to glycan characterization by IM-MS, either through analysis of intact molecular ions or by a fragment-based approach (i.e., collision-induced dissociation, CID) applied pre-IM (separation of glycan fragments) or post-IM (separation of intact glycans) as shown in Fig. 1. Structural information can be gained from IM separation of glycan fragments (i.e., pre-IM CID),

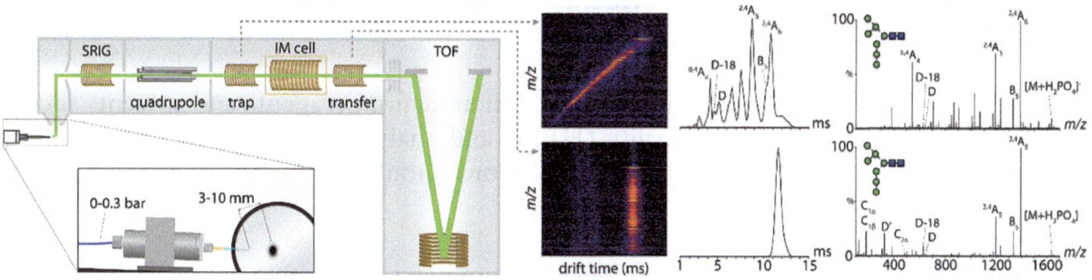

Fig. 1 Schematic of Waters Synapt hybrid Q-IM-TOF mass spectrometer. Setup of direct nanoelectrospray and capillary position with respect to sample cone (inset). Glycan fragmentation can occur in the trap (pre-IM) or transfer (post-IM) region of the instrument. Examples of ion mobility drift plots, arrival time distributions, and MS/MS spectra of a $Man_7GlcNAc_2$ N-glycan fragmented in the trap (top) or transfer (bottom) are shown

in that it facilitates a targeted approach of structural features such as sialic acid linkage [10, 11] or fucosylated epitopes [12] which may be overlooked when investigating the intact glycan. For example, drift times of some ions from glycans containing fucose residues in their antennae are different depending on which antenna they are located [13]. Many isobaric fragment ions can be separated by ion mobility, particularly in negative mode, providing information on the multiplicity of structures that can exist at each *m/z* value. Such information enables greater confidence in their use for structural determination and provides additional structural information. Post-IM CID generates time-aligned fragments, which can be used to extract IM drift time information and uncomplicated identification of glycan isomers. For example, intact $Man_8GlcNAc_2$ N-glycans isomers are easily separated by IM and can be identified by applying CID post-IM and assigning each structure *via* diagnostic fragment ions [14, 15]. Importantly, separation of intact glycan isomers is prominently affected by adduct state, such as different alkali metals and, therefore, ion polarities and sample preparation should be considered when performing identification studies [16, 17].

Overall, each glycan analysis by IM-MS may require testing multiple parameters including sample preparation (adduction) and instrument setup (positive or negative mode), IM settings (N_2 or He drift gas) as well as fragmentation parameters (trap or transfer CID) to achieve optimal results. Here, we outline a step-by-step method to acquire and analyze glycan IM-MS data using a Waters Synapt G2Si mass spectrometer by direct nanoelectrospray infusion. The approach described herein is based on IM-MS analysis of N-glycans, O-glycans, and milk oligosaccharides using nitrogen as the drift gas [5, 12, 13, 15, 18–22]. Similar results may be achieved from other IM-MS instruments (e.g., Agilent 6560 and Bruker timsTOF) (*see* Subheading 2.1).

2 Materials

To minimize sample contamination, reagents and solvents should be high quality (HPLC and analytical grade or greater). Milli-Q water is suitable. Solutions should be freshly prepared prior to analysis.

2.1 Ion Mobility-Mass Spectrometers

1. There are several commercial instruments available for ion mobility-mass spectrometry analysis of oligosaccharides. These include the 6560 IM Q-TOF (Agilent), timsTOF (Bruker), and Synapt (Waters) each having characteristic configurations influencing the type of glycan analysis that can be achieved.
2. Drift-tube (DT) instruments (Agilent 6560) allows for direct CCS measurements without the need for calibration. However, current configurations place the quadrupole and collision cell after the drift tube which restricts IM measurements to intact glycan ions.
3. Trapped ion mobility-based instruments (Bruker timsTOF) have a similar geometry to the Agilent 6560 (i.e., CID after IM) but operate by flowing the IM gas against immobilized ions. These instruments have increased IM resolution compared to commercial DT or traveling wave mass spectrometers.
4. Traveling wave (TW) instruments (Waters Synapt) requires a calibration standard to estimate CCS values, but this instrument has the advantage that glycan fragmentation can occur both pre-, post-, or both pre- and post-IM separation.

2.2 Sample Preparation

1. Water (Milli-Q).
2. Methanol (HPLC or LC-MS grade).
3. Acetonitrile (HPLC or LC-MS grade).
4. Ammonium phosphate ($NH_4H_2PO_4$, ≥98% pure).
5. Ammonium acetate ($NH_4CH_3CO_2$, ≥98% pure).
6. Potassium chloride (KCl, ≥98% pure).
7. Ammonium chloride (NH_4Cl, ≥98% pure).
8. Formic acid (HCO_2H).
9. Acetic acid.
10. Trifluoroacetic acid.
11. Concentrated nitric acid.
12. Nafion® 117 membrane.
13. Convex watch glass.
14. Plastic petri dish with lid.

15. Hypercarb™ SPE Cartridges.
16. ZipTip® C_{18} (0.2 resin bed) pipette tips.
17. Benchtop microcentrifuge.
18. α2–3,6,8,9 neuraminidase.
19. 37 °C incubator.
20. 10 kDa molecular weight cutoff spin filters.

2.3 Glycan Analysis

1. Dextran from *Leuconostoc mesenteroides*, MW 1000.
2. Dextran from *Leuconostoc mesenteroides*, MW 5000.
3. Sodium iodide (2 μg/ml), (NaI, ≥99% pure).
4. Vacuum concentrator.
5. Borosilicate glass capillaries, 1.0 mm OD × 0.78 mm ID, with filament.
6. Flaming/Brown P-1000 micropipette puller.
7. Agar automatic sputter coater.
8. Nanoflow probe tips, long (*see* **Note 1**).
9. Column scribe.
10. 0.5–20 μl Eppendorf GELoader tips.
11. AA stainless steel Dumont tweezers.
12. T2351 stainless steel positioning tweezers, 115 mm.
13. Stereo microscope, 10–30×.

3 Methods

This chapter focuses on IM-MS analysis of N-glycans after their release and purification from glycoproteins. Release, isolation, and purification of glycoconjugates differ considerably and are not described here, but comprehensive protocols are available for isolating N-glycans [23, 24], O-glycans [25, 26], glycosaminoglycans (GAGs) [27, 28], and glycopeptides [29, 30]. The methods described herein are also directly applicable to O-glycans and glycopeptides. Additional considerations should be made when analyzing certain other types of oligosaccharides by IM-MS, such as GAGs (*see* **Note 2**).

3.1 Sample Desalting

A major consideration is purity of the glycan sample, both in terms of salt and noncarbohydrate contaminates. Purification steps are necessary, especially for biologically derived material and can be performed by cation exchange (Nafion membrane) [31] or solid phase extraction (porous graphitized carbon (PGC)-C_{18}) immediately prior to MS.

3.1.1 Desalting—PGC-C_{18} Spin Columns

1. Prepare PGC-C_{18} spin tips. Extract PGC material from Hypercarb SPE cartridge with MeOH (approximately 50 mg/ml PGC slurry). Store excess PCG in MeOH (*see* **Note 3**).
2. Add 15 μl PGC slurry to C_{18} ZipTip and place at the bottom of a 1.5 ml microcentrifuge tube.
3. Wash PGC-C_{18} bed with 25 μl 90% acetonitrile (ACN), 0.5% trifluoroacetic acid (TFA). Centrifuge with benchtop centrifuge (low speed) until solvent has passed through PGC-C_{18} bed. Discard the flow-through. Repeat twice.
4. Equilibrate with 3 × 25 μl 0.5% TFA in Milli-Q water. Centrifuge and discard flow-through between each wash.
5. Apply sample to top of PGC-C_{18} bed and centrifuge for approximately 15 s or until sample has completely passed into the PGC.
6. Wash with 3 × 25 μl 0.5% TFA in water. Centrifuge and discard the flow-through between each wash.
7. Place spin tip in new microcentrifuge tube.
8. Elute with 3 × 25 μl 40% ACN, 0.5% TFA. The sample will collect in the bottom of the microcentrifuge tube.
9. Remove spin tip.
10. Dry sample in a vacuum concentrator.
11. The sample is ready for analysis and transfer to nanospray capillaries (*see* Subheading 3.3).

*3.1.2 Desalting—Nafion Membrane (See **Note 4**)*

1. If necessary, regenerate Nafion® 117 membrane (*see* **Note 5**).
2. Pipette 1–2 ml Milli-Q water on convex watch glass placed in petri dish.
3. Place precut membrane (~5 × 5 cm) on water layer. Do not submerge, but float the membrane on top of the water.
4. Spot sample (2–5 μl) on membrane, cover with Petri dish lid and let sit for 1–12 h. Longer times are necessary for samples with high salt.
5. Carefully remove sample from the membrane.
6. Add water and methanol to 50% (v/v) and a final volume of 9 μl.
7. The sample is ready for analysis and transfer to nanospray capillaries (*see* Subheading 3.3).

3.2 Sialic Acid Removal

Sialylated glycans readily form singly or multiplied charged ions, commonly as $[M-nH]^{n-}$ ions in negative mode, depending on the number of sialic acid residues. The charge is localized to the carboxylate group on sialic acid residues, and collision-induced dissociation (CID) produces spectra of predominantly sialic acid fragment ions (m/z 290 in the case of Neu5Ac). These data do not contain significant structural data and, therefore, additional sample preparation, either removal of sialic acids or modification (e.g., permethylation), is required to produce useful MS/MS spectra for glycan assignments. Considerable information can be gained from a dual approach where both sialylated and a desialylated glycans (e.g., the same sample after all sialic acids are removed) are examined. In doing so, both glycan compositional information (from the nontreated sample) and structural information (linkage and branching) are obtained from the desialytated sample. Sialic acid removal is achieved chemically or enzymatically. Enzymatic treatment can be precise (e.g., trimming α2,3 linked residues only) but is costlier than chemical removal.

3.2.1 Sialic Acid Removal—Acetic Acid

1. Add 5 μl 2% acetic acid (HAc) to small volume of sample (~5–10 μl).
2. Incubate for 60 min at 80 °C.
3. Dry *via* vacuum concentrator.
4. Desalt *via* Nafion® membrane or PGC-C_{18} spin tip.

3.2.2 Sialic Acid Removal—Neuraminidase

1. Add neuraminidase to sample following manufacturers' protocol (*see* **Note 6**).
2. Incubate at 37 °C.
3. Remove neuraminidase *via* 10 kDa molecular weight cutoff (MWCO) centrifuge filters.
4. Dry *via* vacuum concentrator.
5. Desalt *via* Nafion® membrane or PGC-C_{18} spin tip.

3.3 Nano-electrospray of Glycan Ions

Conductive glass nanoflow electrospray capillaries can be purchased but are more expensive long-term than those made in-house. Nanoflow capillaries can be made easily with a micropipette puller and sputter coater [32]. Program values are instrument dependent and fluctuate with new heating filaments (*see* **Note 7**). Inspect needles and make adjustments to ensure a good emitter shape.

3.3.1 Adduct Formation

1. Add 1 μl of the following to generate specific ion adducts (*see* **Note 8**).

Ion	Aqueous solution
$[M+H_2PO_4]^-$	50 mM $NH_4H_2PO_4$
$[M+H]^+$	0.1% HCO_2H
$[M-H]^-$	50 mM $NH_4CH_3CO_2$
$[M+K]^+$	50 mM KCl
$[M+Cl]^-$	50 mM NH_4Cl

2. The sample is ready for transfer to nanospray capillaries.

3.3.2 Sample Loading

1. Pipette sample with GELoader tips (volume depends on the desired acquisition time). Generally, 10 μl is suitable for analyzing complex samples.
2. Hold capillary with positioning tweezers, insert pipette tip, and load sample to the bottom of the nanospray needle. Check that sample is completely at the bottom. Flick the capillary if necessary to pack sample toward the tip.
3. Using a microscope, trim the tip of the needle (for in-house capillaries) using fine tweezers to create an opening approximately 10 μm in diameter (*see* **Note 9**).
4. Place the capillary in the sprayer holder, secure the holder to the sprayer platform, and position emitter tip ~10 mm from the source cone (*see* **Note 10**).
5. Prior to the analysis of glycan samples, ensure that the instrument is mass calibrated either using a dextran homopolymer ladder or NaI solution. Dextran is preferred as it can be used both for mass calibration and CCS calculations (*see* Subheading 3.4).

3.3.3 Nanoelectrospray of Glycan Ions

1. With the instrument set to Mobility TOF mode (positive or negative polarity), turn on all IM gases. All example data and methods outlined in this chapter are based on nitrogen IM gas and argon for trap/transfer CID.
2. Recommended IM-MS instrument settings:

Sample cone	100 V (*see* **Note 11**)
Extraction cone	25 V
Cone gas	40 L/h
Source temperature	80 °C
Trap DC bias	30–60 V (*see* **Note 12**)
Trap collision energy	4–100 V
Transfer collision energy	4–100 V
Wave velocity	450 m/s
Wave height	40 V
Helium gas flow	180 ml/min
Trap gas flow	2 ml/min
IMS gas flow	80 ml/min

3. Apply backing pressure to form a small droplet at the end of the nanoflow emitter (*see* **Note 13**). Increase the pressure until a droplet is visible, but keep pressure to a minimum (0–0.3 bar).
4. Set the capillary voltage to 0.8–1 kV (negative mode) and 1.0–1.5 kV (positive mode). Adjust if required to generate or maintain a stable spray.

5. Watch for a signal and observe its intensity.
6. Optimize signal intensity and stability by adjusting the needle position, backing pressure, capillary voltage, and gas flows (*see* **Note 14**).
7. Record spectrum of intact glycan ions, with a mass range between 50–3500 *m/z*. Adjust mass range accordingly for other glycoconjugates. Set trap/transfer CID collision energies to 4 V to minimize fragmentation of molecular ions.

3.3.4 Tandem MS

1. Quadrupole mass selection of glycan ions should be performed within a 1 *m/z* window and can be adjusted by LM and HM resolution settings.
2. Isolate molecular ion of interest and check mass isolation before CID fragmentation.
3. Increase trap or transfer collision energies and acquire CID spectrum. The required voltage depends on the mass of the precursor glycan ion and should be adjusted until a useful MS/MS spectrum is acquired (*see* **Note 15**).
4. CID in the trap region results in IM separation of glycan fragments. Smaller ions (<200 Da) do not traverse through the IM cell as efficiently as larger fragment ions. Transfer CID generates time-aligned fragment ions and preserves smaller glycan fragment ions typically lost in trap CID experiments.

3.3.5 Extraction of ATDs from Glycan Ions

1. The drift properties (i.e., location in drift plots) of singly, doubly, and triply charged glycan ions are largely constant as shown in Fig. 2.

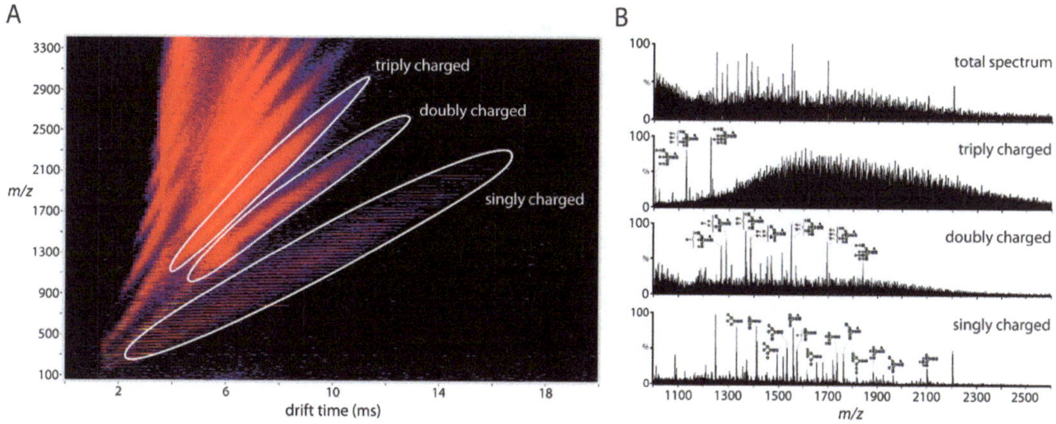

Fig. 2 (**a**) Ion mobility drift plot (or mobiligram) of HIV gp41 N-glycans. Glycan ions of different charge have characteristic drift positions. (**b**) IM extraction of each charge state improves detection of low abundant glycans

2. Using the DriftScope software, open raw IM-MS data file of interest.
3. Using the Selection Tool, select region in the drift plot to extract.
4. Under the File drop-down menu, select *Export to MassLynx* and *Retain Drift Time.*
5. Open IM-extracted data in the MassLynx chromatogram window, and combine all spectra to visualize *m/z* data in Mass Lynx spectrum window.
6. Data are ready for analysis. Examine glycan IM arrival time distributions (ATDs) by extracting *m/z* peaks into the chromatogram window.
7. Refer to the following publications for principles to interpret glycan ion mobility and fragmentation data [33–39].

3.4 Calculating Collision Cross Sections

TWIMS instruments require external calibration to calculate CCSs of unknown ions. It is important to use carbohydrates as calibration standards as the use of other biomolecules will negatively affect CCS accuracy. Dextran homopolymers are suitable calibrants as their absolute CCSs have been established using drift-tube instruments.

3.4.1 Acquiring Dextran Ladder Data

1. Dilute dextran homopolymers MW 1000 and 5000 in Milli-Q water (~mM stock concentration) and mix (1:1 equivalents to begin with). Adjust mixture so both low and high mass ions are detected.
2. Prepare dextran mix to generate $[M+H_2PO_4]^-$ adducts for negative ions or $[M+Na]^+$ for positive ions as described in Subheading 3.3.1.

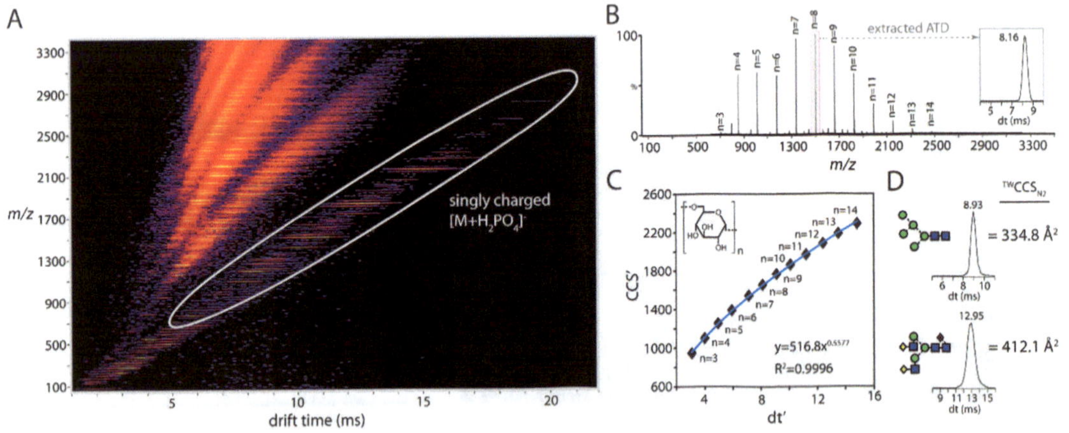

Fig. 3 (**a**) Ion mobility drift plot of dextran phosphate ion ladder. (**b**) IM-extracted spectrum of $[M+H_2PO_4]^-$ ions and associated arrival time distribution of the $n = 8$ dextran oligomer. (**c**) Power fit of dextran ladder CCS′ and d*t*′ values. (**d**) This calibration curve is used to calculate CCSs from ATDs of glycan ions, such as $Man_5GlcNAc_2$ (top) and F(6)A2G2 (bottom)

3. Acquire spectrum making sure low mass dextrans ($n = 3$) and high mass ($n = 14$ or greater) are present (Fig. 3). Adjust instrument tuning parameters if necessary to observe the range of dextran oligomers.

3.4.2 CCS Calculation—Calibration Curve

1. Extract dextran IM data *via* DriftScope (*see* Subheading 3.3.5) and ATD for each dextran ion.
2. Fit each ATD using a Gaussian distribution function to find the accurate drift time (*see* **Note 16**).
3. Set up CCS calibration curve using the freely available Excel spreadsheets and guides from the Pagel lab (logarithmic fit) and Thalassinos lab (power fit) respectively. (https://www.bcp.fu-berlin.de/en/chemie/chemie/forschung/OrgChem/pagel/research/carbohydrates/index.html and http://www.homepages.ucl.ac.uk/~ucbtkth/resources.html) [40].
4. In the *Calibration* tab, enter drift gas mass (28 Da for N_2) and EDC delay coefficient which can be located in the *_extern* file (within the .raw data folder) if it was not recorded during acquisition.
5. Enter *m/z*, charge, drift time, molecular weight (Da), and absolute DT-IMS CCS values (in *lit. CCS in Å²* column) for dextran $[M+H_2PO_4]^-$ ions.

Glucose units (*n*)	*m/z* $[M+H_2PO_4]^-$	CCS (Å²)
3	601.1	214.6
4	763.2	243.1
5	925.2	268.6
6	1087.3	297.4
7	1249.3	318.5
8	1411.4	339.6
9	1573.4	358.2
10	1735.5	377.9
11	1897.6	399.4
12	2059.6	418.7
13	2221.7	438.4
14	2383.7	456.2

6. The CCS′ (a corrected value for charge state and reduced mass) against d*t*′ (corrected drift time for mass-to-charge flight time) is generated and used for calculating CCSs of unknown glycan samples.

7. Measure the drift times of the glycan samples without any changes to IM voltages or gas controls (*see* **Note 17**).
8. Extract and fit ATDs for ions of interest.
9. Plot drift time, *m/z*, molecular weight (Da), and charge (*z*) in *estimated CCS* tab. CCS calibration fits are displayed. Logarithmic or power fits are most appropriate for glycans.
10. Repeat for other intact glycan ions and fragments (trap CID experiments).

3.4.3 CCS Calculation from Waters DriftScope Software

1. Open DriftScope.
2. Select and open a suitable file (e.g., dextran) from which to construct the CCS calibration.
3. Click *Peak Detection*.
4. Click *Detect Peaks* from the drop-down menu (or *Detect Peaks* button directly from the tool bar).
5. In the *Peak Detection Parameters* box that opens, click on *Non-Chromatographic Mode* and set the resolution to approximately the instrument resolution.
6. Click *OK*. The window closes and, after a few seconds, the detected peaks will be marked on the DriftScope plot. If too few peaks are marked, return to the *Peak Detection Parameters* window and reduce the Counts in the *Minimum Intensity Threshold* box. Increase the number if too many peaks are highlighted.
7. From the DriftScope window, click *Peak Detection* and then *Calibrate Peaks* and *New Calibration*.
8. The *CCS Calibration Editor* window opens. Click *File* and then *Open Reference File* and select the file containing the CCS values of the reference compound. The calibration peaks appear in the top panel. For the format of the reference file (*see* **Note 18**).
9. Click *Peaks* > *Auto select*. A small window opens where you can change *m/z tolerance* and *Intensity threshold*. Set as appropriate and click *OK*.
10. The detected calibration peaks will be highlighted in the top and second panel in a different color (which can be changed if required), and the calibration graph appears in the third panel. If too many calibration peaks are missing, return to the *Auto select...* window and adjust the *m/z tolerance* and *Intensity threshold*.
11. Click *Calibrate* (from the *CCS Calibration Editor* window which will still be open) and *Save calibration*. The window closes and the calibration file is saved.

12. Return to the DriftScope window and open the file containing your data.
13. Click *Peak Detection* > *Detect Peaks*. The peaks are detected and highlighted.
14. Click *Peak Detection* > *Calibrate Peaks* > *Apply Calibration*. A small window opens listing the CCS calibration files (named as your original calibration file with a .raw extension. This can be changed if required). Select the required calibration file and click *Open*. The window closes.
15. Make sure that the *Spectrum Display* panel is visible in the DriftScope window and has the focus (click the button with the blue peak in the tool bar) and expand the mass scale by dragging with the mouse over the peak of interest. The centroid of the peak will be shown with a bar.
16. Depress the *Ctrl* key on the keyboard and at the same time move the mouse pointer over the peak displayed in the *Spectrum Display* window. The box that appears shows the mass under the pointer, and when over the peak centroid bar, the peak intensity, drift time, and cross section will appear (if nothing happens, click on the panel to give it focus).

4 Notes

1. Electrospray nanoflow emitters can be purchased. These are more expensive in the short-term than in-house capillaries. Capillaries prepared in-house are sufficient and preferable if many samples are to be analyzed.
2. Sulfate groups on GAGs are labile, influence net charge, sodium adduction and, therefore, gas phase separation by IM requires additional consideration. Care should be taken to minimize IM gas pressures and voltages, and trap DC bias values. Minimizing these values while maintaining ion transmission is required.
3. The top filter must be removed from SPE cartridges to extract PGC solid phase material and can be done with long tweezers. Add MeOH to SPE cartridge and pipette PGC material into a fresh tube.
4. Fluorescent glycan labels (i.e., 2-aminobenzoic acid) are not suitable for Nafion® desalting as they adhere to the membrane.
5. Nafion® membranes are proton-conductive films that need to be periodically regenerated, which is achieved by heating at 80 °C in 69% concentrated nitric acid for 2 h. This should be done in a fume hood. Wash membranes and store in Milli-Q water.

6. Different neuraminidases are available with broad specificity (α2–3,6,8,9 linkages) of Neu5Ac and Neu5Gc (*N*-glycolylneuraminic acid) residues as well as highly specific exoglycosidases that catalyze the removal of α2–3 Neu5Ac residues only.
7. Recommended puller settings (for a P-1000 Sutter instrument) are (1) heat = 530, pull = 0, velocity = 15, time = 250 and (2) heat = 540, pull = 200, velocity = 60, time = 0 with pressure = 200 and ramp = 520; however, these values are initial suggestions.
8. Sodium adducts are generally present without the need to add aqueous NaCl. If needed, add 50 mM NaCl to glycan samples similar to other solutions described above.
9. Overclipping nanospray capillaries will result in unwanted sample discharge or poor desolvation of large droplets. Underclipped capillaries can easily become blocked.
10. It is important that specific tubing is used to connect backing gas flow and the nitrogen source on Synapt instruments when performing direct infusion due to the electronic pressure controller. Detailed plumbing schematic and materials is available from Waters (www.waters.com: *Borosilicate Glass Capillary Sprayer Operator's Guide*, support number: USRM134655272). It is also recommended to periodically check the conductive elastomer between the sprayer holder and glass capillary. Replace when worn.
11. The sample cone voltage will influence the transmission of singly or multiply charged ions. High cone voltage (>100 V) is suitable for singly charged glycan ions. Decreasing this voltage will improve transmission of multiply charged ions.
12. The trap DC bias voltage should not be too high as it will result in unintentional glycan ion fragmentation. The value will vary with glycan mass and typical starting values are between 30 V (milk oligosaccharides and O-glycans) to 60 V (for N-glycans).
13. Keeping backing pressure to a minimum will improve desolvation and prolong sample spray time.
14. If electrospray is not stable and requires continuous adjustment to capillary voltage, consider using a fresh capillary as coating varies, especially for in-house needles, and will influence performance.
15. Constructive MS/MS spectra of glycans contain abundant D-type fragments and cross-ring cleavages (A-type fragments) at the reducing end (shown in Fig. 1). *See* ref. 41 for glycan fragmentation nomenclature. It may be necessary to modulate CID voltages (5 V increments) to generate diagnostic ions. It is

	A	B	C
1	CCS(Lit.)	m/z	z
2	214.6	601.1	1
3	243.1	763.2	1
4	268.6	925.2	1
5	297.4	1087.3	1
6	318.5	1249.3	1
7	399.6	1411.4	1
8	358.2	1573.5	1
9	377.9	1735.5	1
10	399.4	1897.6	1
11	418.7	2059.6	1
12	438.4	2221.7	1
13	456.3	2383.7	1

Fig. 4 CCS reference file obtained in Microsoft Excel (dextran, $[M+H_2PO_4]^-$ ions)

recommended to assess MS/MS data on-the-fly to ensure diagnostic ions are present. *See* refs. 4, 7, 12, 16, 18, 26, 27, 30 for glycan MS/MS interpretation and 24 for suitable CID voltages as a function of glycan mass.

16. An Excel spreadsheet for gaussian fitting of arrival time distributions (GFAT) is available from the Benesch Research Group. http://benesch.chem.ox.ac.uk/resources.html.

17. Changing IM settings after measuring dextran calibrants will result in inaccurate CCS calculations of unknown samples. It is recommended to make an additional dextran calibrant measurement if many samples are being measured on a single day.

18. CCS reference file. Construct the file as depicted in Fig. 4 in Microsoft Excel (shown for dextran, $[M+H_2PO_4]^-$ ions). Save as a .csv file and copy it into the *DriftScope* > *Docs* > *Cal directory*. Rename file if required.

References

1. Struwe WB, Pagel K, Benesch JL et al (2016) GlycoMob: an ion mobility-mass spectrometry collision cross section database for glycomics. Glycoconj J 33:399–404
2. Gray CJ, Thomas B, Upton R et al (2016) Applications of ion mobility mass spectrometry for high throughput, high resolution glycan analysis. Biochim Biophys Acta 1860:1688–1709
3. Hofmann J, Pagel K (2017) Glycan analysis by ion mobility-mass spectrometry. Angew Chem Int Ed Engl 56:8342–8349
4. Manz C, Pagel K (2018) Glycan analysis by ion mobility-mass spectrometry and gas-phase spectroscopy. Curr Opin Chem Biol 42:16–24
5. Struwe WB, Baldauf C, Hofmann J et al (2016) Ion mobility separation of deprotonated oligosaccharide isomers – evidence for gas-phase

charge migration. Chem Commun (Camb) 52:12353–12356
6. Mucha E, Stuckmann A, Marianski M et al (2019) In-depth structural analysis of glycans in the gas phase. Chem Sci 10:1272–1284
7. May JC, Goodwin CR, Lareau NM et al (2014) Conformational ordering of biomolecules in the gas phase: nitrogen collision cross sections measured on a prototype high resolution drift tube ion mobility-mass spectrometer. Anal Chem 86:2107–2116
8. Harvey DJ, Scarff CA, Edgeworth M et al (2013) Travelling wave ion mobility and negative ion fragmentation for the structural determination of N-linked glycans. Electrophoresis 34:2368–2378
9. Harvey DJ, Sobott F, Crispin M et al (2011) Ion mobility mass spectrometry for extracting spectra of N-glycans directly from incubation mixtures following glycan release: application to glycans from engineered glycoforms of intact, folded HIV gp120. J Am Soc Mass Spectrom 22:568–581
10. Guttman M, Lee KK (2016) Site-specific mapping of sialic acid linkage isomers by ion mobility spectrometry. Anal Chem 88:5212–5217
11. Hinneburg H, Hofmann J, Struwe WB et al (2016) Distinguishing N-acetylneuraminic acid linkage isomers on glycopeptides by ion mobility-mass spectrometry. Chem Commun 52:4381–4384
12. Hofmann J, Stuckmann A, Crispin M et al (2017) Identification of Lewis and blood group carbohydrate epitopes by ion mobility-tandem-mass spectrometry fingerprinting. Anal Chem 89:2318–2325
13. Harvey DJ, Struwe WB (2018) Structural studies of fucosylated N-glycans by ion mobility mass spectrometry and collision-induced fragmentation of negative ions. J Am Soc Mass Spectrom 29:1179–1193
14. Harvey DJ, Royle L, Radcliffe CM et al (2008) Structural and quantitative analysis of N-linked glycans by matrix-assisted laser desorption ionization and negative ion nanospray mass spectrometry. Anal Biochem 376:44–60
15. Harvey DJ, Scarff CA, Edgeworth M et al (2016) Travelling-wave ion mobility and negative ion fragmentation of high-mannose N-glycans. J Mass Spectrom 51:219–235
16. Huang YT, Dodds ED (2013) Ion mobility studies of carbohydrates as group I adducts: isomer specific collisional cross section dependence on metal ion radius. Anal Chem 85:9728–9735
17. Struwe WB, Benesch JL, Harvey DJ et al (2015) Collision cross sections of high-mannose N-glycans in commonly observed adduct states – identification of gas-phase conformers unique to [M−H](−) ions. Analyst 14:6799–6803
18. Harvey DJ, Scarff CA, Edgeworth M et al (2016) Travelling-wave ion mobility mass spectrometry and negative ion fragmentation of hybrid and complex N-glycans. J Mass Spectrom 51:1064–1079
19. Harvey DJ, Seabright GE, Vasiljevic S et al (2018) Isomer information from ion mobility separation of high-mannose glycan fragments. J Am Soc Mass Spectrom 29:972–988
20. Harvey DJ, Watanabe Y, Allen JD et al (2018) Collision cross sections and ion mobility separation of fragment ions from complex N-glycans. J Am Soc Mass Spectrom 29:1250–1261
21. Hinneburg H, Hofmann J, Struwe WB et al (2016) Distinguishing N-acetylneuraminic acid linkage isomers on glycopeptides by ion mobility-mass spectrometry. Chem Commun (Camb) 52:4381–4384
22. Hofmann J, Struwe WB, Scarff CA et al (2014) Estimating collision cross sections of negatively charged N-glycans using traveling wave ion mobility-mass spectrometry. Anal Chem 86:10789–10795
23. Bitto D, Harvey DJ, Halldorsson S et al (2015) Determination of N-linked glycosylation in viral glycoproteins by negative ion mass spectrometry and ion mobility. Methods Mol Biol 1331:93–121
24. Royle L, Radcliffe CM, Dwek RA et al (2006) Detailed structural analysis of N-glycans released from glycoproteins in SDS-PAGE gel bands using HPLC combined with exoglycosidase array digestions. Methods Mol Biol 347:125–143
25. Jensen PH, Karlsson NG, Kolarich D et al (2012) Structural analysis of N- and O-glycans released from glycoproteins. Nat Protoc 7:1299–1310
26. Karlsson NG, Schulz BL, Packer NH (2004) Structural determination of neutral O-linked oligosaccharide alditols by negative ion LC-electrospray-MSn. J Am Soc Mass Spectrom 15:659–672
27. Seo Y, Andaya A, Leary JA (2012) Preparation, separation, and conformational analysis of differentially sulfated heparin octasaccharide isomers using ion mobility mass spectrometry. Anal Chem 84:2416–2423
28. Turnbull JE, Miller RL, Ahmed Y et al (2010) Glycomics profiling of heparan sulfate

structure and activity. Methods Enzymol 480:65–85

29. Kolarich D, Jensen PH, Altmann F et al (2012) Determination of site-specific glycan heterogeneity on glycoproteins. Nat Protoc 7:1285–1298
30. Stavenhagen K, Hinneburg H, Kolarich D et al (2017) Site-specific N- and O-glycopeptide analysis using an integrated C18-PGC-LC-ESI-QTOF-MS/MS approach. Methods Mol Biol 1503:109–119
31. Bornsen KO, Mohr MD, Widmer HM (1995) Ion-exchange and purification of carbohydrates on a Nafion(R) membrane as a new sample pretreatment for matrix-assisted laser-desorption ionization mass-spectrometry. Rapid Commun Mass Spectrom 9:1031–1034
32. Hernandez H, Robinson CV (2007) Determining the stoichiometry and interactions of macromolecular assemblies from mass spectrometry. Nat Protoc 2:715–726
33. Harvey DJ (2005) Fragmentation of negative ions from carbohydrates: Part 1. Use of nitrate and other anionic adducts for the production of negative ion electrospray spectra from N-linked carbohydrates. J Am Soc Mass Spectrom 16:622–630
34. Harvey DJ (2005) Fragmentation of negative ions from carbohydrates: Part 2. Fragmentation of high-mannose N-linked glycans. J Am Soc Mass Spectrom 16:631–646
35. Harvey DJ (2005) Fragmentation of negative ions from carbohydrates: Part 3. Fragmentation of hybrid and complex N-linked glycans. J Am Soc Mass Spectrom 16:647–659
36. Harvey DJ, Abrahams JL (2016) Fragmentation and ion mobility properties of negative ions from N-linked carbohydrates: Part 7. Reduced glycans. Rapid Commun Mass Spectrom 30:627–634
37. Harvey DJ, Edgeworth M, Krishna BA et al (2014) Fragmentation of negative ions from N-linked carbohydrates: Part 6. Glycans containing one N-acetylglucosamine in the core. Rapid Commun Mass Spectrom 28:2008–2018
38. Harvey DJ, Jaeken J, Butler M et al (2010) Fragmentation of negative ions from N-linked carbohydrates: Part 4. Fragmentation of complex glycans lacking substitution on the 6-antenna. J Mass Spectrom 45:528–535
39. Harvey DJ, Rudd PM (2011) Fragmentation of negative ions from N-linked carbohydrates: Part 5. Anionic N-linked glycans. Int J Mass Spectrom 305:120–130
40. Thalassinos, K., Grabenauer, M., Slade, S. E., Hilton, G. R., Bowers, M. T. and Scrivens, J. H. (2009) Characterisation of phosphorylated peptides using travelling wave-based and drift cell ion mobility mass spectrometry. Analytical Chemistry 81:248–254.
41. Domon B, Costello CE (1988) A systematic nomenclature for carbohydrate fragmentations in FAB-MS MS spectra of glycoconjugates. Glycoconj J 5:397–409

Part III

Ambient Ionization and Imaging Ion Mobility-Mass Spectrometry

Chapter 14

Desorption Atmospheric Pressure Photoionization Coupled with Ion Mobility-Mass Spectrometry

Tiina J. Kauppila

Abstract

Desorption atmospheric photoionization (DAPPI) is an ambient mass spectrometry (MS) technique that can be used to analyze both polar and nonpolar compounds. Here, the coupling of DAPPI with traveling wave ion mobility-mass spectrometry (TWIM-MS) and application to analysis of food, multivitamin, and pharmaceutical products is described.

Key words Desorption atmospheric pressure photoionization, Ambient mass spectrometry, Dopant, Traveling wave ion mobility spectrometry, Proton transfer, Charge exchange, Electron capture

1 Desorption Atmospheric Pressure Photoionization

Ambient mass spectrometry (MS) is a family of techniques that can be used for the direct ionization of compounds from the surfaces of solid samples. In ambient MS, the sample is typically analyzed as is, without preceding sample preparation or chromatographic separation. The development of ambient mass spectrometry techniques started with the invention of desorption electrospray ionization (DESI) [1] and direct analysis in real time (DART) [2] in 2004 and 2005, respectively. Since then, this topic has attracted a large number of research groups working on the development of new ambient mass spectrometry techniques and applications. A comprehensive review about the variety of ambient MS techniques has been published by Monge et al. and is highly recommended [3].

Desorption atmospheric pressure photoionization (DAPPI) is an ambient MS technique, which utilizes 10 eV photons emitted by a vacuum ultraviolet photoionization lamp for the ionization of the analytes [4]. Heat, provided by a microchip heater (*see* **Note 1**), is used in DAPPI to thermally desorb the analytes from the sampling surface (*see* **Note 2**). The vaporized analytes are then ionized by photon-initiated ionization reactions, similar to those in

Giuseppe Paglia and Giuseppe Astarita (eds.), *Ion Mobility-Mass Spectrometry: Methods and Protocols*, Methods in Molecular Biology, vol. 2084, https://doi.org/10.1007/978-1-0716-0030-6_14,

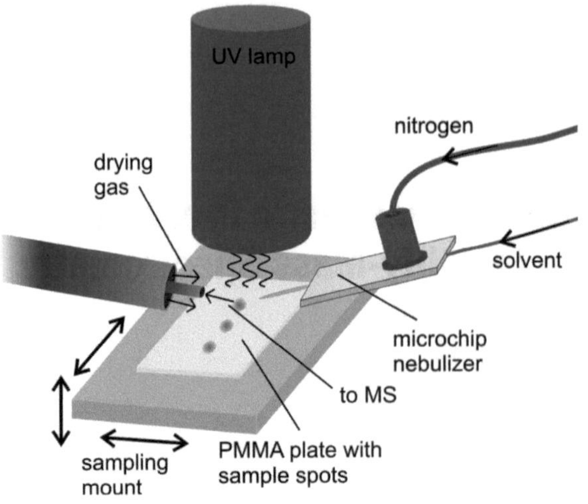

Fig. 1 The setup in desorption atmospheric pressure photoionization. (Reprinted with permission from Haapala M, Pol J, Saarela V, Arvola V, Kotiaho T, Ketola RA, Franssila S, Kauppila TJ, Kostiainen R (2007) Desorption atmospheric pressure photoionization. Anal Chem 79, 7867–7872. Copyright (2007) American Chemical Society)

atmospheric pressure photoionization APPI [5, 6]. Vaporized dopant, a solvent that aids in the ionization process, is supplied to the ionization area through the heated nebulizer microchip (Fig. 1). Toluene and acetone, which both have ionization energies (IEs) below 10 eV, are typical DAPPI dopants (*see* **Note 3**). They are efficiently ionized by the 10 eV photons to form dopant molecular ions (Scheme 1, Reaction 1), which can react further with the analytes through proton transfer or charge exchange reactions (Scheme 1, Reactions 2 and 3). In negative ion mode, the ionization is initiated by electron capture and followed by charge exchange and proton transfer reactions (Table 1, Reactions 4–8).

Table 1 shows the comparison of DAPPI to the most popular and commercially available ambient MS techniques, DESI and DART. Photoionization is more efficient toward neutral and nonpolar compounds than, e.g., electrospray of atmospheric pressure chemical ionization (APCI), and therefore DAPPI widens the group of compounds that can be analyzed with ambient MS toward nonpolar compounds. In comparisons to DESI and DART, DAPPI has been shown to be more sensitive toward neutral and nonpolar compounds [4, 7, 8]. It is also less susceptible to matrix-associated problems than DESI [9]. DAPPI has been applied to the analysis of drugs of abuse in tablet, powder, plant and blotter paper forms [10–13], pesticides from food and plants [14, 15], insect defense compounds from insect body surface [16], steroids, vitamins and other lipids from food or urine [8, 17–19], and atmospheric aerosols from collection filter [20]. In most cases, the analyses have

$$D + h\square \rightarrow D^{+} + e \qquad IE(D) = < 10\ eV \qquad (1)$$

$$D^{+\cdot} + M \rightarrow M^{+\cdot} + D, \qquad \text{if } IE(M) < IE(D) \qquad (2)$$

$$D^{+\cdot} + M \rightarrow [M{+}H]^{+} + [D{-}H]^{\cdot}, \qquad \text{if } PA(M) > PA([D{-}H]^{\cdot}) \qquad (3)$$

$$O_2 + e \rightarrow O_2^{-\cdot} \qquad (4)$$

$$M + e \rightarrow M^{-\cdot}, \qquad \text{if } EA(M) > 0\ eV \qquad (5)$$

$$M + O_2^{-\cdot} \rightarrow M^{-\cdot} + O_2, \qquad \text{if } EA(M) > EA(O_2) \qquad (6)$$

$$M + O_2^{-\cdot} \rightarrow [M{-}H]^{-} + HO_2^{\cdot}, \qquad \text{if } \square_{acid}G(M) < \square_{acid}G\ (HO_2^{+\cdot}) \qquad (7)$$

$$M + O_2^{-\cdot} \rightarrow [M{-}X{+}O]^{-} + OH^{\cdot}, \qquad X = H, X, NO_2 \qquad (8)$$

Scheme 1 Ionization reactions in DAPPI

Table 1
Comparison of DESI, DART, and DAPPI

	DESI	DART	DAPPI
Analyte polarity	• Mid-polar to polar compounds	• Neutral to polar compounds	• Nonpolar to polar compounds
Analyte size	• Up to 60,000 has been reported	• <1000[a]	• <1000[a]
Advantages	• Suitability for thermally labile compounds • High spatial resolution, thus feasible for MS imaging	• Automation and ease of sample introduction, thus higher reproducibility and stability than with most ambient MS techniques	• Wide polarity range • High sensitivity • Better tolerance for matrix compounds than with DESI
Restrictions	• Poor sensitivity toward neutral and nonpolar compounds • Susceptible to matrix compounds	• Only thermally stable compounds	• Only thermally stable compounds • Has not been automated • Not commercially available

[a]This is author's estimation as the analyte size has not been systematically tested

been done without any sample preparation (*see* **Note 4**). Here, the coupling of DAPPI with traveling wave ion mobility-mass spectrometry (TWIM-MS) is reviewed more closely.

2 Coupling Desorption Atmospheric Pressure Photoionization with Traveling Wave Ion Mobility-Mass Spectrometry

Combination of DAPPI with ion mobility spectrometry has been reported in a paper by Räsänen et al. [7]. A self-made DAPPI source was mounted on a Synapt G2 mass spectrometer (Waters Corp., Manchester, UK) with a TWIM-MS option. MicroAPPI (μAPPI) was used in the optimization of the MS parameters prior to the DAPPI experiments (*see* **Note 5**). A VAPUR gas–ion separator tube interface (IonSense, Inc., Saugus, MA) was used to enhance the transfer of ions from the ionization area to the MS. In the VAPUR interface, ions that are produced in the ionization region under atmospheric pressure conditions are suctioned through a ceramic capillary into an enclosure that seals the modified ion spray inlet, while neutrals are removed prior to entering the mass spectrometer by a diaphragm pump.

DAPPI-TWIM-MS was used to analyze almond surface, multivitamin products, and pharmaceutical tablets (*see* **Note 6**). All the samples were analyzed as is, i.e., without prior sample homogenization or extraction. From the almond surface, α-tocopherol was detected by DAPPI-TWIM-MS. The almond was held in front of the heated dopant plume by tweezers (*see* **Note 7**). Figure 2a, b shows the total IM spectrum from the almond surface and the combined mass spectrum from the total IM data of all the ions produced, respectively, from almond surface by positive ion DAPPI. The DAPPI spectra showed the $M^{+\bullet}$ ion of α-tocopherol at *m/z* 430. In addition, an oxidation product $[M-H+O]^+$ at *m/z* 445 was observed. Figure 2c, d shows the single IMS spectrum of the $M^{+\bullet}$ ion of α-tocopherol obtained with DAPPI and the combined mass spectrum of the single IM data. As can be observed, the IM separation prior to MS analysis was observed to greatly reduce the chemical noise in the acquired mass spectrum.

A multivitamin product containing several vitamins, minerals, and excipients was analyzed by DAPPI-TWIM-MS, and 7 out of 13 vitamins reported by the product manufacturer were detected. Figure 3a, b shows the total IM spectrum and the combined mass spectrum from the total IM data, respectively. The main peaks observed in the multivitamin tablet were due to nicotinamide (amide of niacin, vitamin B3, $[M+H]^+$ at *m/z* 123), pyridoxine ((vitamin B6) $[M+H]^+$, and dehydration products $[M+H-H_2O]^+$ and $[M+H-2H_2O]^+$ at *m/z* 170, and 152 and 134, respectively), and α-tocopherol (M^+, oxidation product $[M+O]^+$, and

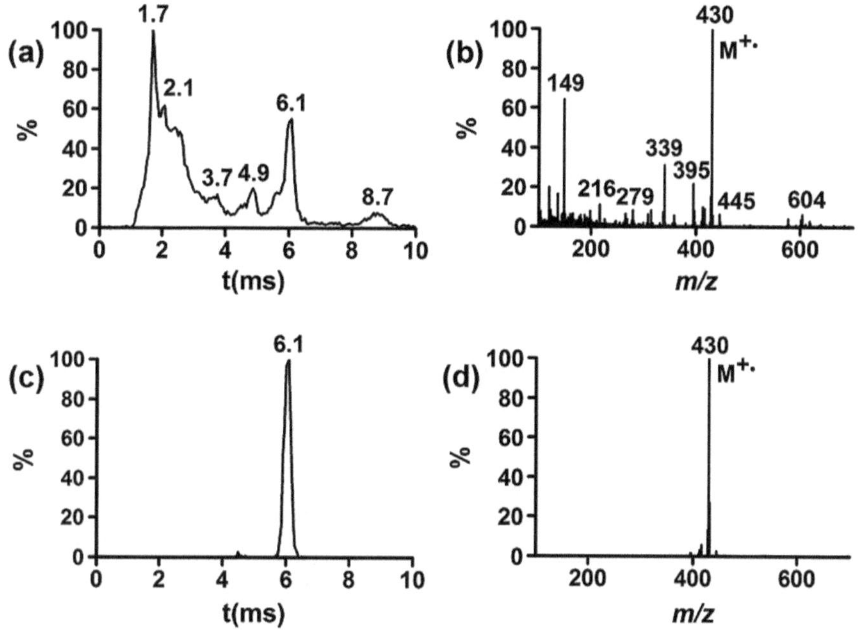

Fig. 2 DAPPI- and DART-TWIM-MS of almond surface: (**a**) DAPPI-TWIM-MS total IM spectrum, (**b**) combined mass spectrum from the total IM data in (**a**), (**c**) single DAPPI-IM spectrum of $M^{+\bullet}$ ion of α-tocopherol at *m*/*z* 430, (**d**) combined mass spectrum from the IM data in (**c**). (Reprinted with permission from Räsänen RM, Dwivedi P, Fernandez FM, Kauppila TJ (2014) Desorption atmospheric pressure photoionization and direct analysis in real time coupled with traveling wave ion mobility-mass spectrometry. Rapid Commun Mass Spectrom 28, 2325–2336. Copyright (2014) John Wiley & Sons)

α-tocopheryl acetate at *m/z* 430, 446, and 472, respectively), as shown in Fig. 3b. The $[M+H]^+$ ion of caffeine was observed at *m/z* 195. Figure 3 shows the single IM spectra of protonated pyridoxine and the hydration products, where the mild tailing observed for the peaks is possibly due to dissociation in the drift region. In addition, less intense peaks that were due to biotin (vitamin B7), thiamine (vitamin B1), and cholecalcipherol (vitamin D3) were observed. The single IM spectra and the corresponding combined mass spectra for these three vitamins are presented in Fig. 3c–h. Biotin (Fig. 3c, d) was detected as $[M+H]^+$ at *m/z* 245, for which the single IM spectrum showed a narrow peak at drift time of 2.4 ms. Thiamine was detected as the M^+ ion at *m/z* 265 with a drift time of 2.7 ms (Fig. 3e, f). The ion at *m/z* 144 (Fig. 3b) was identified as the thiamine fragment ion $[M-C_6H_7N_3]^+$, for which the mobility spectrum is presented in Fig. 3e. Cholecalcipherol was detected as the M^+ ion at *m/z* 384 with a drift time of 5.0 ms (Fig. 3g, h). The single IM spectrum of the cholecalcipherol $M^{+\bullet}$ ion partly overlapped the highly abundant signal of α-tocopherol at *m/z* 430, due to the high concentration of α-tocopherol in the sample (25 mg), and a resulting wide signal in the TWIMS spectrum. Figure 3h shows the combined mass spectrum from the single IM data of the

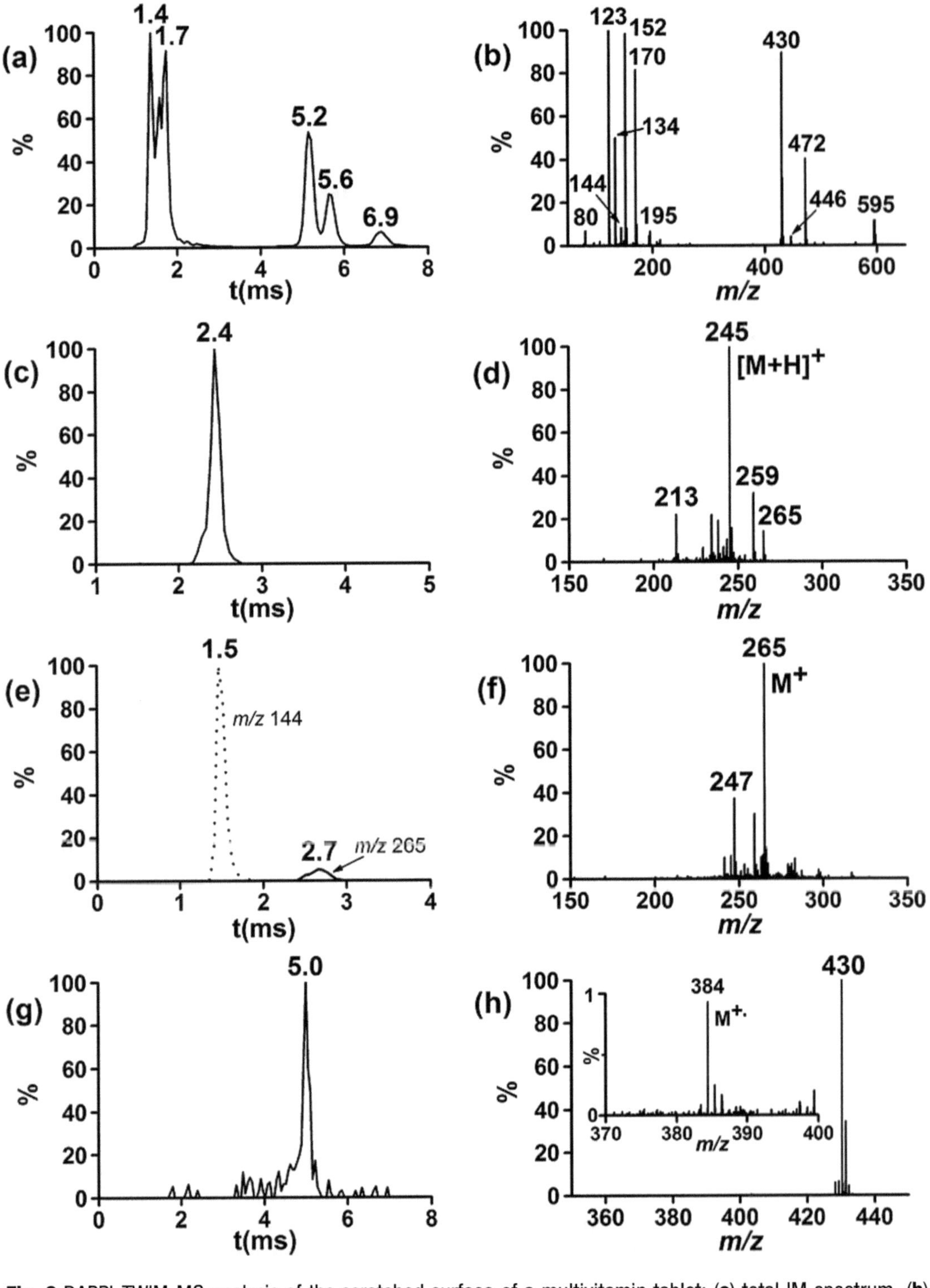

Fig. 3 DAPPI-TWIM-MS analysis of the scratched surface of a multivitamin tablet: (**a**) total IM spectrum, (**b**) combined mass spectrum from the total IM data, where the ion at *m/z* 123 is $[M+H]^+$ of nicotinamide, 170 is $[M+H]^+$ of pyridoxine, 195 is $[M+H]^+$ of caffeine, 430 is $M^{+\bullet}$ of α-tocopherol, and 472 is $[M+H]^+$ of α-tocopheryl acetate, (**c**) single IM spectrum for biotin $[M+H]^+$ at *m/z* 245, (**d**) combined mass spectrum

cholecalcipherol $M^{+\bullet}$ ion, with an insert showing the enlargement of the m/z 370–400 range. The concentration of α-tocopherol in the vitamin product was 1500 times higher than the concentration of the cholecalcipherol (0.0.1 mg), which explains why the intensity of the cholecalcipherol ion peak is low compared to α-tocopherol. The S/N was seven times higher when the ion mobility separation was utilized compared to the case, when it was not, which demonstrates how the mobility separation significantly increased S/N and helped to distinguish low-intensity cholecalcipherol M^+ from other compounds. DAPPI-MS was efficient in the detection of both water- and fat-soluble vitamins, with high intensities. Some vitamins reported in the product were not detected under the experimental conditions of the study. It was, however, speculated that negative ion MS and/or solvent modification could have facilitated the detection of acidic compounds (ascorbic and folic acids), photodegradable compounds (retinol and β-carotene), high molecular weight compounds (cyanacobalamin), and certain water-soluble nitrogen-containing compounds (riboflavin).

Pharmaceutical tablets, namely genuine emergency contraceptive tablets containing levonorgestrel, and falsified tablets, sold as genuine, were analyzed by DAPPI-TWIM-MS. Figure 4a, b shows the total IM spectrum of the ions produced from a genuine pharmaceutical tablet containing levonorgestrel, and the combined mass spectrum from the total IM data, respectively. Levonorgestrel can be observed with high abundance as the $[M+H]^+$ ion at m/z 313. Figure 4c presents the single IM spectrum of the levonorgestrel $[M+H]^+$ ion, showing that the levonorgestrel drift time, 7.9 ms, corresponds to the high-intensity peak in the total IM spectrum. Figure 4d presents the combined mass spectrum from the $[M+H]^+$ levonorgestrel single IM spectrum. The mobility separation was observed to remove the background from the mass spectrum, allowing the detection of the target compound with increased selectivity. The total IM spectrum and the corresponding combined mass spectrum of the potentially falsified tablet are illustrated in Fig. 4e, f, respectively. The DAPPI-TWIM-MS analysis showed no evidence of levonorgestrel in the tablet, but, instead, the

Fig. 3 (continued) from the IM data in (**c**), (**e**) single IM spectrum for the thiamine M^+ ion at m/z 265 (and for thiamine fragment ion at m/z 144 with dashed line), (**f**) combined mass spectrum from the IM data in (**e**), (**g**) single IM spectrum for the cholecalciferol $M^{+\bullet}$ at m/z 384, (**h**) combined mass spectrum from IM data in (**g**) (insert shows enlargement of the spectrum from mass range m/z 370–400). The TWIMS wave velocity was 530 m/s and wave height 35 V. (Reprinted with permission from Räsänen RM, Dwivedi P, Fernandez FM, Kauppila TJ (2014) Desorption atmospheric pressure photoionization and direct analysis in real time coupled with travelling wave ion mobility-mass spectrometry. Rapid Commun Mass Spectrom 28, 2325–2336. Copyright (2014) John Wiley & Sons)

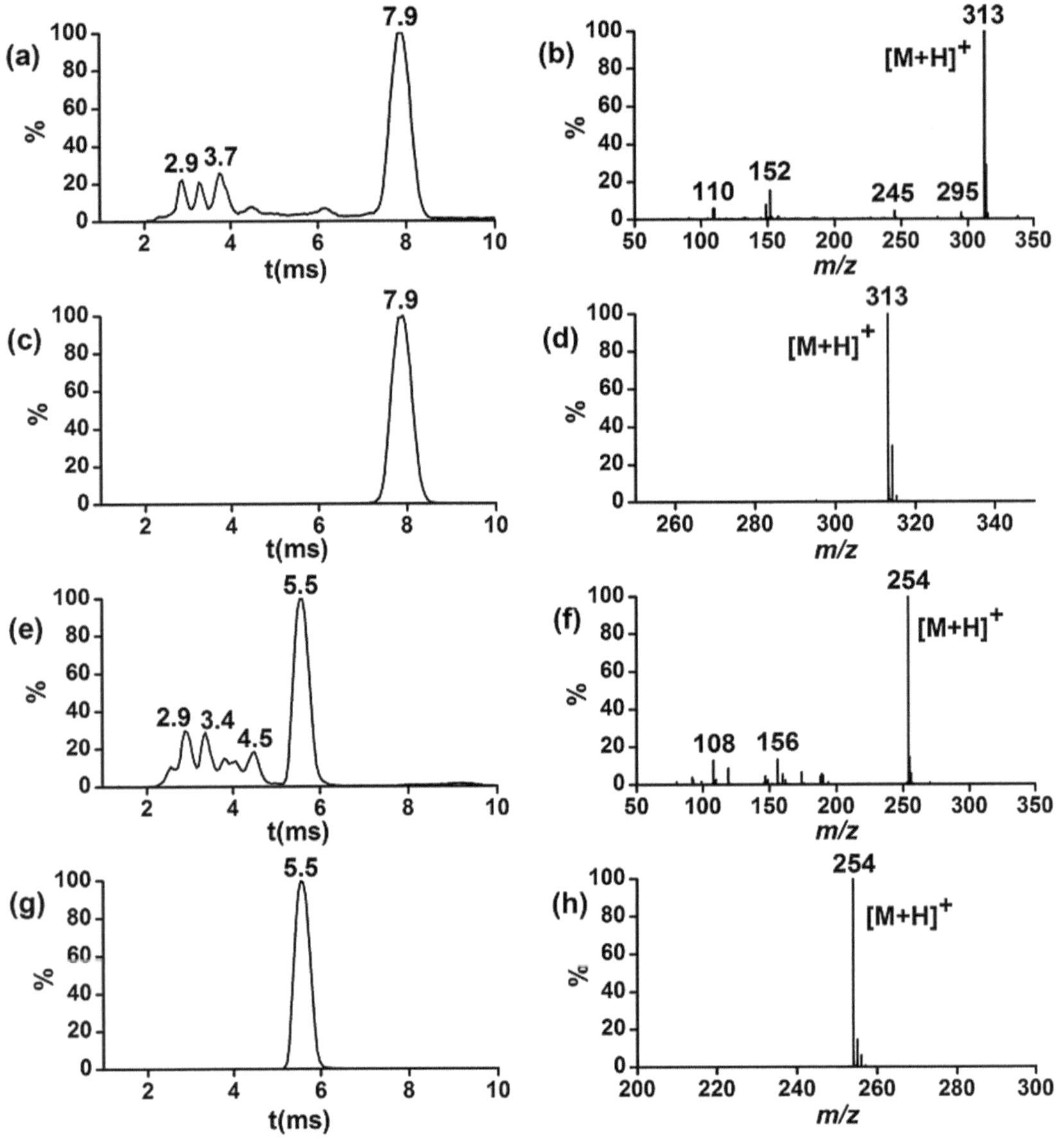

Fig. 4 DAPPI-TWIM-MS analysis from the surface of authentic and potentially falsified levonorgestrel tablets: (**a**) total IM spectrum from the authentic tablet, (**b**) combined mass spectrum from the total IM data in (**a**), (**c**) single IM spectrum of the $[M+H]^+$ of levonorgestrel at *m/z* 313 detected from the authentic tablet, (**d**) combined mass spectrum from the IM data in (**c**), (**e**) total IM spectrum from a potentially falsified tablet, (**f**) combined mass spectrum from the total IM data in (**e**), (**g**) single IM spectrum of the $[M+H]^+$ ion of sulfamethoxazole at *m/z* 254 detected from the falsified tablet, (**h**) combined mass spectrum from the IM data in (**g**). The TWIMS wave velocity was 600 m/s and wave height 24 V. (Reprinted with permission from Räsänen RM, Dwivedi P, Fernandez FM, Kauppila TJ (2014) Desorption atmospheric pressure photoionization and direct analysis in real time coupled with travelling wave ion mobility-mass spectrometry. Rapid Commun Mass Spectrom 28, 2325–2336. Copyright (2014) John Wiley & Sons)

main ingredient was detected at a drift time of 5.5 ms and at *m/z* 254, as shown in the single IM spectrum of the ion at *m/z* 254, and the corresponding combined mass spectrum in Fig. 4g, h, respectively. This species was identified as the $[M+H]^+$ ion of the antibiotic sulfamethoxazole with typical fragments at *m/z* 156 and 108 (Fig. 4f).

3 Conclusions

DAPPI has been successfully combined with TWIM-MS, and it was applied to the direct analysis of compounds from food, multivitamin, and pharmaceutical products. The mobility separation prior to MS analysis was reported to reduce the amount of chemical noise in the mass spectrum and significantly increase the signal-to-noise ratio for the analytes. The mobility separation enabled the detection of species with low signal intensities, e.g., thiamine and cholecalciferol, in the DAPPI-TWIM-MS analysis of multivitamin tablets. The coupling of TWIMS to the DAPPI analysis was reported to increase the spectrum interpretability and the signal-to-noise ratio (S/N), while retaining the speed provided by ambient MS. The combination of DAPPI and TWIM-MS was anticipated to have important applications in, e.g., screening analysis of pharmaceuticals, illicit drugs, toxic chemicals or explosives from tablets, biological matrices, environmental samples, and food products. Especially in cases, where the analyte concentration is close to the detection limit, and the number of matrix components is high, the IM separation can be expected to improve the information content that can be achieved from a particular sample. However, more studies in the coupling of DAPPI with different types of IM separators are still needed to achieve a full picture of the capabilities of DAPPI-IM-MS.

4 Notes

1. A heated tube can be used instead of the microchip-heated nebulizer to supply solvent and gas flow to the ionization area.
2. For small samples (i.e., dried droplets of sample solution on a polymer plate), it is critical that the area heated by the hot solvent and gas plume is in front of the MS inlet. The spot where the heated plume touches the sampling area can be looked up by using a piece of plastic that melts when hit by the heated plume. Once this is known, the sample should be placed in this area.
3. If experiments are to be done with several dopants, it may require a long washing period before the traces of the previous dopant have been removed from the dopant delivery line.

4. Ambient MS techniques can contaminate the MS instrument. If you experience sudden signal loss, it can be due to precipitation of material in the inner surfaces of the MS. Cleaning is then recommended. Another reason for sudden signal loss can be the loss of dopant (empty syringe or stopping of the syringe pump).
5. Optimization of the MS parameters (e.g., ion guiding and voltages and MS^2 parameters) for DAPPI applications is easier to do with a continuous flow technique that supplies a constant analyte solution flow to the ion source, such as APPI or μAPPI.
6. The desorption process takes place very fast. The signal saturates if the sample is held in front of the ionization area for too long.
7. When holding the sample in the ionization area with tweezers: pay attention to not touch high voltages or hot surfaces. The temperature on the surface of the heated nebulizer microchip can be hundreds of degrees.

References

1. Takats Z, Wiseman JM, Gologan B, Cooks RG (2004) Mass spectrometry sampling under ambient conditions with desorption electrospray ionization. Science 306 (5695):471–473. https://doi.org/10.1126/science.1104404
2. Cody RB, Laramee JA, Durst HD (2005) Versatile new ion source for the analysis of materials in open air under ambient conditions. Anal Chem 77(8):2297–2302. https://doi.org/10.1021/ac050162j
3. Monge ME, Harris GA, Dwivedi P, Fernandez FM (2013) Mass spectrometry: recent advances in direct open air surface sampling/ionization. Chem Rev 113(4):2269–2308. https://doi.org/10.1021/cr300309q
4. Haapala M, Pol J, Saarela V, Arvola V, Kotiaho T, Ketola RA, Franssila S, Kauppila TJ, Kostiainen R (2007) Desorption atmospheric pressure photoionization. Anal Chem 79(20):7867–7872. https://doi.org/10.1021/ac071152g
5. Kauppila TJ, Kuuranne T, Meurer EC, Eberlin MN, Kotiaho T, Kostiainen R (2002) Atmospheric pressure photoionization mass spectrometry. Ionization mechanism and the effect of solvent on the ionization of naphthalenes. Anal Chem 74(21):5470–5479. https://doi.org/10.1021/ac025659x
6. Kauppila TJ, Kersten H, Benter T (2014) The ionization mechanisms in direct and dopant-assisted atmospheric pressure photoionization and atmospheric pressure laser ionization. J Am Soc Mass Spectrom 25(11):1870–1881. https://doi.org/10.1007/s13361-014-0988-7
7. Räsänen RM, Dwivedi P, Fernandez FM, Kauppila TJ (2014) Desorption atmospheric pressure photoionization and direct analysis in real time coupled with travelling wave ion mobility mass spectrometry. Rapid Commun Mass Spectrom 28(21):2325–2336. https://doi.org/10.1002/rcm.7028
8. Suni NM, Aalto H, Kauppila TJ, Kotiaho T, Kostiainen R (2012) Analysis of lipids with desorption atmospheric pressure photoionization-mass spectrometry (DAPPI-MS) and desorption electrospray ionization-mass spectrometry (DESI-MS). J Mass Spectrom 47(5):611–619. https://doi.org/10.1002/jms.2992
9. Suni NM, Lindfors P, Laine O, Ostman P, Ojanpera I, Kotiaho T, Kauppila TJ, Kostiainen R (2011) Matrix effect in the analysis of drugs of abuse from urine with desorption atmospheric pressure photoionization-mass spectrometry (DAPPI-MS) and desorption electrospray ionization-mass spectrometry (DESI-MS). Anal Chim Acta 699(1):73–80. https://doi.org/10.1016/j.aca.2011.05.004
10. Kauppila TJ, Arvola V, Haapala M, Pol J, Aalberg L, Saarela V, Franssila S, Kotiaho T, Kostiainen R (2008) Direct analysis of illicit drugs by desorption atmospheric pressure

photolonization. Rapid Commun Mass Spectrom 22(7):979–985. https://doi.org/10.1002/rcm.3461

11. Kauppila TJ, Flink A, Haapala M, Laakkonen UM, Aalberg L, Ketola RA, Kostiainen R (2011) Desorption atmospheric pressure photoionization-mass spectrometry in routine analysis of confiscated drugs. Forensic Sci Int 210(1–3):206–212. https://doi.org/10.1016/j.forsciint.2011.03.018
12. Kauppila TJ, Flink A, Laakkonen UM, Aalberg L, Ketola RA (2013) Direct analysis of cannabis samples by desorption atmospheric pressure photoionization-mass spectrometry. Drug Test Anal 5(3):186–190. https://doi.org/10.1002/dta.1412
13. Luosujarvi L, Laakkonen UM, Kostiainen R, Kotiaho T, Kauppila TJ (2009) Analysis of street market confiscated drugs by desorption atmospheric pressure photoionization and desorption electrospray ionization coupled with mass spectrometry. Rapid Commun Mass Spectrom 23(9):1401–1404. https://doi.org/10.1002/rcm.4005
14. Luosujarvi L, Kanerva S, Saarela V, Franssila S, Kostiainen R, Kotiaho T, Kauppila TJ (2010) Environmental and food analysis by desorption atmospheric pressure photoionization-mass spectrometry. Rapid Commun Mass Spectrom 24(9):1343–1350. https://doi.org/10.1002/rcm.4524
15. Vaikkinen A, Schmidt HS, Kiiski I, Ramo S, Hakala K, Haapala M, Kostiainen R, Kauppila TJ (2015) Analysis of neonicotinoids from plant material by desorption atmospheric pressure photoionization-mass spectrometry. Rapid Commun Mass Spectrom 29(5):424–430. https://doi.org/10.1002/rcm.7123
16. Rejsek J, Vrkoslav V, Hanus R, Vaikkinen A, Haapala M, Kauppila TJ, Kostiainen R, Cvacka J (2015) The detection and mapping of the spatial distribution of insect defense compounds by desorption atmospheric pressure photoionization Orbitrap mass spectrometry. Anal Chim Acta 886:91–97. https://doi.org/10.1016/j.aca.2015.06.007
17. Vaikkinen A, Hannula J, Kiiski I, Kostiainen R, Kauppila TJ (2015) Transmission mode desorption atmospheric pressure photoionization. Rapid Commun Mass Spectrom 29(7):585–592. https://doi.org/10.1002/rcm.7139
18. Vaikkinen A, Rejsek J, Vrkoslav V, Kauppila TJ, Cvacka J, Kostiainen R (2015) Feasibility of desorption atmospheric pressure photoionization and desorption electrospray ionization mass spectrometry to monitor urinary steroid metabolites during pregnancy. Anal Chim Acta 880:84–92. https://doi.org/10.1016/j.aca.2015.03.029
19. Rejsek J, Vrkoslav V, Vaikkinen A, Haapala M, Kauppila TJ, Kostiainen R, Cvacka J (2016) Thin-layer chromatography/desorption atmospheric pressure photoionization orbitrap mass spectrometry of lipids. Anal Chem 88(24):12279–12286. https://doi.org/10.1021/acs.analchem.6b03465
20. Parshintsev J, Vaikkinen A, Lipponen K, Vrkoslav V, Cvacka J, Kostiainen R, Kotiaho T, Hartonen K, Riekkola ML, Kauppila TJ (2015) Desorption atmospheric pressure photoionization high-resolution mass spectrometry: a complementary approach for the chemical analysis of atmospheric aerosols. Rapid Commun Mass Spectrom 29(13):1233–1241. https://doi.org/10.1002/rcm.7219

Chapter 15

Metabolomic Profiling of Adherent Mammalian Cells In Situ by LAESI-MS with Ion Mobility Separation

Sylwia A. Stopka and Akos Vertes

Abstract

Ambient ionization-based mass spectrometry (MS) methods coupled with ion mobility separation (IMS) have emerged as promising approaches for high-throughput in situ analysis for biomedical to environmental applications. These methods are capable of direct profiling and molecular imaging of metabolites, lipids, peptides, and xenobiotics from biological tissues with minimal sample preparation. Furthermore, employing IMS within the workflow improves the molecular coverage, resolves isobaric species, and improves biomolecule identifications through accurate collision cross section measurements. Laser ablation electrospray ionization (LAESI)-MS coupled with IMS has been successful in profiling and molecular imaging of small biomolecules directly from biological tissues and single cells. Herein, we describe a protocol for the direct analysis of adherent mammalian cells with limited perturbations by LAESI-IMS-MS. A benefit of IMS is that within the same LAESI acquisition, the spectral features related to the ESI background, washing buffer, and cell signal can be extracted and isolated separately.

Key words Mass spectrometry, Ion mobility separation, Cell culture, Metabolites, In situ analysis, Laser ablation electrospray ionization, LAESI-IMS-MS

1 Introduction

To uncover the relationship between the genotype and phenotype of a biological system, a comprehensive analysis that encompasses genomics, transcriptomics, proteomics, and metabolomics is needed. Of these omics, metabolomics is the least established due to a variety of challenges, e.g., the wide range of metabolite concentrations, varied compound polarities, and the diverse classes of compounds [1]. Due to their high sensitivity and selectivity, mass spectrometry (MS)-based platforms have been at the forefront in the detection, identification, quantification, and spatiotemporal mapping of biomolecules. The most common MS workflow for metabolite analysis and quantification includes liquid chromatography (LC) combined with stable isotope labeling. For example, the absolute metabolite concentrations (ranging from millimolar to

Giuseppe Paglia and Giuseppe Astarita (eds.), *Ion Mobility-Mass Spectrometry: Methods and Protocols*, Methods in Molecular Biology, vol. 2084, https://doi.org/10.1007/978-1-0716-0030-6_15,

nanomolar) and fluxes of over 150 compounds were determined for immortalized baby mouse kidney cells (IBMK) [2]. However, this analysis requires extensive sample preparation, is considered low-throughput, and does not capture any spatial information of biomolecules within the tissue.

Ambient ionization MS methods provide an alternative approach for metabolite, lipid, peptide, and xenobiotic analysis and imaging from biological tissues with minimal sample preparation and under native conditions. A large number of ambient sampling methods have been introduced including liquid-based, plasma-based, and laser-based approaches [4, 5]. Due to the wide versatility and adaptability of these techniques, applications in biomedical research have covered microbe-host interactions, environmental studies, molecular imaging, and forensics problems [6–9]. Among these, the more established methods include desorption electrospray ionization (DESI), direct analysis in real time (DART), and laser ablation electrospray ionization (LAESI) [10–12]. Due to the high water content found in biological tissues and cells, laser-based approaches that operate in the mid-IR range have been a useful tool for in situ analysis. Here a mid-IR laser beam tuned to the strongest absorbance maximum of water (2940 nm), which corresponds to the symmetric stretching of the O–H bonds, is focused onto the sample resulting in an ablation plume that is ionized by electrospray ionization (ESI). To afford high-throughput and direct sampling capabilities, ambient ionization methods typically omit separation methods, e.g., HPLC, which results in spectral complexity and ion suppression effects.

To retain the benefits of a separation step in combination with direct sampling, ion mobility separation (IMS) can be introduced. This expands the molecular coverage, improves the signal-to-noise ratio, increases confidence in molecular identification through determining the ion collision cross section (CCS), and resolves isobaric species [13–16]. In this gas phase separation, ions are differentiated within milliseconds based on their size, shape, and charge state [17, 18]. The benefits are clearly observed when coupling LAESI-MS with IMS, which have been demonstrated for high-throughput metabolite profiling of living algal cells and plant tissues, identification of isotopologues, determination of turnover rates for small molecules, and separation of compound classes [19, 20].

As many new therapies and diagnostic tools for human disease are based on cellular models, the ability for in situ analysis of adherent mammalian cells is valuable in biology and medicine [21, 22]. To minimize cell manipulation, e.g., lysis and enzymatic digestion that may affect the physiological state and alter biochemical processes, direct sampling is beneficial [23]. Herein, we outline a protocol using LAESI-MS coupled with IMS for the direct analysis of adherent mammalian cells. Separation and isolation of ions related to the ESI background, washing buffer, and cells with limited sample preparation is demonstrated with IMS.

2 Materials

2.1 Reagents and Chemicals

1. Electrospray solution for negative ion mode: methanol:chloroform (2:1, v/v).
2. Dulbecco's Modified Eagle's Medium (DMEM) (ATCC 30-2002) supplemented with 10% fetal bovine serum (FBS) (ATCC 30-2020), 1% MEM nonessential amino acid solution (100×), and 1% penicillin–streptomycin for cell culture.
3. Phosphate-buffered saline (PBS) (ThermoFisher, Waltham, MA) for cell washing and 0.25× trypsin-EDTA for cell detachment.
4. 70% ethanol for sterilization.

2.2 Biological Samples and Materials

1. Cryogenic stock of human neuroblast (SK-N-AS; ATCC CRL-2137) derived from brain and bone marrow (*see* **Note 1**).
2. Humidity-controlled cell incubator set at 5% CO_2 and 37 °C.
3. Automated cell counter.
4. Ethanol-sterilized microscope slides.
5. 100 cm polystyrene petri dish.

2.3 LAESI-IMS-MS

2.3.1 Electrospray

1. The electrospray assembly consists of a conductive union, an insulating mounting bracket, fused silica tubing, a needle port (IDEX Health and Science, Oak Harbor, WA), and a 50 μm ID-tapered metal emitter (MT320-50-5-5; New Objective, Woburn, MA). The ESI assembly is mounted on a three-axis translation stage using optical posts. The ESI emitter tip should be positioned on axis and 12 mm from the mass spectrometer orifice.
2. High voltage (−2.7 kV) is applied to the conductive union by a regulated power supply (PS350; Stanford Research Systems, Sunnyvale, CA) through a high-voltage cable (*see* **Note 2**).
3. A syringe pump is used to supply the ESI solution at a 500 nL/min flowrate from a 500 μL syringe.

2.3.2 Laser Ablation

1. A mid-IR laser source (Opolette 100; Opotek, Carlsbad, CA) used is tuned to 2940 nm wavelength for maximum absorbance. The produced laser pulses are 5 ns in length and are emitted at a repetition rate of 20 Hz. The laser beam is steered with gold-coated mirrors through a 30 mm cage system from the laser cavity to the last optical component (*see* **Note 3**).

2. A 75 mm planoconvex focal length ZnSe lens (Infrared Optical Products, Farmingdale, NY) is used to focus the mid-IR laser beam on the sample. The microscope slide containing the adherent cells is mounted on an *XYZ* stage that is positioned 15 mm below the orifice–ESI axis. The alignment diode can be used to visualize the laser beam on the sample surface.
3. Assure that the laser beam is focused by firing single shots at thermal paper and use a microscope to observe the produced spot size. Adjust the distance between the lens, and the sample until the spot is circular and shows minimum size.

2.3.3 IMS Configuration

1. A homebuilt LAESI source is retrofitted to the front end of a commercial quadrupole time-of-flight mass spectrometer (Synapt G2S; Waters, Milford, MA). This system is equipped with a traveling-wave (T-wave) IMS with a resolving power of $\Omega/\Delta\Omega = 30$.
2. Nitrogen is used as the drift gas and supplied at a flow rate of 90 mL/min, a pressure of 3.35 mbar, and a delay coefficient of 1.41 V (*see* **Note 4**).
3. The height and velocity of the traveling wave were set to 40 V and 650 m/s, respectively.
4. Ion collision cross section (CCS) values were derived using polyalanine oligomers with $n = 3$–14 as the calibrant. This calibration method covered the m/z range from 230 to 1011.

3 Methods

3.1 Sample Preparation

1. Using a glass pen scribe, cut to three parts of similar dimension 75×25 mm^2 microscope slides. Sterilize them with 70% ethanol and UV light on both sides in a sterile environment. Arrange the cut slides next to each other in a 100 cm petri dish covering most of the dish without any overlap. Smaller microscope slides afford shorter measurement times thereby improve reproducibility.
2. Culture the stock SK-N-AS cells in the supplemented DMEM medium in a T75-uncoated culture flask. The medium is aspirated from the flask, washed once with 1× PBS, and filled with 4 mL of prewarmed trypsin (*see* **Note 5**).
3. Place the flask in the incubator for 10 min to allow the cells to detach from the surface. Add prewarmed fresh DMEM medium to the flask to inactivate the trypsin.
4. Use a cell counter to determine the cell density and then dilute to 5×10^5 cells/mL. This cell suspension is then pipetted (500 μL) onto the sterilized cut microscope slides. Place the petri dish in the incubator for ~15 min, so the cells can adhere

to the slides. Prewarmed fresh medium is gently poured into the petri dish without disturbing the slides until the dish is ¾ filled. Place back the petri dish into the incubator and perform LAESI analysis 1 h later.

3.2 LAESI Direct Analysis of Adherent Mammalian Cells

1. Assemble both the ESI and LAESI optical components as mentioned in Subheading 2. Run the ESI for 30 min to assure a stable spray prior to any analysis.
2. Place the mass spectrometer into operational mode and perform the appropriate mass and drift time (DT) calibrations using sodium formate and polyalanine, respectively. LAESI mass spectra should be obtained for both calibration events.
3. Once the LAESI source is optimized, the slides containing the SK-N-AS cells can be analyzed. A slide is removed from the medium with tweezers, gently rinsed with 10× diluted PBS, and placed onto the sample mount for analysis.
4. Start the data acquisition, the first ten scans are of the ESI background followed by firing the laser and rastering the laser beam over the cell surface. Sample related ions can be tracked using their ion chromatogram.
5. End the acquisition by stopping the laser from firing, turning off the ESI high-voltage power supply and the syringe pump. Place the instrument in standby mode.
6. The acquired data are ready to be processed.

3.3 Separation and Classification of Small Molecules from LAESI Data

1. Using the IMS feature, the data sets can be visualized as a DT vs. m/z plot showing the ion intensities on a false color scale using the DriftScope 2.4 software (Waters, Milford, MA). Here the sample-related peaks, PBS ion clusters, and ESI background ions can be differentiated as red, teal, and blue dots, respectively (*see* Fig. 1a). Regions of interest can be extracted individually to produce enhanced mass spectra with reduced spectral interferences for the three different groups (*see* Fig. 1b–d).
2. As each ion appears at a specific DT range based on its size, shape, and charge state, looking at the DT distributions results in separation and improved signal-to-noise ratios. From the LAESI mass spectrum of the SK-N-AS cells, several metabolites can be identified using tandem MS, database searches and the CCS measurements on the order of milliseconds (*see* Fig. 2). Small molecules, e.g., adenine and glutamine have shorter DT values compared to lipids.
3. The CCS calibration is applied after data acquisition using the DriftScope 2.4 software and correlates the polyalanine unit drift times with the CCS values (Table 1). For our example, the related equation is $\Omega_{\text{calibration}} = A(t'_{\text{d}})^B$, in which

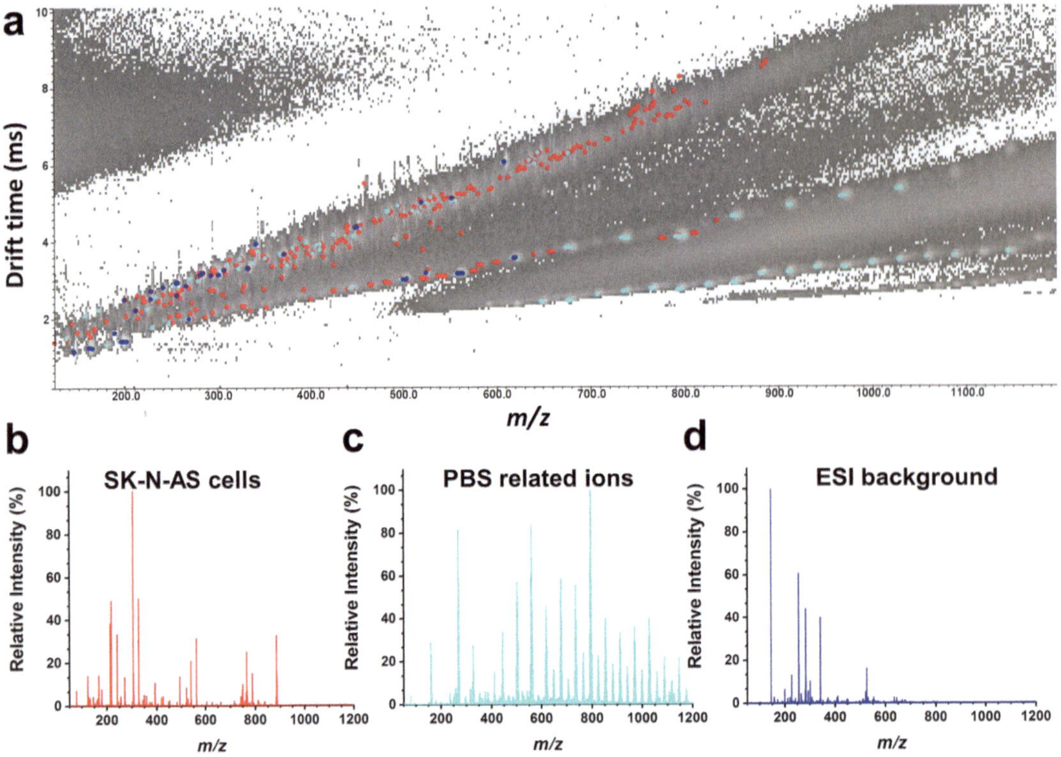

Fig. 1 (**a**) A DT vs. *m/z* map corresponding to a LAESI mass spectrum of adherent SK-N-AS neuroblasts in 10× diluted PBS. Separation of ions related to the cells, PBS, and ESI background within the plot is displayed in red, teal, and blue dots, respectively. To enhance the ion signal and provide spectral separation for each group, three selected regions within the DT vs. *m/z* plot were extracted into individual mass spectra containing the (**b**) cells, (**c**) 10× diluted PBS, and (**d**) ESI-related ions

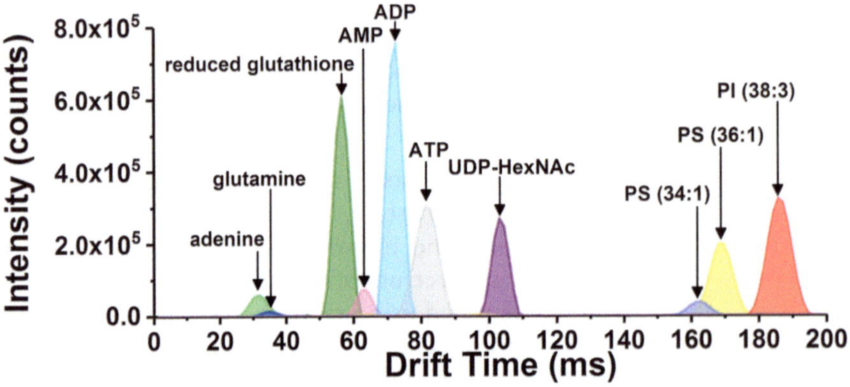

Fig. 2 Measured DT distributions for several metabolites and lipid species from a typical LAESI mass spectrum of SK-N-AS neuroblasts

Table 1
Measured $CCS_{nitrogen}$ and DT values for polyalanine oligomers with $n = 3–14$ in negative ion mode

n	*m/z*	DT (ms)	CCS (N_2) (Å^2)
3	230.117	35	150
4	301.159	41	165
5	372.199	48	179
6	443.237	57	195
7	514.274	65	209
8	585.312	72	223
9	656.348	80	238
10	727.392	89	253
11	798.427	98	267
12	869.471	107	279
13	940.512	117	294
14	1011.547	127	308

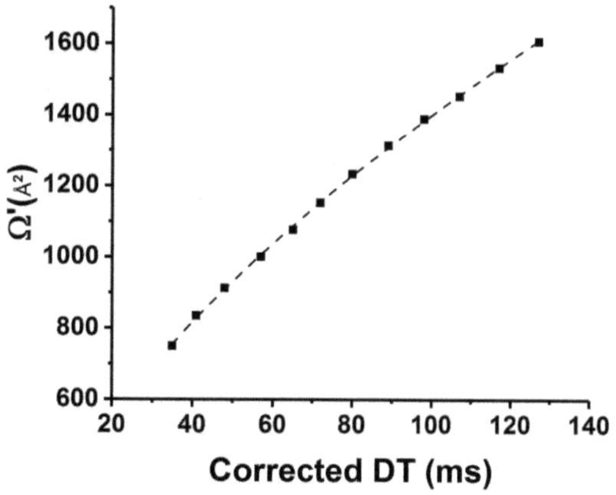

Fig. 3 Polyalanine CCS calibration plot for $n = 3–14$ oligomers using nitrogen buffer gas in negative ion mode. The corrected DT reflects mass-independent flight times

$A = 466.443$, and $B = 0.573$, and R^2 is 0.9996 (*see* Fig. 3). This calibration is then applied to the LAESI DT vs. m/z plots that correlate the detected ion m/z values with the CCS.

4. Extract all CCS measurements and m/z values using the DriftScope 2.4 software.
5. Plotting CCS vs. m/z, some classes of homologous compounds can be revealed based on observed trends, e.g., amino

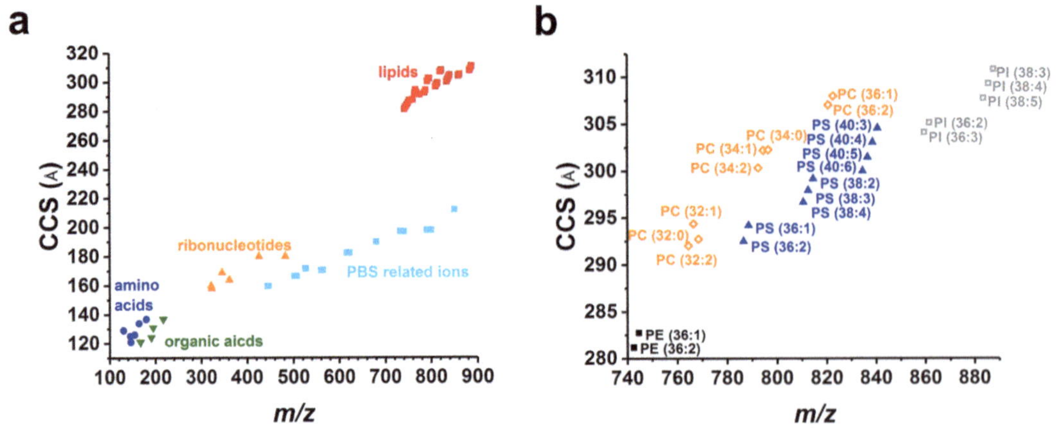

Fig. 4 (**a**) A CCS vs. *m/z* plot that reveals separation and classification of detected metabolites from adherent SK-N-AS cells analyzed by LAESI-IMS-MS. Unique trends were observed for each homologous class of molecules and the PBS-related ions. (**b**) A zoomed area within the lipid region from the CCS vs. *m/z* plot revealed separation of specific lipid classes as well as trends based on degree of saturation. The lipid species phosphatidylethanolamine (PE), phosphatidylserine (PS), and phosphatidylinositol (PI) were detected as $[M-H]^-$ species, whereas the phosphatidylcholine (PC) species were detected as $[M+Cl]^-$ adducts

acid CCS values were within 121–137 Å^2, ribonucleotides ranged between 158 and 180 Å^2, and lipids within 281–310 Å^2 (*see* Fig. 4a).

6. Examining the lipid species in the CCS vs. *m/z* plot, revealed separation based on lipid class and degree of saturation (*see* Fig. 4b). For example, the phosphatidylethanolamine lipids have smaller headgroups than phosphatidylserines and phosphatidylinositols, thus they have a smaller CCS values than the two latter lipid classes.

4 Notes

1. The SK-N-AS cells are biosafety level one (BSL-1), thus all cell handling should be performed in a BLS-1-rated hood. Personnel working with these cells should undergo biohazard training. Additionally biohazard signs should be placed on all equipment and instruments that may come in contact with the cells.
2. Extreme caution should be used when dealing with high-voltage systems. Direct contact to the ESI emitter assembly can result in injury or death. Experimental precautions include shielding, insulting, and limiting the exposure of electrical components. Visual signs indicating high voltage should be displayed, and foot traffic within the surrounding area is limited during experiments. The high voltage should be turned off during removal and placement of samples.

3. The mid-IR laser source is a Class IV laser system, thus precautionary measures must be taken to prevent injury or death. Appropriately rated laser safety goggles and clothing must be worn to avoid laser beam exposure. Beam blocks and enclosure tubes should be used to shield the laser beam throughout the optical path. Proper protection should also be used for the Class II laser alignment diode (635 nm wavelength).
4. Helium, argon, and carbon dioxide can also be used as alternative drift gases and will result in different CCS values.
5. To reduce any environmental shock to the cells, avoid washing with water and store the adherent cells in the incubator between analyses. Analysis should be performed within a couple of minutes of washing the cells with 10× diluted PBS.

Acknowledgments

We thank Lida Parvin for providing a cell stock of SK-N-AS. Research was sponsored by the U.S. Army Research Office and the Defense Advanced Research Projects Agency and was accomplished under Cooperative Agreement Number W911NF-14-2-0020. The views and conclusions contained in this document are those of the authors and should not be interpreted as representing the official policies, either expressed or implied, of the Army Research Office, DARPA, or the U.S. Government.

References

1. Dunn WB, Bailey NJC, Johnson HE (2005) Measuring the metabolome: current analytical technologies. Analyst 130(5):606–625. https://doi.org/10.1039/b418288j
2. Park JO, Rubin SA, Xu YF, Amador-Noguez D, Fan J, Shlomi T, Rabinowitz JD (2016) Metabolite concentrations, fluxes and free energies imply efficient enzyme usage. Nat Chem Biol 12(7):482–489. https://doi.org/10.1038/nchembio.2077
3. Cooks RG, Ouyang Z, Takats Z, Wiseman JM (2006) Ambient mass spectrometry. Science 311(5767):1566–1570. https://doi.org/10.1126/science.1119426
4. Wu CP, Dill AL, Eberlin LS, Cooks RG, Ifa DR (2013) Mass spectrometry imaging under ambient conditions. Mass Spectrom Rev 32(3):218–243. https://doi.org/10.1002/mas.21360
5. Chiang S, Zhang WP, Ouyang Z (2018) Paper spray ionization mass spectrometry: recent advances and clinical applications. Expert Rev Proteomics 15(10):781–789. https://doi.org/10.1080/14789450.2018.1525295
6. Pavlovich MJ, Musselman B, Hall AB (2018) Direct analysis in real time mass sepctrometry (DART-MS) in forensic and security applications. Mass Spectrom Rev 37(2):171–187. https://doi.org/10.1002/mas.21509
7. Alford A (1975) Environmental applications of mass spectrometry. Biomed Mass Spectrom 2(5):229–253. https://doi.org/10.1002/bms.1200020502
8. Ifa DR, Eberlin LS (2016) Ambient ionization mass spectrometry for cancer diagnosis and surgical margin evaluation. Clin Chem 62(1):111–123. https://doi.org/10.1373/clinchem.2014.237172
9. Li H, Balan P, Vertes A (2016) Molecular imaging of growth, metabolism, and antibiotic inhibition in bacterial colonies by laser ablation electrospray ionization mass spectrometry. Angew Chem Int Ed 55(48):15035–15039. https://doi.org/10.1002/anie.201607751

10. Muller T, Oradu S, Ifa DR, Cooks RG, Krautler B (2011) Direct plant tissue analysis and imprint imaging by desorption electrospray ionization mass spectrometry. Anal Chem 83 (14):5754–5761. https://doi.org/10.1021/ac201123t
11. Gross JH (2014) Direct analysis in real time-a critical review on DART-MS. Anal Bioanal Chem 406(1):63–80. https://doi.org/10.1007/s00216-013-7316-0
12. Nemes P, Vertes A (2007) Laser ablation electrospray ionization for atmospheric pressure, in vivo, and imaging mass spectrometry. Anal Chem 79(21):8098–8106. https://doi.org/10.1021/ac071181r
13. Shrestha B, Vertes A (2014) High-throughput cell and tissue analysis with enhanced molecular coverage by laser ablation electrospray ionization mass spectrometry using ion mobility separation. Anal Chem 86(9):4308–4315. https://doi.org/10.1021/ac500007t
14. Gaye MM, Kurulugama R, Clemmer DE (2015) Investigating carbohydrate isomers by IMS-CID-IMS-MS: precursor and fragment ion cross-sections. Analyst 14(20):6922–6932. https://doi.org/10.1039/c5an00840a
15. Paglia G, Kliman M, Claude E, Geromanos S, Astarita G (2015) Applications of ion-mobility mass spectrometry for lipid analysis. Anal Bioanal Chem 407(17):4995–5007. https://doi.org/10.1007/s00216-015-8664-8
16. Chouinard CD, Wei MS, Beekman CR, Kemperman RHJ, Yost RA (2016) Ion mobility in clinical analysis: current progress and future perspectives. Clin Chem 62(1):124–133. https://doi.org/10.1373/clinchem.2015.238840
17. Kliman M, May JC, McLean JA (2011) Lipid analysis and lipidomics by structurally selective ion mobility-mass spectrometry. Biochim Biophys Acta 1811(11):935–945. https://doi.org/10.1016/j.bbalip.2011.05.016
18. Kanu AB, Dwivedi P, Tam M, Matz L, Hill HH (2008) Ion mobility-mass spectrometry. J Mass Spectrom 43(1):1–22. https://doi.org/10.1002/jms.1383
19. Stopka SA, Mansour TR, Shrestha B, Marechal E, Falconet D, Vertes A (2016) Turnover rates in microorganisms by laser ablation electrospray ionization mass spectrometry and pulse-chase analysis. Anal Chim Acta 902:1–7. https://doi.org/10.1016/j.aca.2015.08.047
20. Stopka S, Agtuca B, Khattar R, Anderton C, Koppenaal D, Pasa-Tolic L, Stacey G, Vertes A (2017) Metabolomics of biological nitrogen fixation explored by laser ablation electrospray ionization mass spectrometry combined with fluorescence microscopy. Abstracts of Papers of the American Chemical Society 254
21. De Clercq E (2005) Recent highlights in the development of new antiviral drugs. Curr Opin Microbiol 8(5):552–560. https://doi.org/10.1016/j.mib.2005.08.010
22. Otterbein LE, Bach FH, Alam J, Soares M, Lu HT, Wysk M, Davis RJ, Flavell RA, Choi AMK (2000) Carbon monoxide has anti-inflammatory effects involving the mitogen-activated protein kinase pathway. Nat Med 6 (4):422–428
23. Jacobson RS, Thurston RL, Shrestha B, Vertes A (2015) In situ analysis of small populations of adherent mammalian cells using laser ablation electrospray ionization mass spectrometry in transmission geometry. Anal Chem 87 (24):12130–12136. https://doi.org/10.1021/acs.analchem.5b02971

Chapter 16

Matrix-Assisted Laser Desorption and Desorption Electrospray Ionization Mass Spectrometry Coupled to Ion Mobility

James I. Langridge and Emmanuelle Claude

Abstract

Matrix-Assisted Laser Desorption Ionization (MALDI) and Desorption Electrospray Ionization (DESI) are two complementary ionization techniques that have transformed the field of biomolecular analysis, enabling the measurement of a wide range of biomolecules by mass spectrometry. These techniques have also been applied to imaging mass spectrometry where the spatial localization of molecules is determined. Coupling this with Ion Mobility Spectrometry (IM) allows an additional level of separation and specificity to be obtained. Here, we describe the coupling of the technologies and the practical advantages of these combinations, highlighting specific examples.

Key words MALDI, DESI, Ion mobility, Mass spectrometry, Imaging MS

1 Introduction

The development of MALDI [1] and ESI [2] ionization techniques opened new horizons in molecular analysis for mass spectrometry and has revolutionized the analysis of biological samples, specimens, and systems. Continued development of these techniques along with the associated research and development in mass analyzer technology has enabled the rapid molecular analysis of a wide range of compounds, from proteins and peptides to metabolites and lipids, complementing the sustained advances being made in DNA- and RNA-sequencing technologies.

Both ESI and MALDI ionization were revolutionary developments in the 1990s. MALDI uses a UV-laser absorbing organic matrix to generate ions from a diverse range of samples including peptides, proteins, glycans, xenobiotic drugs, and intact bacteria. It was widely used in the early years for the analysis of peptides and proteins with peptide mass fingerprinting [3] with MALDI being a critical technique in the early proteomics revolution. Many

Giuseppe Paglia and Giuseppe Astarita (eds.), *Ion Mobility-Mass Spectrometry: Methods and Protocols*, Methods in Molecular Biology, vol. 2084, https://doi.org/10.1007/978-1-0716-0030-6_16,

technology improvements to MALDI have been made over the intervening years, successfully enhancing both the speed and sensitivity of the technique, as well as describing the use of the technique for a variety of different applications from bacterial identification [4] to imaging mass spectrometry [5].

Desorption electrospray ionization [6] was the first in a range of ambient ionization technologies that have allowed the direct analysis of samples providing a detailed chemical fingerprint of the composition of samples with minimal or no sample preparation. This is performed at atmospheric pressure and temperature and has been applied to a significant number of analytical challenges and applications. DESI allows mass spectrometry data to be acquired and processed rapidly for near real-time analysis and decision-making.

Secondary ionization mass spectrometry (SIMS) was the first MS technique applied to imaging MS applications, but the approach was transformed by the application of both MALDI and DESI ionization, with these technologies increasingly being applied to a diverse range of imaging mass spectrometry (IMS) studies. These studies determine the spatial localization and distribution of both endogenous and exogenous chemical components, in tissues slices or biological samples, by sequentially sampling the surface in a defined pattern or rastering across the surface. At each sampling point, ionization occurs and a mass spectrum is generated with X, Y coordinates recorded. The MS data are typically recorded over a wide mass range using a time-of-fight mass spectrometer allowing a multitude of analytes to be monitored simultaneously from each position. This approach allows the detailed molecular distribution of a wide range of chemical constituents to be determined in a variety of sample matrices and has found use in applications including lipid profiling [7], neuropeptide and neurotransmitter analysis [8], protein analysis [9], and quantitative studies looking at xenobiotic drug distribution [10].

The use of ion mobility (IM) as an analytical technique coupled with mass spectrometry to detect and separate biomolecules has found application in proteomics [11, 12], glycomics [13], and metabolomics [14]. IM separates gas phase ions based upon their mobility in a buffer gas in an analogous fashion to liquid phase capillary electrophoresis. The separation is related to an ion's rotationally averaged collisional cross section, (dependent upon the ions size and shape in the gas phase) as well as m/z and charge. The basis for separation by traditional drift-tube ion mobility at a low electric limit can be derived from the Mason–Champ equation,

$$K = \frac{3}{16}\left(\frac{2\pi}{\mu k_B T}\right)^{1/2}\frac{ze}{N\Omega}$$

where K = drift velocity (vd)/electric field strength (E), μ is the reduced mass of the ion, k_B is the Boltzmann constant, T is the temperature of the drift gas, z is the charge state of the analyte ion, e is the charge on an electron, N is the number density of the drift gas, and Ω = rotationally averaged collisional cross section.

The hyphenation of ion mobility spectrometry (IM) with MS is often referred to as ion mobility-mass spectrometry (IM-MS). The most common mass analyzer coupled with IMS comprises a time-of-flight (TOF) instrument, due to the high sampling rate (kHz) of the mass analyzer complementing and maintaining the fidelity of the millisecond timeframe of the IM separation, although other mass detection systems have been described. Several different methods of ion mobility separation are currently used in combination with MS, including drift time ion mobility spectrometry (DTIMS), aspiration ion mobility spectrometry (AIMS), differential-mobility spectrometry (DMS), also called field-asymmetric waveform ion mobility spectrometry (FAIMS), trapped ion mobility spectrometry (TIMS), and traveling wave ion mobility spectrometry (TWIMS). A description of these methods is beyond the scope of this paper since they have been reviewed in detail elsewhere [15, 16].

1.1 Traveling Wave Ion Mobility-Mass Spectrometry

The principle of TWIMS is briefly discussed, with extensive details previously described [17, 18]. Ions are formed by electrospray ionization (ESI) in the source and subsequently pass through a quadrupole for mass selection before trapping and injection into the ion mobility cell. Unlike other instruments, which use a uniform electric field across the cell for ion mobility experiments, so called drift tube IMS, this device uses traveling wave (T-wave) technology. The T-wave cell consists of a stacked-ring radio frequency (RF) ion guide which incorporates a repeating sequence of transient DC voltages applied to the ring electrodes. These voltage pulses result in a traveling electric field that propels ions through the background gas present in the mobility cell. The total time taken for an ion to drift through the cell depends on its mobility as well as the wave period/height and the gas pressure. Ions with high mobility are better able to keep up with traveling waves and are pushed more quickly through the cell. Ions with low mobility crest over the waves more often and have to wait for subsequent waves to push them forward, resulting in longer drift times. To measure an arrival time distribution (ATD), ions are gated into the mobility cell using an up-front stacked-ring RF device that traps ions before releasing them into the IMS cell. The oa-TOF pulses in an asynchronous manner, sending ions that have exited the mobility cell into the TOF mass analyzer and finally to the detector.

In this chapter, the application of MALDI and DESI combined with IM for use in imaging mass spectrometry is discussed with specific experimental considerations and application examples that demonstrate the benefits of integrating IM into the analytical schema in terms of increased separation power and charge state separation that enhances sensitivity.

2 Materials

All solutions were prepared using ultrapure water (18 MΩ-cm at 25 °C) and analytical grade reagents with a purity of ≥99.5%.

Reagents should be disposed of in compliance with health and safety regulations and in accordance with COSHH guidelines.

1. Ethyl alcohol, 200 proof, ACS reagent ≥99.5%.
2. Embedding mediums; 1% carboxymethylcellulose (CMC) or 5% gelatine.
3. Ammonium bicarbonate, ≥99.5%.
4. Trypsin Gold, Mass Spectrometry Grade.
5. Octanoyl-*N*-Methylglucamide (MEGA-8), Anagrade.
6. Octyl-α/β-glucoside 10 mM solution (OcGlu).
7. α-Cyano-4-hydroxycinnamic acid (CHCA), suitable for MALDI-TOF (MS grade).
8. Aniline, ACS reagent ≥99.5%.
9. Acetonitrile, for HPLC, ≥99.9%.
10. Methanol, for HPLC, ≥99.9%.
11. Chloroform ≥99.5%.
12. Trifluoroacetic acid, 99%.

3 Methods

3.1 Optimizing Traveling Wave Ion Mobility on the Mass Spectrometer

As described above, in traveling wave ion mobility devices (TWIMS), ions are accumulated in a trapping region within a stacked-ring ion guide, prior to being released into an ion mobility cell. Ions are then pushed through the mobility cell by a series of DC traveling waves superimposed on a confining RF field applied to the stacked-ring ion guide. The wave height and velocity are set such that waves overtake the ions, and the low mobility ions fall back and 'roll over' more waves than ions with high motilities. The transit time through the mobility cell is therefore no longer a simple proportional relationship to the collisional cross section, compared to drift tube ion mobility. A full description of the theoretical description and ion optical treatment of traveling wave ion mobility is outside the scope of this chapter. However, it has been determined that despite the relationship between drift time and cross section while not being linear, with appropriate calibration, it is possible to determine collision cross section values assuming that gas pressure, traveling wave height, and velocity are kept constant. The section below describes how to set up and optimize TWIMS ion mobility on a SYNAPT mass spectrometer for both DESI and MALDI MS.

1. The mass spectrometer should be tuned appropriately, with detector setup performed and baseline sensitivity and mass resolution checked.
2. Ion mobility should be optimized upon a suitable sample covering the m/z and drift time range expected from the samples to be measured in the experiment.
3. Ion mobility and TWIMS starting parameters are listed (*see* **Note 1**).
4. The TWIMS cell parameters should be optimized to maximize the separation space, ensuring that all ions elute over the appropriate m/z and mobility arrival time (drift time) range for the experiment. This is important to avoid wrap-around, where ions of low mobility are present in the high mobility region of the subsequent release.
5. A generic starting point for T-WAVE velocity optimization is 800 m/s for the V-mode of TOF operation over an m/z range of 50–1200. A typical optimization experiment is presented in Fig. 1. Here in the MALDI ionization mode, the effect of optimizing the traveling wave velocity and the effect on the ion mobility separation are shown. Poorly optimized variable wave velocities where the IM drift time vs m/z space is poorly utilized with ions eluting too quickly (a) or the ions take too long to exit the device and wrap round to the next IMS cycle (b) are presented. Optimized traveling wave velocities, for both fixed wave velocity (540 m/s) and variable wave velocity (1000–300 m/s) where the drift time vs m/z space is optimized appropriately are shown in (c) and (d), respectively.
6. It should be noted that calibration of the TWIMS cell is only required to obtain absolute collision cross section (CCS) measurements. In the case that ion mobility is used purely as an additional separation stage, or for relative measurement and comparison of drift times, it is important that constant temperature, gas pressure, and traveling wave voltage are maintained.
7. In all cases, it is advisable to switch the instrument into ion mobility mode for at least 1 h to ensure stabilization of pressure in the ion mobility cell.
8. Calibration of CCS and m/z should be performed using a suitable multipoint calibrant covering the m/z range of the experiment. For electrospray experiments, infusion of the

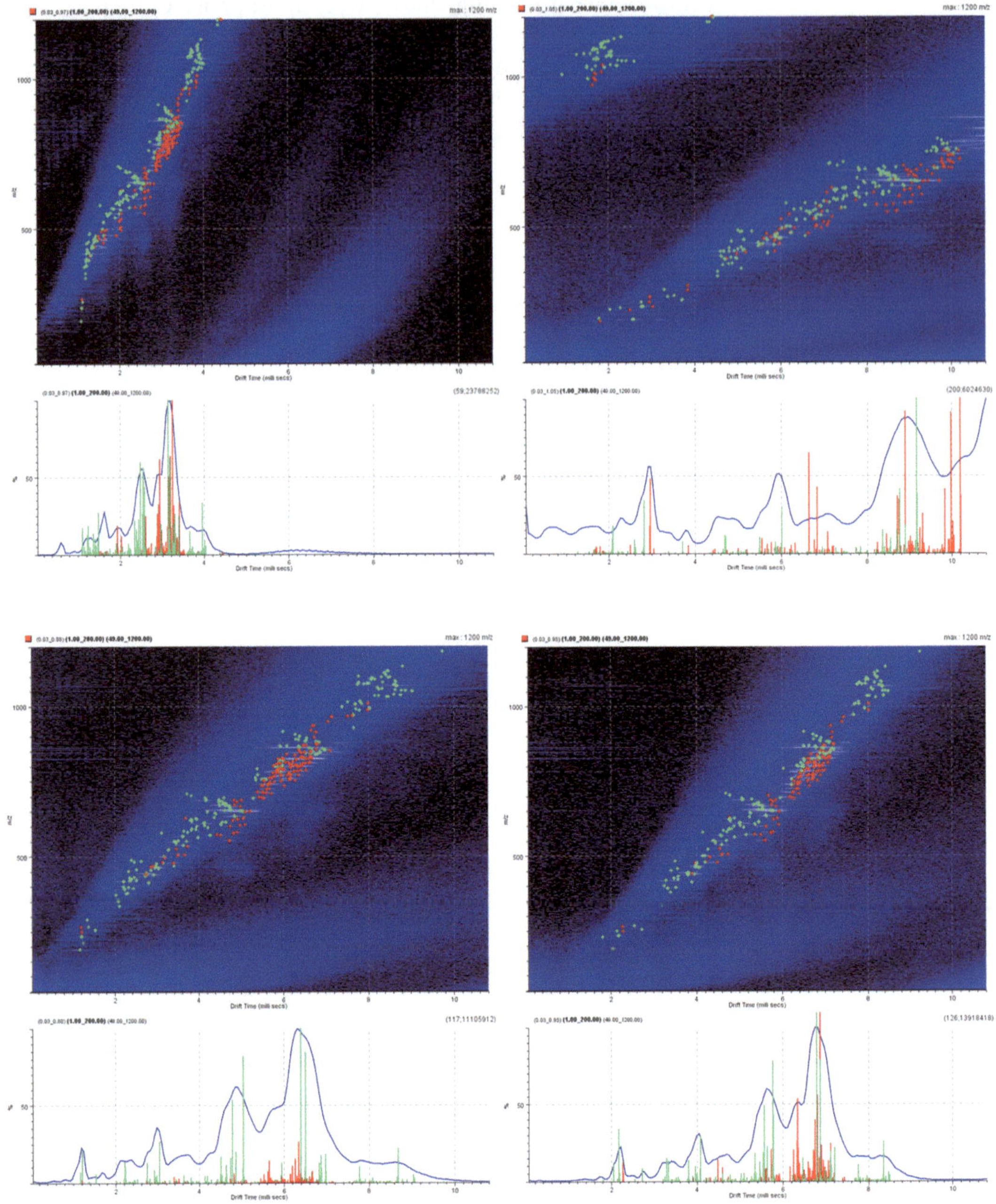

Fig. 1 The effect of T-WAVE velocity on the ion mobility separation, shown for a MALDI experiment. The nested drift time vs *m*/*z* plot is displayed for lipids (red) and background matrix ions (green). (**a**) Variable wave velocity 500–1000 m/s (top left), (**b**) Variable wave velocity 1000–750 m/s (top right), (**c**) Fixed wave velocity 540 m/s (bottom left) and (**d**) variable wave velocity 1000–300 m/s (bottom right)

sample can be performed manually or via the IntelliStart system. In this example, the MajorMix sample has been used (*see* **Note 2**).

9. For MALDI IM calibration, 1 μL of 10 mg/mL polyalanine in methanol should be deposited on a target plate followed by 1 μL of a suitable matrix solution, for example, 9-AA.
10. The CCS area calibration can then again be performed using Intellistart, which automatically generates the constants for the CSS area calculation.
11. T-WAVE IM calibration result should result in RSD values <1.5% for the calibration to be accepted. This will enable routine CCS measurements from analytes with <2% accuracy.

3.2 Imaging Mass Spectrometry

Imaging mass spectrometry has developed rapidly due to the ability to provide spatial localization of molecular information across a wide range of substrates and samples. The development of more sensitive and selective technologies and workflows has enabled the expansion of tissue-based research and holds significant potential to impact the area of histopathology, supplementing digital pathological analysis. Imaging mass spectrometry was initially performed using secondary ionization MS as an ionization source, but the majority of the work and rapid expansion of the approach occurred through the use of MALDI. More recently the uses of alternative, ambient ionization techniques, such as DESI, have allowed metabolites and lipids to be studied with high sensitivity and specificity. The advantages of the ambient DESI approach are that minimal sample preparation is required, no matrix is needed, and in addition the technique can be operated with parameters and solvents to be nondestructive to the tissue surface [19]. The lack of matrix provides inherent advantages, such as a lack of matrix-related background ions, while the nondestructive nature of DESI allows previously analyzed tissue sections to be studied by other imaging modalities both MS and non-MS approaches. For example, matrix could be applied, and the section could be studied by MALDI MS, or the section could easily be subjected to chemical staining, such as Hematoxylin and Eosin (H&E) staining.

MALDI has been extensively optimized and applied to a broad range of samples and applications. The workflow for acquisition of imaging mass spectrometry data using MALDI is shown in Fig. 2. The workflow starts with sectioning of the tissue sample, followed by fixing to the substrate and deposition of matrix onto the tissue section. The experiment is then setup by taking an optical image of the tissue section and coregistering with the MS acquisition system. Once the experiment parameters are defined, the sample can be placed into the MALDI interface for data acquisition and finally data processing and visualization.

The sections below describe the steps required to obtain imaging mass spectrometry data using MALDI and DESI with ion mobility separation.

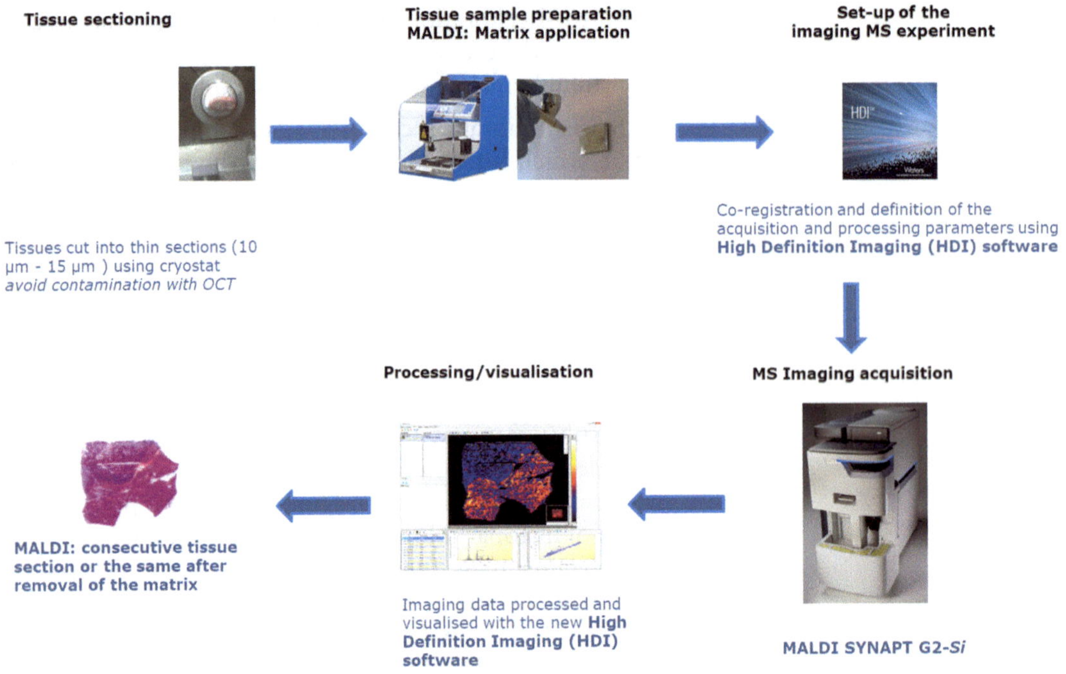

Fig. 2 Workflow for imaging mass spectrometry using MALDI with ion mobility

3.2.1 Sample Sectioning

Sectioning of the tissue sample is critical for good results by both DESI and MALDI techniques, and cryosections must be used for imaging mass spectrometry. The embedding material used is important to generate useable imaging MS results and should not include traditional polymer-based material, such as OCT, that causes contamination of the mass spectrometer. OCT can be used to attach the sample to the holder but should not be in direct contact with the sample that will be sectioned. The use of carboxymethyl cellulose (CMC) is generally recommended combined with gelatin [20]. This can vary depending upon the tissue being sectioned but a starting embedding material would be 5% CMC + 10% gelatin.

1. Place the slide inside the cryostat and wait until it has reached the appropriate temperature.
2. Fresh frozen tissue can be added directly to the cryostat. Embedding material such as OCT should be avoided, or minimized where possible, as this will result in a background polymeric signal in the mass spectrometer. Embedding material such as CMC or gelatin is recommended for imaging mass spectrometry.
3. Thickness required for MALDI imaging mass spectrometry is 10–15 µM. If spray preparation is to be performed then thinner sections are more appropriate <12 µM. For DESI tissue, thickness is not as critical as for MALDI.

4. Cut tissue sections and transfer tissue section onto the cold slide.
5. Warm the slide while still inside cryostat to avoid condensation.
6. Remove slide from cryostat briefly treat for 15 min in a desiccator prior to storage at −80 °C (or before use for DESI or application of the appropriate MALDI matrix).

3.2.2 MALDI Optimization

To perform a MALDI imaging, experiment on a sectioned tissue sample includes the following steps:

1. Choosing the appropriate matrix and applying to the sectioned tissue.
2. Setting up of the MS system to acquire data.
3. Processing and visualization.

3.2.3 Matrix Selection and Deposition

Matrix is an essential component of the MALDI process, and selection of matrix, as well as the method of deposition onto the sample, is critical to obtain good imaging mass spectrometry data [21]. The choice of MALDI matrix is critical for succesful results and has been reviewed extensively elswhere.

The device and method for matrix applications should be carefully determined and the choice may be dictated by personal preference or availability, but this strongly dictates the quality of results. Application of the MALDI matrix can be achieved using a variety of methods that can be divided into sublimation, spraying, or spotting techniques. Sublimation results in the addition of dry matrix and has advantages in both the crystal size and uniformity. Spotting and spraying are more commonly used wet matrix application techniques. A typical workflow for the matrix application and for an imaging mass spectrometry experiment using a spraying device is outlined below and is based upon modification of a previously published protocol [22]:

1. The SunCollect system (www.SunChrom.de) should be flushed with the solvent solution.
2. The system should be flushed through with the solvent that will be used for the MALDI matrix.
3. Adjust the gas pressure to an appropriate value (a default value is 2 bar).
4. The sprayer needs to deposit a smooth, homogenous stream of solvent prior to setting up the device for the tissue section.
5. Identify the location of the tissue section with the control software; in the case of the SunCollect device, this involves the recording of the coordinates from the top left and bottom right of the section for spraying.
6. Select a suitable matrix based upon the type of molecule being analyzed (*see* Table 1).

Table 1
Selected T-WAVE IM tuning parameters

	Positive	Negative
Source		
Capillary	3.2	2.5
Cone	40	40
Source	80 °C	80 °C
Quadrupole (4 K)		
LM	4.7	4.7
HM	15	15
Prefilter	2	2
T-WAVE parameters		
StepWave 1 and 2		
Wave velocity (m/s)	300	300
Wave height (V)	15	15
TriWave IMS		
Wave velocity (m/s)	650	300
Wave height (V)	40	15
IMS ramping		
Mobility trapping		
Trap height (V)	15	15
Extract height (V)	0	0
Release time (μs)	500	500
TriWave transfer		
Wave velocity (m/s)	220	220
Wave height (V)	4	4
TriWave DC		
Entrance (V)	3	3
Bias	45	45
IMS entrance	20	20
Helium cell (V)	50	50
Helium exit (V)	−20	−20

7. For peptides, CHCA was used as a matrix for the MALDI analysis.
8. Matrix solutions were made at 10 mg/mL in ACN:water:TFA (1:1:0.1 in volume).
9. The matrix solution should then be flushed through the sprayer until the matrix is eluting from the spray tip.
10. Set the matrix spraying as follows: In medium speed mode: ten layers of matrix should be applied using the continual spraying mode at medium speed, with the spraying needle at a distance of 2 cm from the tissue surface. The first two layers can be subsequently sprayed at a flow rate of 10 μL/min, layers 3–8 at 15 μL/min, and the final two layers at 20 μL/min.
11. Start the spray sequence and assess the quality of matrix deposition.

3.2.4 MALDI TWIMS Data Acquisition and Processing

1. The appropriate polarity (positive/negative) should be selected based upon the molecules being studied. (If lipids are to be investigated, then negative mode is most informative).
2. Ensure that the system has been set up for ion mobility (as detailed in Subheading 1.1), and that gas pressures have stabilized in the ion mobility mode.
3. Ensure the TOF system has been mass calibrated in the mode of operation as well as the same mass range.
4. Set the appropriate mass range for data acquisition. For lipids/drug molecules, this would be *m/z* 50–600 or 50–1200 and for peptides *m/z* 50–2000.
5. MS parameters such as the type of experiment (MS, MS/MS, in TOF mode only or with ion mobility separation (IM)), mass range, polarity, and collision energy are defined also in HDI. Processing parameters can also be set prior to the acquisition.
6. Save the experiment parameters and create the worksheet to be loaded into MassLynx.
7. Start the sample list acquisition using MassLynx.
8. Load the processed data into HDI software for visualization of the imaging experiment.

A typical example of a result obtained from the MALDI TWIMS analysis of an on-tissue tryptic digest of a rat brain sagittal section is shown in Fig. 3. In this example, data were obtained from the analysis of the tissue digest by MALDI on a SYNAPT G2-Si at 250 μM pixel resolution with a 1 kHz Nd-Yag solid-state laser. The upper pane in Fig. 3 shows the nested drift time (ms) versus *m/z* plot for the tissue section. Without IM separation as seen in Fig. 3a, the level of information content acquired from the data is severely limited by the preferential ionization of compounds on tissue.

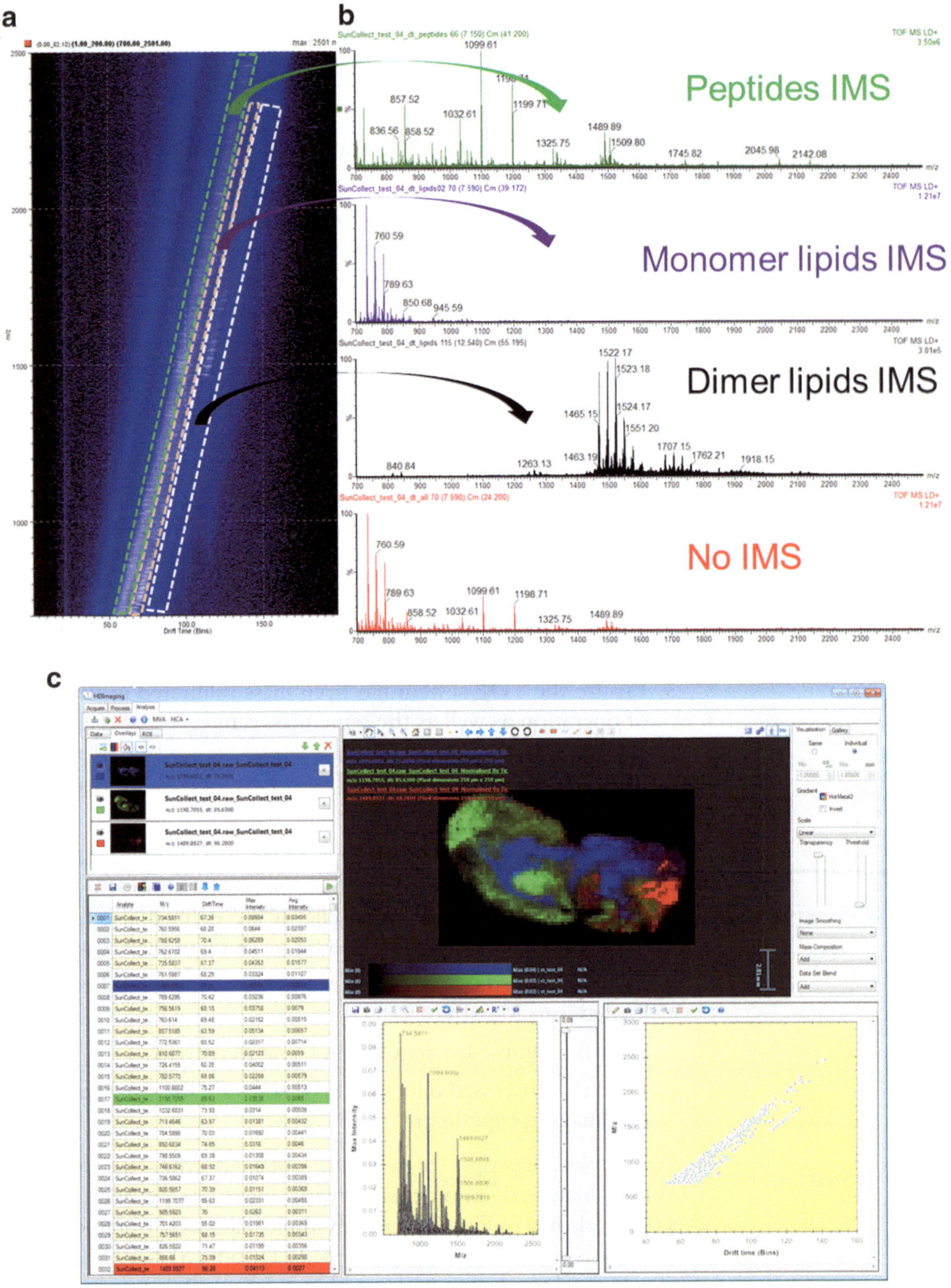

Fig. 3 MALDI TWIMS analysis of an on-tissue tryptic digest of a rat brain sagittal section. Data were acquired in MALDI mode on a SYNAPT G2-Si at 250 μM pixel resolution. (**a**) Drift time vs *m/z* plot. (**b**) Selected spectra showing the effect of TWIMS separation on the lipids. (**c**) Spatial distribution of selected lipid species

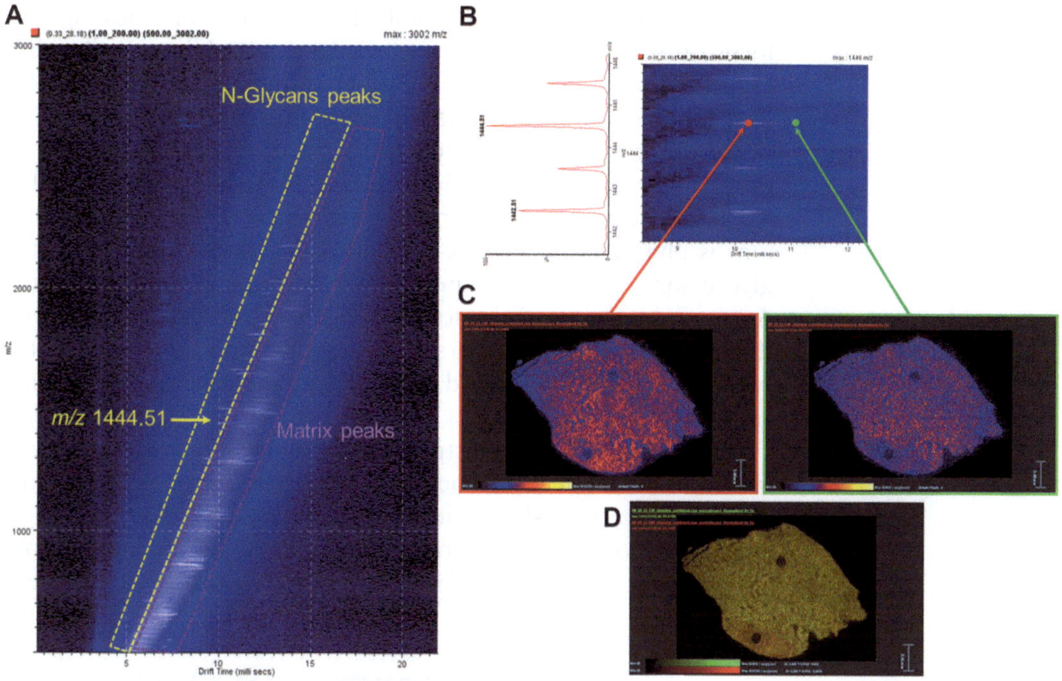

Fig. 4 MALDI imaging TWIMS analysis of released N-linked glycans from tumor bearing formalin-fixed paraffin-embedded (FFPE) colorectal tissue. Glycans were released using a previously published procedure [19]. (**a**) Drift time vs *m/z* plot for the data showing the glycan and matrix ion trend lines; (**b**) zoom section of IMS drift time versus *m/z* plot that corresponds to *m/z* 1444.5 showing two peaks can be observed, one an N-glycan, confirmed by MS/MS as Hex4 dHex1 HexNac3 while the other is a matrix-related ion; (**c**) spatial distribution of these two separate ions where the glycan and matrix-related species have different spatial localization; (**d**) RGB overlay of the two ions

Close examination of the IM separation shows three distinct series nested by their drift time and *m/z* relationship. The series correspond to different molecular species, all singly charged. The most mobile series, with the shortest IM drift time are related to tryptic peptides, with their elution ordered by *m/z*. The ion series with the intermediate drift time correspond to abundant singly charged lipids and lipid-related ions. The lowest mobility ion series contains dimers of lipids. Figure 3c shows the spatial distribution of three distinct peptides, *m/z* 1099.6, *m/z* 1198.7, and *m/z* 1489.9 that localize and correlate with different areas of the rodent brain.

A further example is shown in Fig. 4 for the MALDI imaging analysis of released N-linked glycans from tumor-bearing formalin-fixed paraffin-embedded (FFPE) colorectal tissue. The glycans were released using a previously published procedure [19] and analyzed using MALDI imaging analysis. When the IM data were explored in the drift time and *m/z* space, it can be clearly seen that two trend lines were observed. The fastest trend line that with the highest ion mobility (Fig. 4a) corresponds to the N-linked glycans and the slowest nested trend line to the CHCA matrix molecules. Specific focus on specific regions shows broader ion mobility profiles than expected for a single species. Figure 4b shows a zoom section of IM drift time versus m/z plot that corresponds to m/z 1444.5. Two peaks can be observed, one that is an N-glycan that could be confirmed by MS/MS as Hex4 dHex1 HexNac3 while the other is a matrix-related ion. The spatial distribution of these two separate ions is illustrated in Fig. 4c where it can clearly be seen that the glycan and matrix-related species have different spatial localization. As expected, the matrix is more homogeneous across the tissue with the glycan more localized to specific regions. An RGB overlay of the two ions is shown in Fig. 4d where the glycan and background ion are shown.

The advantage of an instrument geometry where a collision cell for collision-induced dissociation (CID) after the ion mobility separation stage is shown in Fig. 5a. Here, CID has been used after ion mobility separation (transfer T-WAVE fragmentation) allowing the separation of isobaric species (precursors) that can

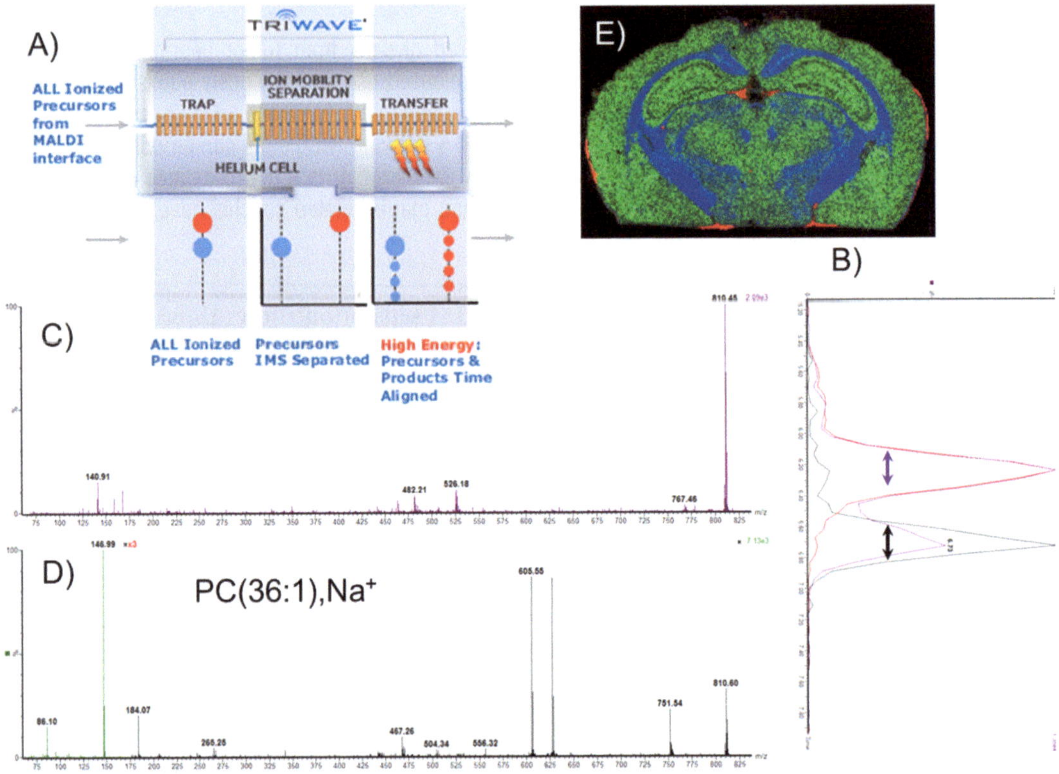

Fig. 5 MALDI imaging data from the analysis of lipids from a sagittal Mouse brain tissue section at 45 μM spatial resolution in positive ion mode, showing the specificity obtained by TWIMS combined with post IM fragmentation. (**a**) Schematic of the post IM experiment, (**b**) IM drift time separation of two lipids, (**c**, **d**) fragment ion spectra of IM-separated lipid and background ion, (**e**) spatial distribution of the two ions

then be subjected to fragmentation. In this case, MALDI imaging mass spectrometric analysis using ion mobility of lipids is shown from a sagittal Mouse brain tissue section acquired at 45 μM spatial resolution in positive ion mode. Figure 5b shows the ion mobility drift time separation of two isobaric ions which upon collision-induced dissociation (post IMS) gives rise to the spectra shown in (c) and (d). The fragmentation spectra show significant differences in the associated fragment ions, with the spectrum shown in (c) as an unidentified species, while the spectrum shown in (d) can be identified and confirmed as the Na+ adduct of a phosphatidylcholine (36:1).

3.2.5 DESI and TWIMS Ion Mobility

DESI-MS has been extensively used for the direct analysis of a wide range of biological samples with an increasing role in applications to imaging mass spectrometry. In DESI, a nebulized electrospray probe directs solvent toward the target surface, where analytes are desorbed from the surface at atmospheric pressure. The desorbed ions are then sampled via an ion collecting inlet capillary into the source of the mass spectrometer. To obtain mass spectral images, data are collected in a serial fashion rastering the probe across the surface, thus maintaining the spatial resolution of the molecules desorbed from the surface. This MS data are then subsequently processed and displayed, showing both the mass and intensity of ions within the spatially defined matrix.

A schematic of a DESI ion source appears in Fig. 6. The spatial resolution of the DESI image depends upon the electrospray spray size, and therefore the spray impacts upon the surface, as well as the speed at which the sample stage is moved under the spray (as well as the MS data acquisition rate).

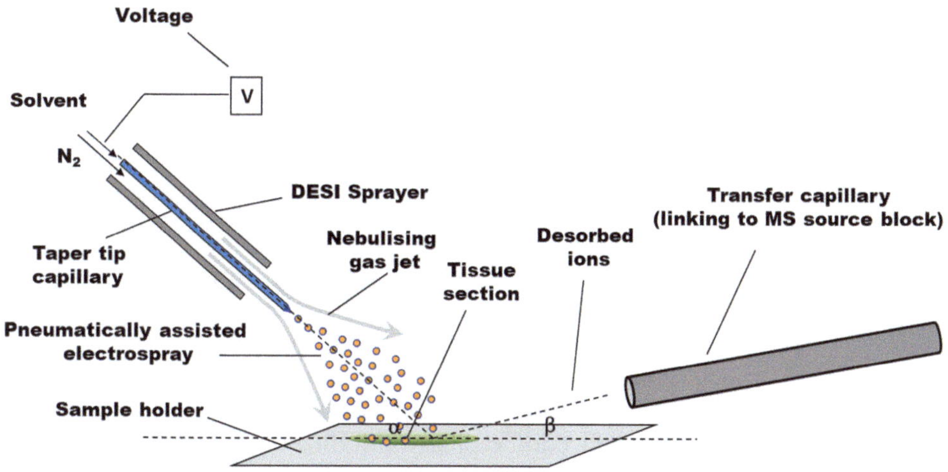

Fig. 6 Schematic of the desorption electrospray (DESI) probe and source assembly

Optimizing the DESI ion source is critical to obtain strong signal intensity of the analyte ions of interest, but importantly also in obtaining the required stability of signal for lengthy imaging mass spectrometry experiments. Ideally, signal optimization should be performed using reference samples as close as possible to the sample of interest. For instance, for the analysis of lipids on tissue sections, then the signal should ideally be optimized on serial sections. Specific details for setup of DESI for imaging are:

1. Tune the mass spectrometer to achieve the desired performance.
2. Calibrate the mass spectrometer using a suitable multipoint calibrant covering the *m/z* range of the experiment (if not already performed).
3. Fit the DESI source to the mass spectrometer.
4. Select the appropriate electrospray solvent for DESI imaging. This is typically >95% methanol with a weak acid modifier (0.1% acid). However, this is highly sample and surface dependent and should be optimized.
5. Select the appropriate flow rate through the DESI sprayer. This is typically set to 1–2 μL/min with a nebulizer gas flow rate of 7 bar. It should be noted that DESI spot size and hence spatial resolution will be defined by spray flow rate and nebulizer gas pressure.
6. The DESI ion source (Prosolia, USA) parameters should as a first step be optimized by analyzing a spot drawn using a fine, red marker containing the rhodamine 6G dye (*m/z* 443) on a glass microscope slide.
7. An important aspect of the DESI is the pneumatically assisted spray, and it is critical that this should be optimized for each set of samples.
8. The spray should be optimized for each polarity, by monitoring both the signal intensity and stability using reconstructed mass chromatograms.
9. The DESI pneumatically assisted electrospray probe should be assembled to ensure that the ESI emitter protrudes from the nebulizer by a few mm. The position should then be optimized to obtain a good spray and signal intensity.
10. The DESI probe source should be spatially aligned with the desorption surface. Adjust the electrospray incidence angle (α), angle of ion collecting inlet capillary (β), distance between emitter and sample surface, the distance between the electrospray emitter and MS inlet capillary, and the distance between ion collection capillary and the sample surface (Fig. 6).

11. An optical image of the specimen on the slide should be acquired using a regular flatbed scanner. The optical image is initially coregistered with the software and used to define an area of DESI-MS imaging.
12. The translation stage used in imaging is controlled by the manufacturer's software, such as HDI Imaging software (www.waters.com). Stages are programmed to raster in successive rows creating a two-dimensional image.
13. Select the area of imaging, define pixel sizes (equal value for x and y is recommended), and input desired stage scan speed (e.g., 100 μm per second).
14. To obtain data in ion mobility mode, ensure that the scan speed is not set lower than 0.3 s per spectrum; the recommended setting is 1 scan per second.
15. Acquire IM-MS imaging data. After the DESI imaging data acquisition concludes, put the mass the spectrometer in Standby mode, stop the nitrogen gas flow, and turn off syringe pump.
16. Process mass spectral data. Ions can be rendered on the basis of m/z ratio and drift times. Other analyses include hierarchal clustering of ions, which groups related ions, and spatial correlation analysis, which shows all other ions with similar or opposite spatial distribution.

A typical example from the DESI imaging mass spectrometric analysis of gangliosides from murine brain sample is shown in Fig. 7. These data are acquired in negative ion mode, and in contrast to the MALDI imaging data, DESI results in distinct series of multiply charged species. The ion mobility separation allows the gas phase separation of the multiply charged ions from singly charged ions resulting in nested m/z vs drift time groups. As shown in Fig. 7a, the m/z vs drift time plot shows both single deprotonated, doubly and triply deprotonated trendlines in the data. This extra separation power and peak capacity allows extra information to be discerned from the data compared to data without IM separation, as presented in Fig. 7b, c. Data without IM separation (Fig. 7b) show a congested spectrum with numerous overlapping ions, whereas the data shown in Fig. 7c present a clearer MS picture, with charge state separation providing specific information on lysophospholipids (singly charged), glycerophosholipids (doubly charged), and ceramide-based lipids and gangliosides (triply charged). The highlighted peaks shown in Fig. 7c have distinct spatial distributions as shown and yet are completely obscured in the MS data acquired without IM, showing the extra information provided by the incorporation of the ion mobility separation stage.

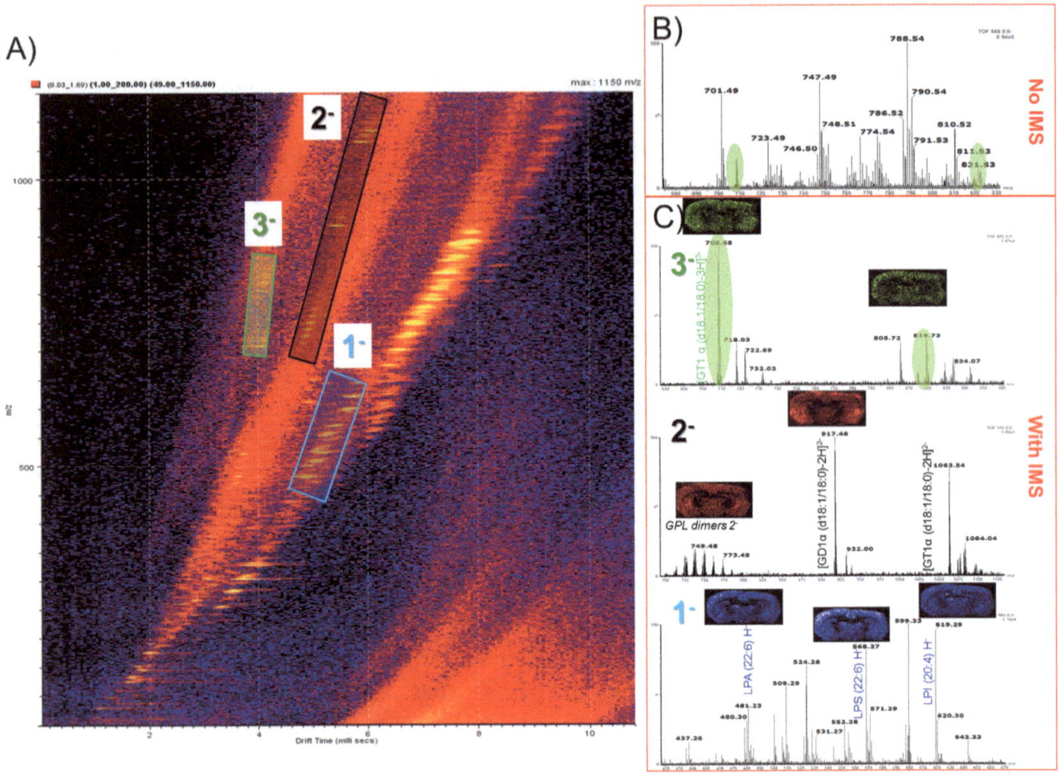

Fig. 7 DESI imaging TWIMS analysis of gangliosides from a murine brain sample: (**a**) drift time vs *m*/*z* distribution for the gangliosides showing singly, doubly, and triply deprotonated species, (**b**) complex mass spectrum obtained without IMS separation, (**c**) mass spectra obtained from the different ion mobility vs *m*/*z* nested distributions highlighting the specific information obtained on lysophospholipids (singly charged), glycerophospholipids (doubly charged), and ceramide-based lipids and gangliosides (triply charged)

4 Notes

1. Critical user parameters affecting the separation of ion are the T-WAVE height (voltage) and velocity settings, as well as the ion mobility gas cell pressure. In practice, altering the traveling wave velocity allows the separation to be optimized using either a fixed traveling wave velocity, or alternatively the velocity can be ramped with drift time. In most cases, a voltage ramp is recommended to maximize the ion mobility separation space.
2. A working MajorMix calibration solution is made from mixing the stock Major Mix (Table 2) with an LC-MS QC cal mix (Table 3) and diluting as detailed in Table 4.

Table 2
MajorMix

Reference compound	Concentration	Volume (μL)	Notes
Succinic acid	1 mg/mL	10	Only for neg ion
Salicylic acid	100 ng/μL	10	Only for neg ion
Theophylline	100 ng/μL	10	Only for neg ion
Pantothenic acid	10 ng/μL	10	Only for neg ion
Stearic acid	100 ng/μL	20	Only for neg ion
Perfluoroheptanoic acid	10 ng/μL	10	Only for neg ion
Perfluorooctanoic acid	10 ng/μL	10	Only for neg ion
Perfluorodecanoic acid	10 ng/μL	10	Only for neg ion
Perfluorododecanoic acid	10 ng/μL	10	Only for neg ion
Polyalanine	3 mg/mL	10	
Ultramark 1621	1:1000 dil (v/v)	10	
Formic acid	10%	10	
1:1 ACN:water		860	
Total		*1000*	

Table 3
LC-MS QC ref. standard

Name	Exact mass $[M+H]^+$	Exact mass $[M-H]^-$	Conc (μg/mL)
Acetaminophen	152.0712	150.0555	10
Caffeine	195.0882		1.5
Sulfaguanidine	215.0603	213.0446	5
Sulfadimethoxine	311.0814	309.0658	1
Val-Tyr-Val	380.2185	378.2029	2.5
Verapamil	455.2910		0.2
Terfenadine	472.3216		0.2
Leucine-enkephalin	556.2771	554.2615	2.5
Reserpine	609.2812		0.6

Table 4
Working MajorMix calibration solution

	Volume (μL)	
Major mix	100	Table 2
LC-MS QC ref. standard	50	Table 3
(ACH:H_2O) 1% formic (1:1)	850	
Total	*1000*	

References

1. Hillenkamp F, Karas M, Beavis RC, Chait BT (1991) Matrix-assisted laser desorption/ionization mass spectrometry of biopolymers. Anal Chem 63(24):1193A–1203A
2. Yamashita M, Fenn JB (1984) Electrospray ion source. Another variation on the free-jet theme. J Phys Chem 88(20):4451–4459
3. Henzel WJ, Billeci TM, Stults JT, Wong SC, Grimley C, Watanabe C (1993) Identifying proteins from two-dimensional gels by molecular mass searching of peptide fragments in protein sequence databases. Proc Natl Acad Sci U S A 90(11):5011–5015
4. Cain TC, Lubman DM, Weber WJ Jr (1994) Differentiation of bacteria using protein profiles from matrix-assisted laser desorption/ionization time-of-flight mass spectrometry. Rapid Commun Mass Spectrom 8:1026–1030
5. Caprioli RM, Farmer TB, Gile J (1997) Molecular imaging of biological samples: localization of peptides and proteins using MALDI-TOF MS. Anal Chem 69:4751
6. Takáts Z, Wiseman JM, Gologan B, Cooks RG (2004) Mass spectrometry sampling under ambient conditions with desorption electrospray ionization. Science 306(5695):471–473
7. Ifa DR, Wiseman JM, Song QY, Cooks RG (2007) Development of capabilities for imaging mass spectrometry under ambient conditions with desorption electrospray ionization (DESI). Int J Mass Spectrom 259:8–15
8. Kallback P, Shariatgorji M, Nilsson A, Andren PE (2012) Novel mass spectrometry imaging software assisting labeled normalization and quantitation of drugs and neuropeptides directly in tissue sections. J Proteome 75:4941–4951
9. Chaurand P, Schwartz SA, Caprioli RM (2002) Imaging mass spectrometry: a new tool to investigate the spatial organization of peptides and proteins in mammalian tissue sections. Curr Opin Chem Biol 6:676–681
10. Goodwin RJ, Mackay CL, Nilsson A, Harrison DJ, Farde L, Andren PE et al (2011) Qualitative and quantitative MALDI imaging of the positron emission tomography ligands raclopride (a D2 dopamine antagonist) and SCH 23390 (a D1 dopamine antagonist) in rat brain tissue sections using a solvent-free dry matrix application method. Anal Chem 83:9694–9701
11. Ibrahim YM, Prior DC, Baker ES, Smith RD, Belov ME (2010) Characterization of an ion mobility-multiplexed collision induced dissociation-tandem time-of-flight mass spectrometry approach. Int J Mass Spectrom 293 (1–3):34–44
12. Rodriguez-Suarez E, Hughes C, Gethings L, Giles K, Wildgoose J, Stapels M, Fadgen K, Geromanos S,Vissers JPC, Elortza F, Langridge JI (2013) An Ion Mobility Assisted Data Independent LC-MS Strategy for the Analysis of Complex Biological Samples. Current Analytical Chemistry 9(2):199–211.
13. Vakhrushev SY, Langridge J, Campuzano I, Hughes C, Peter-Katalinić J (2008) Ion mobility mass spectrometry analysis of human glycourinome. Anal Chem 80(7):2506–2513
14. Paglia G, Williams JP, Menikarachchi L, Thompson JW, Tyldesley-Worster R, Halldórsson S, Rolfsson O, Moseley A, Grant D, Langridge J, Palsson BO, Astarita G (2014) Ion mobility derived collision cross sections to support metabolomics applications. Anal Chem 86(8):3985–3993
15. May JC, McLean JA (2015) Ion mobility-mass spectrometry: time-dispersive instrumentation. Anal Chem 87(3):1422–1143
16. Ewing MA, Glover MS, Clemmer DE (2016) Hybrid ion mobility and mass spectrometry as a separation tool. J Chromatogr A 1439:3–25
17. Pringle SD, Giles K, Wildgoose JL, Williams JP, Slade SE, Thalassinos K, Bateman RH, Bowers MT, Scrivens J (2007) An investigation of the mobility separation of some peptide and protein ions using a new hybrid quadrupole/travelling wave IMS/OA-ToF instrument. Int J Mass Spectrom 261:1–12
18. Giles K, Pringle SD, Worthington KR, Little D, Wildgoose JL, Bateman RH (2004) Applications of a travelling wave-based radio-frequency-only stacked ring ion guide. Rapid Commun Mass Spectrom 18 (20):2401–2414
19. Powers TW, Neely BA, Shao Y, Tang H, Troyer DA, Mehta AS, Haab BB, Drake RR (2014) MALDI imaging mass spectrometry profiling of N-glycans in formalin-fixed paraffin embedded clinical tissue blocks and tissue microarrays. PLoS One 9:9
20. Nelson KA, Daniels GJ, Fournie JW, Hemmer MJ (2013) Optimization of whole-body zebrafish sectioning methods for mass spectrometry imaging. J Biomol Tech 24(3):119–127
21. Shariatgorji M, Svenningsson P, Andrén PE (2014) Mass spectrometry imaging, an emerging technology in neuropsychopharmacology. Neuropsychopharmacology 39:34–49

22. Djidja MC, Francese S, Loadman PM, Sutton CW, Scriven P, Claude E, Snel MF, Franck J, Salzet M, Clench MR (2009) Detergent addition to tryptic digests and ion mobility separation prior to MS/MS improves peptide yield and protein identification for in situ proteomic investigation of frozen and formalin-fixed paraffin-embedded adenocarcinoma tissue sections. Proteomics 9(10):2750–2763

Part IV

Bioinformatic Solutions for Analyzing IM-MS Data and Deriving CCS Values

Chapter 17

The Use of LipidIMMS Analyzer for Lipid Identification in Ion Mobility-Mass Spectrometry-Based Untargeted Lipidomics

Xi Chen, Zhiwei Zhou, and Zheng-Jiang Zhu

Abstract

Untargeted lipidomics aims to comprehensively measure and characterize all lipid species in biological systems. Ion mobility-mass spectrometry (IM-MS) has showed a great potential for untargeted lipidomic analysis. Coupling with liquid chromatography and data-independent tandem MS techniques, acquired IM-MS data set contains four-dimensional information for lipid identification, including *m/z* of MS1 ion, retention time (RT), collision cross section (CCS), and MS/MS spectra. In this protocol, we introduced a data processing workflow using an integrative web server, namely, LipidIMMS Analyzer, to support accurate lipid identification. The protocol demonstrated the integration of all four dimensional information to achieve unambiguous identifications of lipids in complex biological samples.

Key words Untargeted lipidomics, Ion mobility-mass spectrometry, LipidIMMS Analyzer, Lipid identification, Collision cross section

1 Introduction

Lipids play many vital roles in cell biology, such as energy storage, cell constituent components, and cell signal transduction [1–3]. Dysregulations in lipid metabolism are closely related to the occurrence and development of many diseases, including cardiovascular diseases, neurodegenerative diseases, and cancer [1, 2, 4]. Thus, the comprehensive and quantitative analysis of lipid species, called untargeted lipidomics, plays a vital role in lipid metabolism study. However, due to the large number and structural diversity of lipids, separation and identification of lipid structures presents a grand challenge. Recently, ion mobility-mass spectrometry (IM-MS) has showed a great application potential for untargeted lipidomics [5–7]. IM-MS separated structurally different lipid ions rapidly in gas phase through many collisions occurring between lipid ions and inert buffer gas under an electric field [8, 9]. The physicochemical property obtained from ion

Giuseppe Paglia and Giuseppe Astarita (eds.), *Ion Mobility-Mass Spectrometry: Methods and Protocols*, Methods in Molecular Biology, vol. 2084, https://doi.org/10.1007/978-1-0716-0030-6_17,

mobility termed as collision cross section (CCS) provided another dimensional information to characterize and distinguish different lipid structures including lipid isomers [5, 10]. In addition, coupling IM-MS with liquid chromatography (LC) and data-independent tandem MS techniques (i.e., all-ion-fragmentation, AIF) enables to simultaneously characterize and measure lipid ions with four-dimensional information, including m/z of MS1, retention time (RT), collision cross section (CCS), and MS/MS spectra [10, 11]. The combination of all four dimensional information could support an unambiguous identification of lipids in complex biological samples.

However, IM-MS-based lipidomics is still significantly restricted by lacking software tools for analyzing multidimensional information toward lipid identification. To address this challenge, our lab developed a freely available webserver, namely, LipidIMMS Analyzer, to support the lipid identification in IM-MS-based lipidomics [12]. It provides an interactive interface to help users to integrate four-dimensional information (i.e., m/z, RT, CCS, and MS/MS spectra) for lipid identification. Moreover, it also supports different data acquisition approaches (e.g., LC-IM-MS and direct infusion-IM-MS) and different IM-MS instruments (e.g., Agilent and Waters). In this protocol, we systemically summarized the details of data analysis using LipidIMMS Analyzer, including data preparation from different IM-MS platforms, performing lipid identification, and result interpretation.

2 Materials

2.1 Reagents and Equipments

1. Differential Ion Mobility System Suitability Lipidomix (DIMS Kit, Avanti Lipids, No. 330708) (*see* **Note 1**).
2. Extracted lipid samples from biofluids, cells, and animal tissues in autosampler vials [9, 13, 14].
3. LC system: Ultra-performance liquid chromatography (UPLC, 1290 series, Agilent Technologies); ACQUITY I-Class UPLC (Waters).
4. IM-MS system: Agilent 6560 drift-tube IM-MS with Agilent Mass Hunter Version B.08.00; Waters Synapt G2-Si traveling wave IM-MS with MassLynx 4.1.
5. A personal computer with an Internet connection and a web browser like Chrome, Firefox, or Safari.

2.2 Equipment Setup

1. MS instrument parameters for data acquisition: please refer to the previous protocols [10, 13, 15].
2. Software tools for data preparation:

Agilent:

- PNNL PreProcessor (https://omics.pnl.gov/software/pnnl-preprocessor).
- Agilent IM-MS Reprocessor (Version B.08.00 or later).
- Agilent Mass Profiler (Version B.08.00 or later).
- Agilent IM-MS Browser (Version B.08.00 or later).

Waters:

- Progenesis QI (Version 2.4 or later).

3. Data formatting: Microsoft office Excel 2010.
4. Data processing webserver: LipidIMMS Analyzer (http://imms.zhulab.cn/LipidIMMS/).

3 Methods

3.1 Data Preparation

3.1.1 Overview

LipidIMMS Analyzer supports both Agilent (Subheading 3.1.2) and Waters instruments (Subheading 3.1.3). It requires three types of files for the analysis:

- A MS1 peak table (.csv format): The MS1 peak table is a list of peaks with *m/z*, retention time (RT), collision cross-section (CCS), and peak intensities.
- MS/MS data files (.mgf/.msp/.cef format): The MS2 data files are the MS/MS spectra for MS1 peaks.
- An RT calibration table (.csv format, only required for RT match): The RT calibration table is a table used for RT calibration.

We recommend the users to use vendors' software for peak picking and generating the MS1 table peak and MS/MS data files.

3.1.2 Agilent DTIM-MS

1. Raw data preprocessing (Single-field-based data acquisition).
   - Postcalibrate raw data in IM-MS Reprocessor software from Agilent.
   - Smooth and repair saturated peaks using PNNL PreProcessor [16]. After processing, two new data files were generated and named as "FILENAME MX_DX_CX" and "FILENAME MX_DX_CX_SR." For parameters, we recommend:
     - Moving average smoothing number of points: checked.
     - *m/z*: none.
     - drift: 3.
     - chromatography/infusion: 3.
     - Repair saturation for points above abundance limit (%): checked, 40.

   - Repair fragmentation spectra (high CE): checked.
   - Export frame metadata (original data file): checked.
- Calculate the calibration coefficient (T_{fix} and β) of calibrates (Agilent tune mixture solution) using Agilent IM-MS Browser, and save calibration coefficient to all saturation repaired data files.
- Open files in Agilent Mass Profiler, and select the appropriate parameters to peak detection and alignment. In the demo data (*see* **Note 2**), we use the following parameters:
   - *Feature Finding/Loading*:

     Measure of abundance-Max ion volume.

     Ion intensity $\geq$ 100 count.

     Isotope model: Common organic molecules.

     Limit charge states to a range of 1–1.
   - *Alignment and Normalization*:

     RT tolerance = ±(0.0% + 0.3 min).

     Mass tolerance = ±(15 ppm + 0.2 mDa).
   - *Statistics and Filters*:

     Missing sample treatment: Assign 0 abundance.

     Feature filter: *Q*-score $\geq$ 70.0.

     Sample occurrence: Frequency $\geq$ 50% in at least one group.

2. Export the MS1 peak table in Agilent Mass Profiler, and modify the peak table as the following format (*see* **Note 3**):
   - The first column is set as the mass-to-charge ratio ("mz").
   - The second column is set as the retention time ("rt").
   - The third column is set as the collision cross-section ("ccs").
   - Other columns are set as peak abundances in each sample.
3. After the peak detection using Mass Profiler, open data files using Agilent IM-MS Browser (*see* **Note 4**). Then, export MS/MS spectra of all features in Mascot Generic Format (*.mgf). We highly suggest only using MS/MS data files from the pooled quality control samples. For the version of Agilent Mass Profiler B.08.01 (B153, Beat Release), it supports to directly export MS/MS spectra for each detected feature as CEF format instead of using Agilent IM-MS Browser. After the raw data processing, users could export MS/MS spectra in the tab "File," then select "Export Each Sample to CEF" (*see* **Note 5**). Users can choose either of them.
4. Prepare the RT calibration table. These RTs need to be manually recorded by extracting ion chromatography (EIC) using

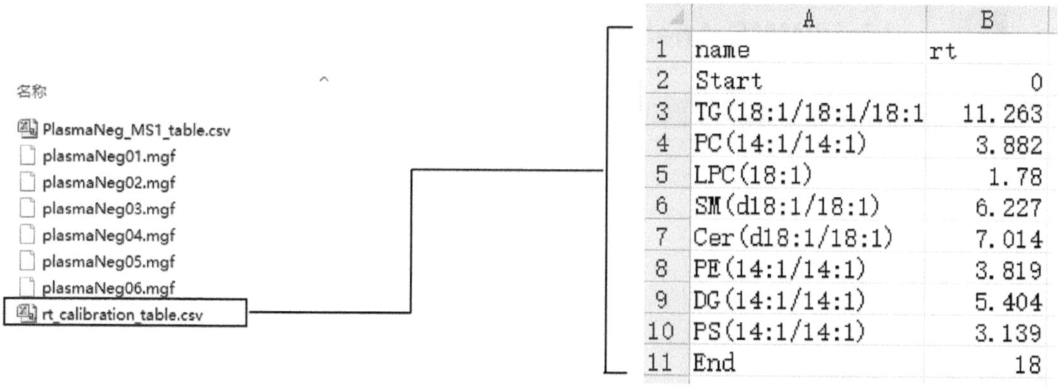

Fig. 1 The screenshot of the imported RT calibration table

vendor software, e.g., Agilent MassHunter Qualititive software, Waters MassLynx. The RT calibration table is prepared in a specific format (.csv file). The first column is "name" of the lipids in DIMS Kit (*see* **Note 6**). The second column "rt" is the experimental retention times of lipids. An example of RT calibration table is given in Fig. 1.

3.1.3 Waters TWIM-MS

1. Please refer to Progensis QI tutorial for the raw data processing [17]. Some critical parameters of data processing used in our demo data set were as follows (*see* **Note 7**):

   Alignment reference: Assess all runs in the experiment for suitability.
   Sensitivity: Automatic.
   Adducts: M+H-H_2O, M+H, M+NH_4, M+Na.

2. Users could export the MS1 peak table in the step of "Review Compounds." The exported .csv table is required to be modified as the following format (*see* **Note 8**):
   - The first column is set as the mass-to-charge ratio ("mz").
   - The second column is set as the retention time ("rt").
   - The third column is set as the collision cross section ("ccs").
   - The fourth column is the compound name provided by Progenesis QI ("compound").
   - Other columns are peak abundances of MS1 peaks in each sample.
3. The MS/MS spectra in Progensis QI were exported in the "Review Compounds—Export fragment database." The MS/MS spectra are saved as the format of "Mass Spectral Database (∗.msp)." Only Mass Spectral Database (∗.msp) is supported for Waters TWIM-MS data.

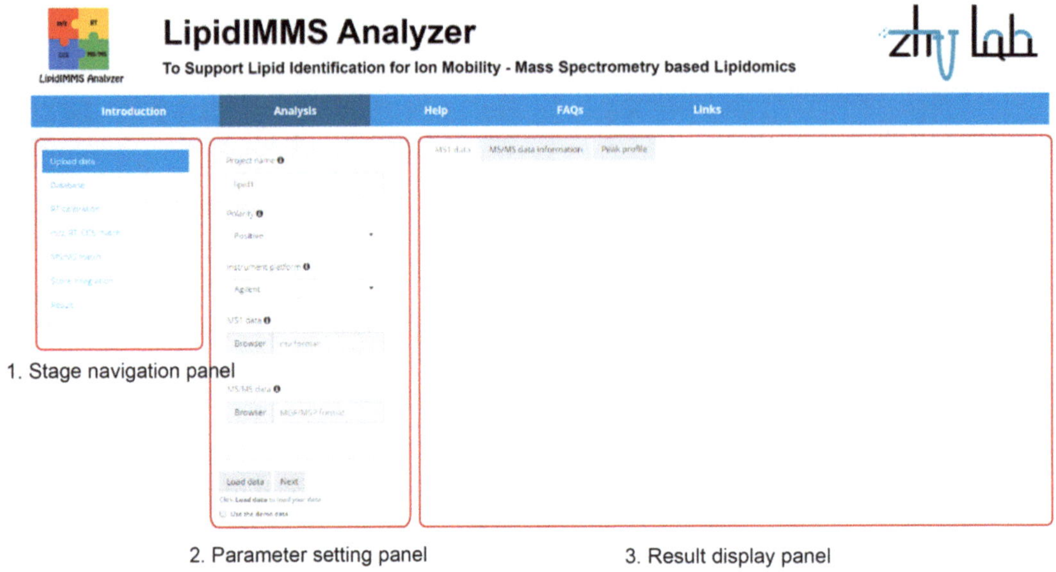

Fig. 2 The layout of LipidIMMS Analyzer webserver

4. Prepare the RT calibration table. The requirement of the RT calibration table for Waters data is same as Agilent data (Subheading 3.1.2, **step 4**).

3.2 Data Analysis with LipidIMMS Analyzer

The LipidIMMS Analyzer provides an interactive interface to help users to analyze the data from IM-MS. It consists of three parts for each page (Fig. 2), including: "Stage navigation panel," "Parameter setting panel," and "Result display panel." Users just click the first choice in "stage navigation panel" to start the analysis. In each stage, users could set parameters in the "parameter setting panel," and results will be quickly returned in the "result display panel." Generally, it contains seven steps:

1. Upload data.
   - Input basic information of the project, including "Project name," "Polarity," and "Instrument platform."
   - Upload MS1 data (MS1 peak table) and MS/MS data (*see* **Note 9**).
   - Click the "Load data."
   - View results in the "MS1 data," "MS/MS data information," and "Peak profile" tabs. "MS1 data" displays the uploaded MS1 peak table; "MS/MS data information" displays the statistical information of the uploaded MS/MS data files; "Peak profile" displays the plot of peak profile, including "m/z vs. RT" and "m/z vs. CCS" plots.
   - Click the "Next" to proceed to next step.

2. Database.
   - Select specific lipid classes for identification. All lipid classes are chosen by default.
   - Click the "Submit."
   - View and check the lipid classes in "Database information" tab. The abbreviations of lipid classes are listed in the "Abbreviations" tab.
   - Click the "Next" to proceed to next step.
3. RT calibration (*see* **Note 10**).
   - Select "Column type," and upload the RT calibration table.
   - Click the "Submit."
   - View and check the uploaded table in "RT calibration table" tab. The "RT calibration plot" tab display the fitted curve of calibration.
   - Click the "Next" to proceed to next step.
4. m/z, RT, CCS match (*see* **Note 11**).
   - Input m/z tolerance for match.
   - Check or uncheck "RT match"; input the minimum and maximum tolerances for RT match score.
   - Check or uncheck "CCS match"; input the minimum and maximum tolerances for CCS match score.
   - Click the "Submit."
   - View match result in "Match result" tab. In the table, lipid candidates will be displayed both in "lipid species" and "lipid molecular species" (Subheading 3.3). In addition, m/z error, rt. error, rt. score, CCS error, and CCS score will also be displayed. Only top ten candidates for each peak are given in the result table. One can select "All" entries and download the result table in either .csv or Excel format.
   - Click the "Next" to proceed to next step.
5. MS/MS match.
   - Check or uncheck the "MS/MS match" (*see* **Note 12**).
   - Set the mass range for MS/MS spectra.
   - Input the values of "absolute intensity cutoff," "relative intensity cutoff," and "MS/MS score cutoff."
   - Click the "Submit."
   - View match result in "Match result" tab. In this table, lipid candidates are both displayed in "lipid species" and "lipid molecular species." MS/MS spectral match scores ("msms

score") are calculated. Please refer to Subheading 3.3 for more details about the result.

- Click the "Next" to proceed to next step.

6. Score Integration.
   - Input the values of "RT score weight," "CCS score weight," and "MS/MS score weight," respectively, to calculate the composite score; input the value of "Composite score cut-off" (*see* **Note 13**).
   - Click the "Submit."
   - View match results in "Match result" tab. In this table, composite score is calculated, and lipid candidates are given if the score is larger than the cutoff score.
   - Click the "Next" to proceed to next step.
7. Result.

   In result page, it includes three tabs: "Summary," "Result table," and "Detail."
   - Summary: It summarizes the analysis result, and users could download a zip file of analysis result. The zip file includes an html report and an identification table.
   - Result table: Lipid identifications for each feature are listed in the result table, and users could browse them in the webserver. All identified lipids are displayed as both "lipid species" and "lipid molecular species" levels. Please refer to Subheading 3.3 for the definition of these levels. For each feature, the table only displays the top ten lipid candidates ranked by their composite scores. Users could select a row of interested feature, and view all identification information in the "Detail" tab.
   - Detail: This tab displays all candidates for the users' selected feature. It contains all identification information, including *m/z* error, CCS error, MS/MS score, composite score, and so on. If users want to visualize the MS/MS spectral match plot, they could select the specific row of one candidate, and click the "show MS/MS plot."

   We will interpret the details of the analysis result in the Subheading 3.3.

3.3 Interpretation of Analysis Result

The result zip file could be downloaded in the "Summary" tab in the final "Result" step. The downloaded zip file contains two files: "analysis report" and "result table."

1. Analysis report: This report is created in the html format, and users could open it using common browsers like Chrome, Firefox, Safari, etc. It includes four components: "Introduction," "Parameters of Analysis," "Result," and "Plots."

- Introduction: It records the version of LipidIMMS Analyzer in the analysis.
- Parameters of Analysis: It displays a table to record the parameters of the analysis.
- Result: It summarizes the total numbers of identifications, feature numbers, and lipid candidates in the analysis.
- Plots: It displays several figures in the analysis, including peak profile and RT calibration plot.

2. Result table: The result table is generated in a csv format. It includes all identification results in the MS1 peak table. The explanation of the columns are listed as follows:
   - Number: The feature or peak number.
   - Feature: The name of each peak, it defined by *m/z*, RT, and CCS. For example, M803T356C287 represents a feature with *m/z* 802.559, RT 356 s, and CCS 286.6 $Å_2$.
   - Lipid species: The lipid candidates defined in "lipid species" level, and the order is ranked by their composite scores. The definition of "lipid species" could be seen in refs. 12, 18. It refers to lipid subclass and characterizes the lipids by the number of carbons and double bonds. For example, PC (36:2) represents that a glycerophosphatidylcholine has 36 carbons and 2 double bonds.
   - Lipid molecular species: The lipid candidates are defined in "lipid molecular species" level, and the order is ranked by their composite scores. The definition of "lipid molecular species" could be seen in ref. 12. It characterizes the lipids in fatty acyl position level. For example, PC(20:4/16:0) represents that it has two acyl chains with 20:4 and 16:0 fatty acyls in sn-1 and sn-2, respectively.
   - Adducts: The adduct forms of lipid candidates.
   - *mz* error: The error between experimental *m/z* value and *m/z* value in the library. The unit is ppm.
   - rt. error: The error between experimental rt. value and rt. value in the library. The unit is second (s).
   - rt. score: The match score of RT match using a trapezoidal function [12]. The score ranges from 0 to 1, referring to from no match to a perfect match.
   - ccs error: The error between experimental CCS value and the CCS value in the library. The unit is percentage (%).
   - ccs score: The match score of CCS match using a trapezoidal function [12]. The score ranges from 0 to 1, referring to from *no match* to a *perfect match*.

- msms score: The match score of MS/MS similarity between experimental MS/MS spectrum and the predicted MS/MS spectrum. It is calculated by the reverse dot-product function [12]. The score ranges from 0 to 1, referring to from *no match* to a *perfect match.*
- Composite score: The composite score is calculated by a linear combination of scores (i.e., rt. score, CCS score and msms score) [12]. The score ranges from 0 to 1, representing the confidence level.

4 Notes

1. DIMS Kit is used to apply the RT match in different LC conditions and correct RT drift in different experiments. An RT calibration method was developed using a list of 20 lipids to recalibrate all RT values in the database (Tables 1 and 2). Users could select some of these lipids for the calibration. For convenience, users could directly use a commercially available lipid mixture (Differential Ion Mobility System Suitability Lipidomix, DIMS Kit, Avanti Lipids), which contains ten different lipids in the calibration list (the first ten lipids in the Tables 1 and 2).
2. Demo data are available in http://imms.zhulab.cn/LipidIMMS/.
3. In MS1 peak table of Agilent, the order and names of the first three columns must be "mz," "rt," and "ccs." The unit of retention time must be minute. The csv file must be separated by comma.
4. The MS/MS data files have to be exported one by one due to the limitation of IM-MS browser software. An update will be provided by Agilent to export all MS/MS data files together. We recommend exporting the top 100 fragment ions for each feature. Please set this parameter in the "Method–Find peaks in mass spectrum–Maximum peak count."
5. This function was only available for Mass Profiler B.08.01 (B153, Beat Release) or latter. It was recommended to put an additional parameter file "FragmentDriftTimeOffset.txt" (Fig. 3) in the "C:\temp" folder of your computer to correct the small shift of drift time between the precursor and fragment ions.
6. In RT calibration table, the "Start" and "End" represents the start and end points of gradient, which must be filled. The name of lipids must be consistent with the template. If some lipids were not detected in your system, please remove the rows of missing lipids. The unit of RT must be minute.

Table 1
The retention times of lipids for RT calibration on a reverse phase column

No.	Lipid name	$Adduct^{pos}$	m/z^{pos}	$Adduct^{neg}$	m/z^{neg}	RT(s)
1	TG(18:1/18:1/18:1)	$[M+NH_4]^+$	902.8171	n.a.	n.a.	659
2	SM(d18:1/18:1)	$[M+H]^+$	729.5905	$[M+HCOO]^-$	773.5809	332
3	Cer(d18:1/18:1)	$[M+H]^+$	564.5350	$[M+HCOO]^-$	608.5254	384
4	LPC(18:1)	$[M+H]^+$	522.3554	$[M+HCOO]^-$	566.3458	86
5	DG(14:1/14:1)	$[M+NH_4]^+$	526.4472	n.a.	n.a.	283
6	PC(14:1/14:1)	$[M+H]^+$	674.4755	$[M+HCOO]^-$	718.466	193
7	PS(14:1/14:1)	$[M+H]^+$	676.4184	$[M-H]^-$	674.4099	142
8	PG(14:1/14:1)	$[M+Na]^+$	685.4051	$[M-H]^-$	661.4081	135
9	PE(14:1/14:1)	$[M+H]^+$	632.4286	$[M-H]^-$	630.4135	190
10	PI(14:1/14:1)	$[M+Na]^+$	773.4211	$[M-H]^-$	749.4241	122
11	TG(14:1/14:1/14:1)	$[M+NH_4]^+$	734.6299	n.a.	n.a.	516
12	SM(d18:1/12:0)	$[M+H]^+$	647.5123	$[M+HCOO]^-$	691.5027	218
13	Cer(d18:1/12:0)	$[M+H]^+$	482.4568	$[M+HCOO]^-$	526.4472	254
14	LPC(24:0)	$[M+H]^+$	608.4650	$[M+HCOO]^-$	652.4554	201
15	DG(17:1/17:1)	$[M+NH_4]^+$	610.5411	n.a.	n.a.	415
16	PC(17:0/17:0)	$[M+H]^+$	762.6008	$[M+HCOO]^-$	806.5912	394
17	PS(17:0/17:0)	$[M+H]^+$	764.5436	$[M-H]^-$	762.5291	344
18	PG(17:0/17:0)	$[M+Na]^+$	773.5303	$[M-H]^-$	749.5338	363
19	PE(17:0/17:0)	$[M+H]^+$	720.5538	$[M-H]^-$	718.5392	382
20	PI(16:0/18:1)	$[M+Na]^+$	859.5307	$[M-H]^-$	835.5342	305

7. The accuracy of CCS values from TWIM-MS depends on the structural similarity between the analytes and the calibrant ions. It is highly recommended to use the lipid calibrants for the calculation of CCS values [19].
8. In MS1 peak table of Waters, the order and names of the first four columns must be "mz," "rt," "ccs," and "compound." The unit of retention time must be minute. The csv file must be separated by comma. The only difference of the MS1 peak tables between Agilent and Waters is the "compound" column.
9. If no MS/MS data files are available, one can only upload the MS1 data.
10. If users want to skip the RT calibration, please directly click the "Next." RT calibration is required for the following RT match.

Table 2
The retention times of lipids for RT calibration on a HILIC column

No.	Lipid name	Adduct^{pos}	m/z^{pos}	Adduct^{neg}	m/z^{neg}	RT(s)
1	TG(18:1/18:1/18:1)	$[M+NH_4]^+$	902.8171	n.a.	n.a.	159
2	SM(d18:1/18:1)	$[M+H]^+$	729.5905	$[M+CH_3COO]^-$	787.5971	434
3	Cer(d18:1/18:1)	$[M+H]^+$	564.5350	$[M-H]^-$	562.5205	32
4	LPC(18:1)	$[M+H]^+$	522.3554	$[M+CH_3COO]^-$	580.3620	470
5	DG(14:1/14:1)	$[M+NH_4]^+$	526.4472	n.a.	n.a.	38
6	PC(14:1/14:1)	$[M+H]^+$	674.4755	$[M+CH_3COO]^-$	732.4821	404
7	PS(14:1/14:1)	$[M+H]^+$	676.4184	$[M-H]^-$	674.4099	366
8	PG(14:1/14:1)	$[M+Na]^+$	685.4051	$[M-H]^-$	661.4081	150
9	PE(14:1/14:1)	$[M+H]^+$	632.4286	$[M-H]^-$	630.4135	359
10	PI(14:1/14:1)	$[M+Na]^+$	773.4211	$[M-H]^-$	749.4241	223
11	TG(14:1/14:1/14:1)	$[M+NH_4]^+$	734.6299	n.a.	n.a.	157
12	SM(d18:1/12:0)	$[M+H]^+$	647.5123	$[M+CH_3COO]^-$	705.5188	436
13	Cer(d18:1/12:0)	$[M+H]^+$	482.4568	$[M-H]^-$	480.4422	36
14	LPC(24:0)	$[M+H]^+$	608.4650	$[M+CH_3COO]^-$	666.4716	461
15	DG(17:1/17:1)	$[M+NH_4]^+$	610.5411	n.a.	n.a.	34
16	PC(17:0/17:0)	$[M+H]^+$	762.6008	$[M+CH_3COO]^-$	820.6073	402
17	PS(17:0/17:0)	[M+H]+	764.5436	$[M-H]^-$	762.5291	353
18	PG(17:0/17:0)	$[M+Na]^+$	773.5303	$[M-H]^-$	749.5338	123
19	PE(17:0/17:0)	$[M+H]^+$	720.5538	$[M-H]^-$	718.5392	331
20	PI(16:0/18:1)	$[M+Na]^+$	859.5307	$[M-H]^-$	835.5342	223

11. In this step, accurate mass match is required. If you want to skip the "RT match" and/or "CCS match," please uncheck the "RT match" and/or "CCS match." If the "RT match" and/or "CCS match" is unchecked, the inputted parameters would be invalid.

12. If you want to skip the "MS/MS match," please uncheck the "MS/MS match." Also, if uncheck the "MS/MS match," the inputted parameters would be invalid.

13. The sum of three weights should be equal to 1. If any match is unchecked, the corresponding score weight should be 0. For example, if the RT match is unchecked, the RT score weight should be 0. If no match was performed, please set all weights to 0. Then, click the "Submit."

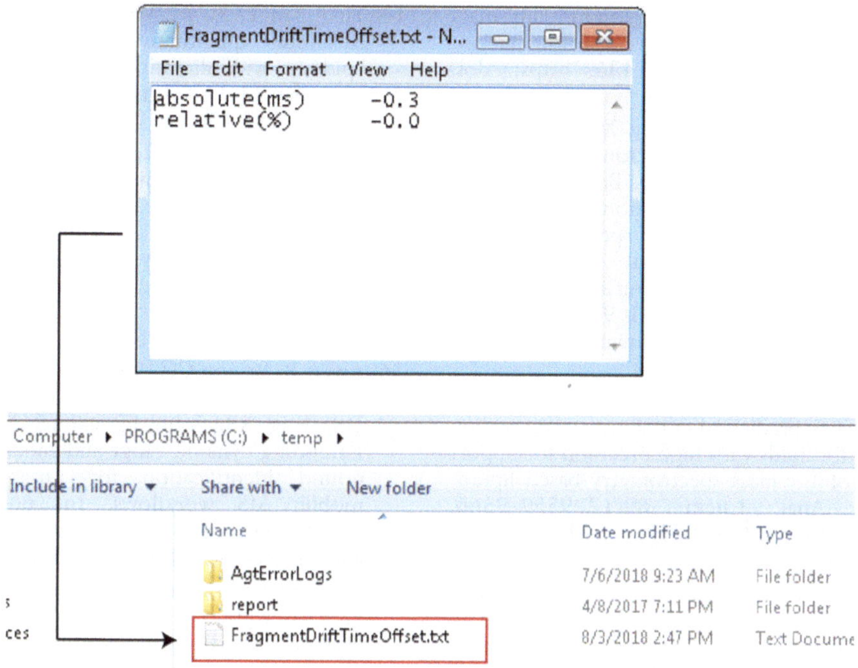

Fig. 3 The addition of a "FragmentDriftTimeOffset.txt" file in "C:\temp" folder

Acknowledgments

The work has been supported by National Key R&D Program of China (2018YFA0800902), National Natural Science Foundation of China (Grants 21575151, 31971356) and Chinese Academy of Sciences Major Facility-based Open Research Program. Z.-J. Z. is supported by Thousand Youth Talents Program.

References

1. Wenk MR (2010) Lipidomics: new tools and applications. Cell 143(6):888–895. https://doi.org/10.1016/j.cell.2010.11.033
2. Han X (2016) Lipidomics for studying metabolism. Nat Rev Endocrinol 12(11):668–679. https://doi.org/10.1038/nrendo.2016.98
3. Harayama T, Riezman H (2018) Understanding the diversity of membrane lipid composition. Nat Rev Mol Cell Biol 19(5):281–296. https://doi.org/10.1038/nrm.2017.138
4. Rohrig F, Schulze A (2016) The multifaceted roles of fatty acid synthesis in cancer. Nat Rev Cancer 16(11):732–749. https://doi.org/10.1038/nrc.2016.89
5. Hinz C, Liggi S, Griffin JL (2018) The potential of ion mobility mass spectrometry for high-throughput and high-resolution lipidomics. Curr Opin Chem Biol 42:42–50. https://doi.org/10.1016/j.cbpa.2017.10.018
6. Paglia G, Kliman M, Claude E, Geromanos S, Astarita G (2015) Applications of ion-mobility mass spectrometry for lipid analysis. Anal Bioanal Chem 407(17):4995–5007. https://doi.org/10.1007/s00216-015-8664-8
7. Zheng X, Smith RD, Baker ES (2018) Recent advances in lipid separations and structural elucidation using mass spectrometry combined with ion mobility spectrometry, ion-molecule reactions and fragmentation approaches. Curr Opin Chem Biol 42:111–118. https://doi.org/10.1016/j.cbpa.2017.11.009

8. May JC, McLean JA (2015) Ion mobility-mass spectrometry: time-dispersive instrumentation. Anal Chem 87(3):1422–1436. https://doi.org/10.1021/ac504720m
9. Zheng X, Wojcik R, Zhang X, Ibrahim YM, Burnum-Johnson KE, Orton DJ, Monroe ME, Moore RJ, Smith RD, Baker ES (2017) Coupling front-end separations, ion mobility spectrometry, and mass spectrometry for enhanced multidimensional biological and environmental analyses. Annu Rev Anal Chem (Palo Alto, Calif) 10(1):71–92. https://doi.org/10.1146/annurev-anchem-061516-045212
10. Zhou Z, Tu J, Xiong X, Shen X, Zhu ZJ (2017) LipidCCS: prediction of collision cross-section values for lipids with high precision to support ion mobility-mass spectrometry-based lipidomics. Anal Chem 89(17):9559–9566. https://doi.org/10.1021/acs.analchem.7b02625
11. Paglia G, Angel P, Williams JP, Richardson K, Olivos HJ, Thompson JW, Menikarachchi L, Lai S, Walsh C, Moseley A, Plumb RS, Grant DF, Palsson BO, Langridge J, Geromanos S, Astarita G (2015) Ion mobility-derived collision cross section as an additional measure for lipid fingerprinting and identification. Anal Chem 87(2):1137–1144. https://doi.org/10.1021/ac503715v
12. Zhou Z, Shen X, Chen X, Tu J, Xiong X, Zhu ZJ (2019) LipidIMMS Analyzer: integrating multi-dimensional information to support lipid identification in ion mobility-mass spectrometry based lipidomics. Bioinformatics 35(4):698–700. https://doi.org/10.1093/bioinformatics/bty661
13. Paglia G, Astarita G (2017) Metabolomics and lipidomics using traveling-wave ion mobility mass spectrometry. Nat Protoc 12(4):797–813. https://doi.org/10.1038/nprot.2017.013
14. Tu J, Yin Y, Xu M, Wang R, Zhu ZJ (2018) Absolute quantitative lipidomics reveals lipidome-wide alterations in aging brain. Metabolomics 14(1):5. https://doi.org/10.1007/s11306-017-1304-x
15. Stow SM, Causon T, Zheng X, Kurulugama RT, Mairinger T, May JC, Rennie EE, Baker ES, Smith RD, McLean JA, Stephan H, John CF (2017) An interlaboratory evaluation of drift tube ion mobility–mass spectrometry collision cross section measurements. Anal Chem 89(17):9048–9055. https://doi.org/10.1021/acs.analchem.7b01729
16. Aivett B, Bryson CG, Joon YL, Kent JB, Jennifer EK, Matthew EM, Thomas OM, Richard DS, Erin SB, John F, Samuel HP (2018) Extending dynamic range and enhancing compound identification for untargeted ion mobility-MS workflows. In: 66th annual ASMS conference on mass spectrometry and allied topics, San Diego, CA, June 2–6, 2018
17. http://www.nonlinear.com/progenesis/qi/v2.3/user-guide/, Nonlinear Dynamics, Waters
18. Hartler J, Triebl A, Ziegl A, Trotzmuller M, Rechberger GN, Zeleznik OA, Zierler KA, Torta F, Cazenave-Gassiot A, Wenk MR, Fauland A, Wheelock CE, Armando AM, Quehenberger O, Zhang Q, Wakelam MJO, Haemmerle G, Spener F, Kofeler HC, Thallinger GG (2017) Deciphering lipid structures based on platform-independent decision rules. Nat Methods 14(12):1171–1174. https://doi.org/10.1038/nmeth.4470
19. Hines KM, May JC, McLean JA, Xu L (2016) Evaluation of collision cross section calibrants for structural analysis of lipids by traveling wave ion mobility-mass spectrometry. Anal Chem 88(14):7329–7336. https://doi.org/10.1021/acs.analchem.6b01728

Chapter 18

MolFind^2: A Protocol for Acquiring and Integrating MS^3 Data to Improve In Silico Chemical Structure Elucidation for Metabolomics

Milinda A. Samaraweera, Dennis W. Hill, and David F. Grant

Abstract

Structure elucidation of metabolites (<1000 Da) in biofluids is extremely challenging due to the diversity and complexity of chemical structure space. Generally, due to lack of reference tandem mass data (MS^2), in silico fragmenters are used to rank candidates acquired from chemical databases as a function on how well they explain an experimental collision-induced dissociation spectrum. However, multistage fragmentation data (i.e., MS^3) have not been adequately utilized in current metabolomics structure elucidation pipelines. To address this shortcoming, here we describe an experimental (nontargeted direct infusion ion mobility-mass spectrometry-based) and computational workflow to acquire and utilize multistage mass (MS^3) spectrometry data for database-assisted structure elucidation.

Key words Multistage mass spectrometry, Ion mobility, Direct infusion, Metabolomics, Structure elucidation, Small molecules

1 Introduction

Metabolomics is the study of small molecules (<1000 Da) involved in biochemical processes that take place in cells of living organisms [1]. Nontargeted metabolomics is the unbiased investigation of all the metabolites contained in cells, body fluids, and tissues of an organism (the metabolome). As the composition of the metabolome is sensitive to perturbations of the genome, transcriptome, proteome, epigenome, microbiome, and exposome [2]; metabolomics is routinely used in a wide range of biomedical applications such as biomarker discovery, customized drug treatments, and for understanding disease mechanisms [3–5].

In nontargeted metabolomics, identification and quantification of metabolites still remains a major limitation despite the recent increase in popularity and the number of applications of the method [6]. Generally, investigators rely on analytical liquid

Giuseppe Paglia and Giuseppe Astarita (eds.), *Ion Mobility-Mass Spectrometry: Methods and Protocols*, Methods in Molecular Biology, vol. 2084, https://doi.org/10.1007/978-1-0716-0030-6_18,

chromatography with electrospray ionization mass spectrometry (LC-ESI-MS) for detecting unknowns [3]. LC-ESI-MS facilitates separation and detection of compounds with a wide range of polarities and thermolabilities in the native state [7]. Moreover, because of recent developments in small particle diameter packing material with high-peak resolution and LC systems capable of operating at ultra-high pressures (UPLC), UPLC-ESI-MS has grown in popularity among metabolomics investigators for separation, identification, and quantification of biomolecules in complex biological samples [8].

Metabolomics investigations typically require extensive study design and examination of multibatch samples followed by a variety of potential statistical analyses [9]. As UPLC/ESI/MS^2-based screening methods are generally slow and susceptible to batch-to-batch variation in experimental conditions (i.e., stationary phase deterioration, changes in concentration of mobile phase additives), the consistency and reproducibility of the data generated can deteriorate over time [9, 10]. Even though this could be minimized using quality control (QC) and statistical methods, due to the wide range of matrices and types of compounds (polarity, charge, elemental composition) in biological samples (i.e., urine/blood) [11], the absence of a one-size-fits-all (i.e., reverse phase, normal phase) LC/MS method remains a key disadvantage [12].

Direct infusion mass spectrometry (DIMS) based on ESI ionization and high-resolution time-of-flight (TOF) and Fourier transform ion cyclotron resonance (FT-ICR) MS have emerged as viable substitutes for nontargeted LC/MS-based profiling and are slowly gaining in popularity among metabolomics researchers [9, 13, 14]. In DIMS, complex samples (i.e., urine) are directly introduced to the mass spectrometer, and the sum of the *m/z* signal intensities are used for statistical analysis to investigate sample to sample variation of analytes [15, 16]. Advantages of DIMS include minimizing the requirement for chromatography of complex mixtures, facilitating simultaneous acquisition of all the metabolites contained in the sample (usually less than 5 min), quantifying metabolite concentrations merely by direct comparison of signal intensities among closely related (matrix) samples and acquisition of MS^2 fragmentation data using data-dependent acquisition (DDA) for structure elucidation [17]. To illustrate, in a recent study by Holčapek et al. compared DIMS vs. several other LC/MS methods in a nontargeted lipidomic analysis. They determined that, in terms of comprehensiveness and validity of the results, DIMS provided the most inclusive results, identifying several lipid classes that had not been detected by any of the other LC/MS techniques tested [18].

Nevertheless, two of the key drawbacks of the DIMS approach (separation and detection of isobaric compounds and ion suppression due to competitive ionization of the components in the

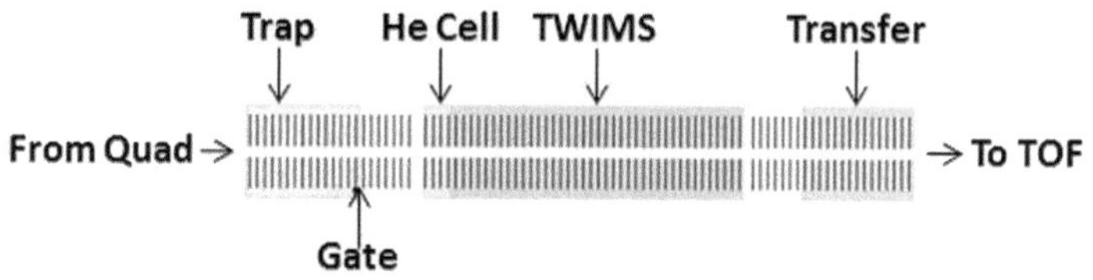

Fig. 1 Major components of the Waters Synapt G2-S*i* TRIWAVE (not to scale) region: Trap collision cell, He cell, TWIMS cell, and Transfer collision cell

matrix) have been recently alleviated by the incorporation of ion mobility spectrometry (IMS) and utilization of chip-based nanospray emitters [19, 20]. IMS, specifically, adds an additional dimension of separation, thus increasing the number of detectable compounds in a mixture [21, 22]. In IMS, ions are separated according to their size, shape, and charge as they drift through an inert gas (i.e., He, N_2) under the influence of an electric field. Currently, four types of IMS devices exist: differential, drift time, aspiration, and traveling wave [23]. The functioning principles of these methods are discussed elsewhere [24].

The Synapt G2-S*i* (Synapt) instrument is one of the latest versions (second generation) of IMS instruments based on traveling wave (TW) technology (TWIMS) [25, 26]. The IMS region of the system (Fig. 1) consists of four TW cells named: Trap, He, TWIMS, and Transfer cells. The heart of the Synapt IMS system is the TWIMS cell. It operates based on nonuniform, dynamic electric pulses to drive ions against a neutral buffer gas, maintained at pressures of 1.0–3.0 mbar. As the ions travel (under the influence of the electric field), they interact/collide with the buffer gas and slow down depending on the structure, charge, and mass; hence mobility separation occurs. The Trap cell accumulates and release ion packets into the TWIMS cell while at the same time functions as a collision cell. Similarly, the Transfer cell transports the mobility-separated ion packets to the TOF mass analyzer, and also can function as a collision cell. The He cell is used to slow down the ions prior to injection into the TWIMS cell, preventing any further fragmentation of the precursor ions [27].

In generic mass spectrometers, analytes are first ionized (i.e., ESI) and then sorted based on their mass-to-charge ratio by a mass analyzer (i.e., TOF, FT-ICR) before finally generating a mass spectrum based on intensity vs. m/z (mass-to-charge ratio). Over the past half-century, tandem mass spectrometry (MS^2), a technique used within mass spectrometry, has become a popular approach for chemical structure elucidation [28]. In detail, in MS^2, first, a precursor ion (based on m/z) of interest is selected; then, the selected ion is fragmented to produce primary fragment ions by colliding it with a neutral gas (i.e., CID). Finally a MS^2 spectrum is generated which provides important aspects of the chemical structure of the selected precursor ion.

However, the unique design of the Synapt, having one TW collision cell before and one TW collision cell after the TWIMS cell, enables acquisition of pseudo-MS^3 spectra of primary fragment ions generated in the Trap cell; this is referred to as time aligned parallel (TAP) fragmentation [27]. In a TAP experiment, a precursor ion with a predetermined *m/z* is first selected by using a quadrupole mass filter; then, the selected precursor is subjected to CID fragmentation in the Trap cell generating primary fragment ions; next, a gate at the end of the Trap cell (Fig. 1) collects and releases packets of ions (a mixture of precursor and primary fragment ions) to the TWIMS cell. The injected ion mixture is then separated in the IMS cell and passed onto the Transfer cell, where they are further fragmented to produce secondary fragment ions. Finally, association of secondary fragment ions to specific drift times of primary fragment ions is carried out to produce a pseudo-MS^3 spectrum. The total cycle span for a TAP fragmentation experiment is 10.75 ms or 200 drift-time bins; thus, each drift bin is equivalent to 53.75 μs.

In typical metabolomics experiments, MS^2 CID spectra are used to identify the chemical structures of unknown metabolites by comparing the experimental spectra against reference spectra available on mass spectral databases (i.e., MassBank, NIST, Metlin) [29–31]. However, it has been estimated that, on average, only 1.8% of MS^2 spectra produced in untargeted metabolomics experiments can be successfully annotated to reference spectra, rendering the vast majority of mass spectral data collected useless (i.e., dark matter) [32]. To overcome this dilemma, in the last decade, a plethora of computational fragmentation prediction software (i.e., MetFrag, CFM:ID, CSI:FingerId) have been developed to compare and rank predicted CID spectra (based on the candidate chemical structure) against the experimental data generated [32, 33]. The advantages and disadvantages of these various software packages have been extensively discussed elsewhere [29, 34, 35].

Even though MS^2 data can be useful for characterization of a molecular structure, multistage fragmentation (specially MS^3) generates fragments of fragment ions which provide a wealth of information that is vital to identify the substructures and ultimately the overall structure of an unknown [36]. Apart from the in silico fragmenter MAGMa [36], little work has been done to develop experimental protocols and to devise computational workflows to use MS^3 data in nontargeted metabolomics.

To address this shortcoming, here we describe a TAP-based experimental workflow to acquire MS^3 data and a computational workflow to utilize the generated MS^3 data for identification of substructures and ultimately improving the performance of in silico fragmenters.

2 Materials

Always use analytical grade reagents mixed in ultra-pure solvents (LCMS grade or 18 MΩ water) to prepare reference standards. Use disposable glassware and labware to limit cross contamination. Some chemicals maybe toxic; use with caution.

2.1 TWIMS Calibration Mix

1. Prepare 10 mL of 5 μmol/mL stock solutions of quaternary ammonium standards (tetramethylammonium bromide, tetraethylammonium bromide, tetrapropylammonium bromide, tetrabutylammonium bromide, tetrapentylammonium bromide, tetrahexylammonium bromide, tetraheptylammonium bromide, tetraoctylammonium bromide, tetrakis(decyl)ammonium bromide, and tetradodecylammonium bromide) in methanol.
2. Prepare the Traveling Wave Ion Mobility Spectrometry (TWIMS) calibration mixture by adding the volume of each quaternary ammonium stock solution indicated in Table 1 to a 25 mL Nalgene bottle and diluting with 20 mL of 0.1% formic acid in H_2O:Acetonitrile (1:1) (*v*:*v*).

2.2 Test Analyte Mixture

1. Add an appropriate amount of the test analyte mixture to a suitable solvent in a Nalgene bottle. Use this mixture for direct infusion analysis. Make sure the detector responses of the analytes of interest are within the instruments' Linear Dynamic Range (LDR).

Table 1
Volume required from each of the quaternary ammonium stock solution to prepare the TWIMS calibration mix

Compound	Monoisotpic molecular weight	Volume from stock solution (μL)	Concentration in the calibration mix (nmol/mL)
Tetramethylammonium ion	74.0964	4980.00	996.00
Tetraethylammonium ion	130.1590	10.00	2.00
Tetrapropylammonium ion	186.2216	2.50	0.50
Tetrabutylammonium ion	242.2842	2.50	0.50
Tetrapentylammonium ion	298.3468	2.50	0.50
Tetrahexylammonium ion	354.4094	2.50	0.50
Tetraheptylammonium ion	410.4720	2.50	0.50
Tetraoctylammonium ion	466.5350	2.50	0.50
Tetrakis(decyl)ammonium ion	578.6599	5.00	1.00
Tetradodecylammonium ion	690.7850	10.00	2.00

3 Methods

Use ultra-high Pure (UHP)-grade N_2 as the collision gas to limit fragmentation of labile precursor ions at low collision energies. Perform a mass calibration in the range 20–1000 m/z. Turn on the Mobility mode and wait at least 1 h for the instrument to equilibrate before optimization, calibration, and data acquisition.

3.1 Optimization of the Synapt System Parameters

1. Connect the Nalgene bottle containing the TWIMS reference mixture to the Synapt built-in syringe system and start infusing at a flow rate of 5 μL/min.
2. Set the following ESI+ source conditions and optimize if needed:
   - Polarity: ES+.
   - Capillary (kV): 2.0.
   - Source temperature (°C): 100.
   - Sampling cone (V): 40.
   - Source offset (V): 80.
   - Source gas flow (mL/min): 0.
   - Desolvation temperature (°C): 200.
   - Cone gas flow (L/h): 0.
   - Desolvation gas flow (L/h): 800.
   - Nebuliser gas flow (Bar): 6.5.
   - LM resolution: 16.
   - HM resolution: 16.
   - Trap CE (V): 2.
   - Transfer CE (V): 4.
   - Trap gas (Nitrogen) flow (mL/min): 2.00.
   - Helium cell gas flow (mL/min): 180.00.
   - IMS Gas (Nitrogen) Flow (mL/min): 90.00.
   - Start mass: 20.000.
   - End mass: 1000.000.
3. Use the default lock mass calibrant, Leucine Enkephalin, as the internal standard to correct for any drift in the mass axis. Use the following source conditions for the acquisition of the lock mass spectra:
   - LockSpray Infusion Flow Rate (μL/min): 5.00.
   - LockSpray Capillary (kV): 3.0.
4. Adjust the Trap Bias to 38.0 V to limit fragmentation of labile compounds.

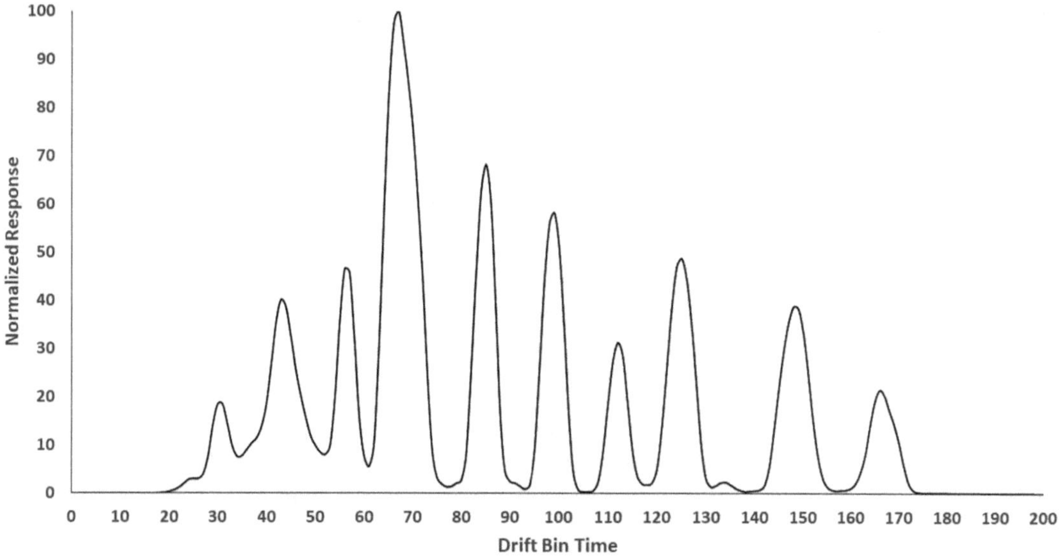

Fig. 2 TWIMS-MS mobilogram after optimization

5. Use the parameters given below as a starting point to optimize the TWIMS drift cell. The goal is to adjust the traveling wave velocities and heights, such that the peak corresponding to tetramethylammonium elutes around 20 drift bin units and the peak corresponding to tetradodecylammonium elutes near 180 drift bin units in the mobilogram (Fig. 2).
   - Variable wave height enabled: FALSE.
   - Variable wave velocity enabled: TRUE.
   - Wave velocity ramp type: Linear.
   - Wave velocity start (m/s): 2000.
   - Wave velocity end (m/s): 190.
   - Wave velocity using full IMS: TRUE.
   - Wave velocity ramp (%): 100.
   - IMSRFOffset: 180.
   - Variable wave height enabled: FALSE.
   - IMS wave height (V): 33.
   - Drift-time bins: 0.
   - Mobility trapping release time (μs): 500.
   - Mobility trap height (V): 15.
   - Mobility extract height (V): 0.
   - Trag Gate LUT table enabled: FALSE.
   - Using drift-time trimming: TRUE.
   - Using mobility delay after trap release: TRUE.
   - IMS wave delay (μs): 1000.

3.2 Traveling Wave IM Collision Cross Section Calibration

In order to obtain the collision cross section (CCS) value of the test compounds, a calibration must be carried out using reference standards with known CCS values (i.e., CCS measured in a drift-tube instrument using Nitrogen as the drift gas). The following procedure describes how to achieve this task.

1. Set the instrument parameters to values optimized in Subheading 3.1.
2. Create a CCS calibration reference file using the CCS values provided in Table 2.
3. Use the IntelliStart software in Masslynx v4.1 (Waters) to perform the CCS calibration.
4. After the calibration report has been generated by MassLynx, make sure all ten reference compounds have been detected, and the root mean square error is less than 5%.

3.3 Time-Aligned Parallel (TAP) Fragmentation

Seven TAP experiments (using seven MS method files in MassLynx v4.1) must be carried out to isolate each of the test compounds to generate the primary and subsequent secondary fragment ions required for the structure elucidation (*see* **Note 1**). An additional experiment must be carried out to obtain the CCS values. The

Table 2
CCS values of quaternary ammonium standards used for the calibration

Compound	*m/z*	CCS ($Å^2$)
Tetramethylammonium ion	74.0964	107.4[a]
Tetraethylammonium ion	130.1590	122.2[a]
Tetrapropylammonium ion	186.2216	143.3[b]
Tetrabutylammonium ion	242.2842	167.7[b]
Tetrapentylammonium ion	298.3468	191.0[b]
Tetrahexylammonium ion	354.4094	212.8[b]
Tetraheptylammonium ion	410.4720	236.1[b]
Tetraoctylammonium ion	466.5350	256.9[b]
Tetrakis(decyl)ammonium ion	578.6599	291.9[b]
Tetradodecylammonium ion	690.7850	320.2[b]

[a]Campuzano, I.D.G.; Bush, M.F.; Robinson, C.V.; Beaumont, C.; Richardson, K.; Kim, H.; Kim, H.I. Structural characterization of drug-like compounds by ion mobility mass spectrometry: comparison of theoretical and experimentally derived nitrogen collision cross-sections. Anal. Chem. 2012, 84, 1026–1033 (*see* Table 1 and Supporting Information Tables S1 and S2)

[b]Hines, Kelly M., et al. Evaluation of collision cross section calibrants for structural analysis of lipids by traveling wave ion mobility-mass spectrometry. Anal. Chem. 88, 14 (2016): 7329–7336 (see supporting information S9)

following section describes how to setup the MassLynx MS method files for these experiments.

1. Set the instrument parameters to the values optimized in Subheading 3.1.
2. Create seven separate MS method files using MassLynx software having trap collision energies sector ramped at: 2, 2–10, 10–20, 20–30, 30–40, 40–50, and 50–60 V, respectively. Set a scan rate of 1.0 s and data acquisition time of 3.0 min. Set the *m/z* range to 20–1000. Set the *m/z* of the mass of the precursor ion to be isolated in the quadrupole.
3. In each TAP method file, add 13 mobility functions to generate secondary fragment ions by ramping the transfer cell collision energies (CE) at 0–2, 2–4, 4–6, 6–8, 8–10, 10–12, 12–14, 14–16, 16–25, 16–35, 16–45, 16–55, and 16–65 V at fixed-trap CE.
4. Create an addition MS method file, having a single mobility function, setting the trap, and transfer CE at 2 and 4 V, respectively. Set data acquisition time to 1 min. The rest of the parameters are identical to the ones used in the previous steps. CCS values can be calculated using the DriftScope v2.8 (Waters) software.

3.4 Spectral Processing

The following section describes the steps to compile the CID spectra at MS^2 and MS^3 levels using MassLynx v4.1 (Waters) and DriftScope v2.8 (Waters) software.

1. Using the MassLynx v4.1 (Waters) software, generate the CID spectra of the isolated precursor ions in each of the seven TAP experiments (trap at 2, 2–10, etc.) at a fixed transfer cell ramp of 0–2 V (to minimize generation of secondary fragment ions).
2. Calculate the survival yield (% of the precursor ion intensity divided by the total ion current) of the precursor ion in each of the generated CID spectra.
3. Average the generated CID spectra that have survival yields (SY) of the precursor ion greater than zero using the approach describe by Wolf et al. [34] This represents a MS^2 spectra that is generally used for structure elucidation purpose.
4. Select the TAP experiment where the precursor ion survival yield is closest to 50% (SY50).
5. Use the DriftScope v2.8 (Waters) software to calculate the drift-time bin value of each of the primary ions observed (SY50).
6. Create a list of primary fragment ions (having *m/z*, intensity, and drift-time bin values) excluding any ions with drift-time bin differences less than or equal to ±2 drift-time bin units.

7. Next, obtain m/z, intensity drift-time bin values of ions generated at each of the transfer CE values (i.e., 2–4, 4–6, ..., 15–65 V) in the TAP experiment selected in **step 4**.
8. Assign the secondary fragment ions to primary fragment ions (with drift-time bin values within ±1 drift-time bin of the primary ion) to construct the pseudo MS^3 spectrum of a primary fragment ion at each transfer cell CE ramp.
9. Finally, using the same approach described in **step 3**, construct an averaged MS^3 spectra for each of the primary fragment ions from **step 6** where the SY of the primary fragment ion is greater than zero.

3.5 Acquisition of Candidate Structures from the PubChem Database and Ranking

The following section describes acquisition and ranking of candidate compounds mined from PubChem database and processing using in silico fragmentation prediction software MassFrontier (Thermo-fisher) and MetFrag [34, 37]. The PubChem database was selected to acquire candidate sets as it is the largest (approximately 90 million compounds) publicly available, general purpose compound database (https://pubchemdocs.ncbi.nlm.nih.gov/rdf-statistics).

1. Download candidate structures from the PubChem database that match the experimental mono isotopic molecular weight (MIMW) of the precursor ion within a relative mass error of ±15 ppm (*see* **Note 2**).
2. Eliminate compounds in candidate sets that contain salts, disconnections, heavy isotopes, an overall charge, only carbon and hydrogen, or are duplicate stereoisomers.
3. Using the batch mode of the software MassFrontier (version 7.0.5.9), generate candidate structures for the primary fragment ions using the following parameters for each of the candidate structures: consider both database and general fragmentation rules to generate fragment ions, ionization method = protonation, maximum number of reaction steps = 5, reaction limit = 10,000, and mass range = 30–1000 Da.
4. Download the command line version of MetFrag (from http://c-ruttkies.github.io/MetFrag/) to rank the candidate set using the experimental MS^2 and MS^3 generated in Subheading 3.4. Set the relative and absolute mass errors at 15.0 ppm and 0.01 Da to annotate fragments against peaks in the respective CID spectra. Keep the rest of the parameters as default.
5. The averaged CID spectra from **step 3** in Subheading 3.4 can be used as a pseudo MS^2 spectrum (as little or no secondary fragment ions are generated in the transfer region) to generate a score/number of matches for each of the candidates using the program MetFrag. Save the output of the MetFrag as a .csv file.

6. For each of the primary fragment ions with MS^3 data available from **step 9** in Subheading 3.4, use MetFrag to rank putative fragment ions generated by MassFrontier in **step 3**. Save the output of the MetFrag as a csv file.
7. Using the following formula, calculate an overall score ($MolFnd^2score$) and rank candidates based on highest value to the lowest. Candidates ranked at the top have a higher probability of being similar to the structure of the unknown compound.

$$\text{MolFnd}^2\text{score} = \text{ScoreMS}^2 + \sum_{i=1}^{\substack{\text{number of primary}\\ \text{fragment ions}\\ \text{with MS}^3\text{ data}}} \text{ScoreMS}^3$$

where $ScoreMS^2$ is the number of matches found by MetFrag to the peaks in the MS^2 CID spectrum of the candidate structure of interest. $ScoreMS^3$ is the number of matches found by MetFrag in the MS^3 CID spectrum using the primary fragment ion structure generated (within <15 ppm or <0.01 Da of the m/z of the primary fragment ion with MS^3 data) by MassFrontier (by fragmenting the candidate structure). If multiple primary fragment structures are generated by MassFrontier, use the structure with maximum number of matches to the MS^3 spectrum.

4 Notes

1. The resolution of the Synapt TWIM system is approximately 40 (CCS/ΔCCS); thus, it is challenging to separate primary fragment ions with similar CCS values.
2. If the unknown compound is not contained in the PubChem database, it will not be correctly identified.

References

1. Fiehn O, Robertson D, Griffin J, van der Werf M, Nikolau B, Morrison N, Sumner LW, Goodacre R, Hardy NW, Taylor C, Fostel J, Kristal B, Kaddurah-Daouk R, Mendes P, van Ommen B, Lindon JC, Sansone S-A (2007) The metabolomics standards initiative (MSI). Metabolomics 3(3):175–178
2. Lu Y-F, Goldstein DB, Angrist M, Cavalleri G (2014) Personalized medicine and human genetic diversity. Cold Spring Harb Perspect Med 4(9):a008581
3. Fuhrer T, Zamboni N (2015) High-throughput discovery metabolomics. Curr Opin Biotechnol 31:73–78
4. Hamburg MA, Collins FS (2010) The path to personalized medicine. N Engl J Med 363:301–304
5. Wishart DS (2016) Emerging applications of metabolomics in drug discovery and precision medicine. Nat Rev Drug Discov 15 (7):473–484

6. Trivedi DK, Hollywood KA, Goodacre R (2017) Metabolomics for the masses: the future of metabolomics in a personalized world. New Horiz Transl Med 3(6):294–305
7. Soga T, Ohashi Y, Ueno Y, Naraoka H, Tomita M, Nishioka T (2003) Quantitative metabolome analysis using capillary electrophoresis mass spectrometry. J Proteome Res 2 (5):488–494
8. Want EJ, Wilson ID, Gika H, Theodoridis G, Plumb RS, Shockcor J, Holmes E, Nicholson JK (2010) Global metabolic profiling procedures for urine using UPLC–MS. Nat Protoc 5(6):1005–1018
9. Kirwan JA, Weber RJ, Broadhurst DI, Viant MR (2014) Direct infusion mass spectrometry metabolomics dataset: a benchmark for data processing and quality control. Sci Data 1:140012
10. Sangster TP, Wingate JE, Burton L, Teichert F, Wilson ID (2007) Investigation of analytical variation in metabonomic analysis using liquid chromatography/mass spectrometry. Rapid Commun Mass Spectrom 21(18):2965–2970
11. Dunn WB, Broadhurst D, Begley P, Zelena E, Francis-McIntyre S, Anderson N, Brown M, Knowles JD, Halsall A, Haselden JN (2011) Procedures for large-scale metabolic profiling of serum and plasma using gas chromatography and liquid chromatography coupled to mass spectrometry. Nat Protoc 6(7):1060–1083
12. Bouatra S, Aziat F, Mandal R, Guo AC, Wilson MR, Knox C, Bjorndahl TC, Krishnamurthy R, Saleem F, Liu P (2013) The human urine metabolome. PLoS One 8(9):e73076
13. Draper J, Lloyd AJ, Goodacre R, Beckmann M (2013) Flow infusion electrospray ionisation mass spectrometry for high throughput, non-targeted metabolite fingerprinting: a review. Metabolomics 9(1):4–29
14. Koulman A, Tapper BA, Fraser K, Cao M, Lane GA, Rasmussen S (2007) High-throughput direct-infusion ion trap mass spectrometry: a new method for metabolomics. Rapid Commun Mass Spectrom 21(3):421–428
15. Giavalisco P, Hummel J, Lisec J, Inostroza AC, Catchpole G, Willmitzer L (2008) High-resolution direct infusion-based mass spectrometry in combination with whole 13C metabolome isotope labeling allows unambiguous assignment of chemical sum formulas. Anal Chem 80(24):9417–9425
16. Castrillo JI, Hayes A, Mohammed S, Gaskell SJ, Oliver SG (2003) An optimized protocol for metabolome analysis in yeast using direct infusion electrospray mass spectrometry. Phytochemistry 62(6):929–937
17. Brown SC, Kruppa G, Dasseux JL (2005) Metabolomics applications of FT-ICR mass spectrometry. Mass Spectrom Rev 24 (2):223–231
18. Lísa M, Cífková E, Khalikova M, Ovčačíková M, Holčapek M (2017) Lipidomic analysis of biological samples: comparison of liquid chromatography, supercritical fluid chromatography and direct infusion mass spectrometry methods. J Chromatogr A 1525:96–108
19. Bedair M, Sumner LW (2008) Current and emerging mass-spectrometry technologies for metabolomics. TrAC Trends Anal Chem 27 (3):238–250
20. Paglia G, Astarita G (2017) Metabolomics and lipidomics using traveling-wave ion mobility mass spectrometry. Nat Protoc 12(4):797–813
21. Collins D, Lee M (2002) Developments in ion mobility spectrometry-mass spectrometry. Anal Bioanal Chem 372(1):66–73
22. Dwivedi P, Wu C, Matz LM, Clowers BH, Siems WF, Hill HH (2006) Gas-phase chiral separations by ion mobility spectrometry. Anal Chem 78(24):8200–8206
23. Eiceman GA (1991) Advances in ion mobility spectrometry: 1980–1990. Crit Rev Anal Chem 22(1–2):471–490
24. Cumeras R, Figueras E, Davis C, Baumbach JI, Gracia I (2015) Review on ion mobility spectrometry. Part 1: Current instrumentation. Analyst 140(5):1376–1390
25. Shvartsburg AA, Smith RD (2008) Fundamentals of traveling wave ion mobility spectrometry. Anal Chem 80(24):9689–9699
26. Giles K, Pringle SD, Worthington KR, Little D, Wildgoose JL, Bateman RH (2004) Applications of a travelling wave-based radio frequency-only stacked ring ion guide. Rapid Commun Mass Spectrom 18(20):2401–2414
27. Merenbloom SI, Flick TG, Williams ER (2012) How hot are your ions in TWAVE ion mobility spectrometry? J Am Soc Mass Spectrom 23 (3):553–562
28. McLafferty FW (1981) Tandem mass spectrometry. Science 214:280–287
29. Allen F, Greiner R, Wishart D (2015) Competitive fragmentation modeling of ESI-MS/MS spectra for putative metabolite identification. Metabolomics 11(1):98–110
30. Horai H, Arita M, Kanaya S, Nihei Y, Ikeda T, Suwa K, Ojima Y, Tanaka K, Tanaka S, Aoshima K (2010) MassBank: a public repository for sharing mass spectral data for life sciences. J Mass Spectrom 45(7):703–714
31. Smith CA, O'Maille G, Want EJ, Qin C, Trauger SA, Brandon TR, Custodio DE, Abagyan R, Siuzdak G (2005) METLIN: a

metabolite mass spectral database. Ther Drug Monit 27(6):747–751
32. da Silva RR, Dorrestein PC, Quinn RA (2015) Illuminating the dark matter in metabolomics. Proc Natl Acad Sci U S A 112 (41):12549–12550
33. Blaženović I, Kind T, Torbašinović H, Obrenović S, Mehta SS, Tsugawa H, Wermuth T, Schauer N, Jahn M, Biedendieck R (2017) Comprehensive comparison of in silico MS/MS fragmentation tools of the CASMI contest: database boosting is needed to achieve 93% accuracy. J Chem 9(1):32
34. Wolf S, Schmidt S, Müller-Hannemann M, Neumann S (2010) In silico fragmentation for computer assisted identification of metabolite mass spectra. BMC Bioinformatics 11(1):148
35. Dührkop K, Shen H, Meusel M, Rousu J, Böcker S (2015) Searching molecular structure databases with tandem mass spectra using CSI: FingerID. Proc Natl Acad Sci U S A 112 (41):12580–12585
36. Ridder L, van der Hooft JJ, Verhoeven S (2014) Automatic compound annotation from mass spectrometry data using MAGMa. Mass Spectrom (Tokyo) 3(Spec Iss 2):S0033
37. Bolton EE, Wang Y, Thiessen PA, Bryant SH (2008) PubChem: integrated platform of small molecules and biological activities. Annu Rep Comput Chem 4:217–241

Chapter 19

Collision Cross Section Calculations Using HPCCS

Gabriel Heerdt, Leandro Zanotto, Paulo C. T. Souza, Guido Araujo, and Munir S. Skaf

Abstract

A technical overview of the High Performance Collision Cross Section (HPCCS) software for accurate and efficient calculations of collision cross sections for molecular ions ranging from small organic molecules to large protein complexes is presented. The program uses helium or nitrogen as buffer gas with considerable gains in computer time compared to publicly available codes under the Trajectory Method approximation. HPCCS is freely available under the Academic Use License at https://github.com/cepid-cces/hpccs.

Key words Ion mobility, Mass spectrometry, Collision cross section, HPCCS, Trajectory method

1 Introduction

Biomolecules' function and properties are directly connected to their structures, and full knowledge about their three-dimensional spatial arrangement is required to properly comprehend their behavior. There are, currently, numerous biophysical techniques capable of providing insights into the conformational structure of biopolymers. Among them, X-ray crystallography and NMR spectroscopy are the most popular ones, but technical issues like crystallization and structure size often make these experiments a challenging endeavor [1–6].

In the last few years, Ion Mobility-Mass Spectrometry (IM-MS) has become a promising tool in the field of structural determination of large biomolecular systems. The IM-MS, based on the gas phase ion separation, is a fairly rapid experiment and requires no previous purification or crystallization of the target compound. The possibility to transfer biomolecules from solution into the gas phase by soft ionization methods such as electrospray ionization [7] offers an exciting opportunity to separate and study analytes by mass spectrometry and related techniques [8–10]. Accurate measurements of the time-of-flight of ionized analytes inside a

Giuseppe Paglia and Giuseppe Astarita (eds.), *Ion Mobility-Mass Spectrometry: Methods and Protocols*, Methods in Molecular Biology, vol. 2084, https://doi.org/10.1007/978-1-0716-0030-6_19,

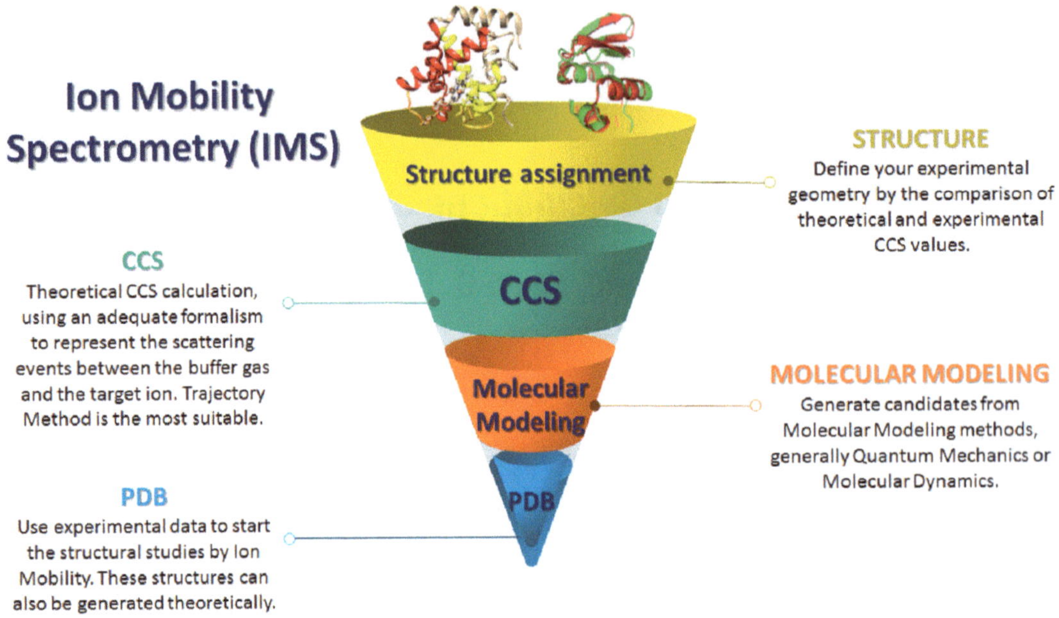

Fig. 1 Workflow for structure assignment from Ion Mobility-Mass Spectrometry experiments. These steps represent the theoretical efforts for data interpretation, considering you have experimental mobilities and/or CCS

gas filled drift tube can be obtained in an IM-MS experiment which is, in turn, closely related to the shape and spatial conformation of the analyte. Ion mobility-mass spectrometry-coupled experiments have, hence, emerged as a powerful aid to the study of molecular structures and conformations.

Preserving native-like geometry of noncovalent assemblies during IM-MS experiments requires very specific soft conditions that aim at minimizing structural perturbations. The analytes are typically electrosprayed from neutral aqueous solution and drift-tube voltages are controlled to avoid inadequate desolvation [11, 12] or collisional heating [13–15]. However, the major challenge to obtain structural information from IM-MS experiments is data interpretation. For any given geometry, the rotationally averaged collision cross section (CCS) can be accurately calculated, with an appropriate algorithm, for subsequent comparison with experimental values derived from time-of-flight or inverse mobility measurements (Fig. 1).

Typically, the experimental structure assignment from IM-MS studies follows these five steps:

1. Obtain experimental Collision Cross Sections (CCS), generally from ion mobilities values that can be converted into CCS;
2. Use experimental data, such as the three dimensional structures from the Protein Data Bank, as starting point for the target structure;

3. Apply Molecular Modeling methods to obtain candidate geometries for each molecular ion of interest. Quantum Chemistry and or Molecular Dynamics are commonly used to achieve that;
4. Compute CCS for the selected structures, with an appropriate treatment of the collisions between the molecular ion and the buffer gas;
5. Directly compare calculated CCSs with the experimental ones, defining the experimental structure from the stable candidates.

During the last two decades, Mobcal [16, 17] has been the most widely used program for theoretical CCS calculations. Developed in 1996, Mobcal is a sequential code for computing the rotationally averaged cross section from an input atomic coordinate file for the analyte at three different levels of theory for the ion-buffer gas collisions: The projection approximation (PA) [18], the exact hard sphere scattering (EHSS) [17, 19], and the trajectory method (TM) [17]. Since Mobcal, efforts have been dedicated to design new algorithms capable of calculating CCS for biomolecules more rapidly and computationally less costly. Both PA and EHSS methods allow CCS calculations for systems containing a large number of atoms, but results can be inaccurate [20, 21] by more than 30%, since, unlike TM, these methods neglect details about the interactions between the gas and molecular ion, as well as the roughness of the molecular ion surface.

The current parallel processing features available in modern computers enable the application of new algorithms and software for scientific areas. High-Performance Computing (HPC) can leverage on modern many core-computing clusters and parallelization algorithms to considerably improve the performance of parallel computational methods. HPC technology focuses on developing parallel processing algorithms and systems by incorporating both administration and parallel computational methods. Making use of these tools, the High Performance Collision Cross Section (HPCCS) software, published in 2018 by Zanotto et al. [22], represents an important contribution for IM-MS data analysis. HPCCS is capable of performing accurate and efficient CCS calculations for a large variety of molecular ions, ranging from small organic molecules to large protein complexes, containing tens to hundreds of thousands of atoms. Focused only on the Trajectory Method, HPCCS was improved with new features that enable application speedups using HPC. Large biomolecules can now be simulated with HPCCS within much shorter times and high accuracy, which helps the interpretation of experimental IM-MS data, allowing the experimental structural assignment.

In this chapter, we present the mains technical and methodological aspects associated with calculating CCS using the HPCCS code, available as free software under the Academic Use License at https://github.com/cepid-cces/hpccs.

2 Theoretical Background: The Trajectory Method

To obtain structural information from ion mobility measurements, CCSs can be calculated for conformers obtained from experiments or generated by molecular modeling methods and compared with experimental CCS results. The collision cross section gives an orientationally averaged result and the mobility, K, of a molecular ion in a low pressure buffer gas, which, within the linear response regime, can be calculated from [17, 23]:

$$K = \frac{\sqrt{18\pi}}{16}\left[\frac{1}{m} + \frac{1}{m_{\mathrm{B}}}\right]^{1/2} \frac{ze}{(k_{\mathrm{B}}T)^{1/2}} \frac{1}{\Omega_{\mathrm{avg}}^{(1,1)}} \frac{1}{N} \tag{1}$$

where m is the molecular ion mass, m_{B} and N are the buffer gas mass and density, respectively, ze is the ionic charge, and $\Omega_{\mathrm{avg}}^{(1,1)}$ is the orientationally averaged collision integral given by Eq. 2 below. Since the $\Omega_{\mathrm{avg}}^{(1,1)}$ integral is related to the scattering angles χ, depicted in Fig. 2, the orientationally averaged collision cross section can be accurately determined by integrating overall possible collision geometries [17, 23]:

$$\Omega_{\mathrm{avg}}^{(1,1)} = \frac{1}{8\pi^2}\int_0^{2\pi} d\theta \int_0^{\pi} d\phi \, \sin\phi \int_0^{2\pi} d\gamma \frac{\pi}{8}\left(\frac{\mu}{k_{\mathrm{B}}T}\right)^3 \int_0^{\infty} dg \, e^{-\frac{\mu g^2}{2k_{\mathrm{B}}T}} g^5 \int_0^{\infty} db 2b(1 - \cos\chi(\theta, \phi, \gamma, g, b)) \tag{2}$$

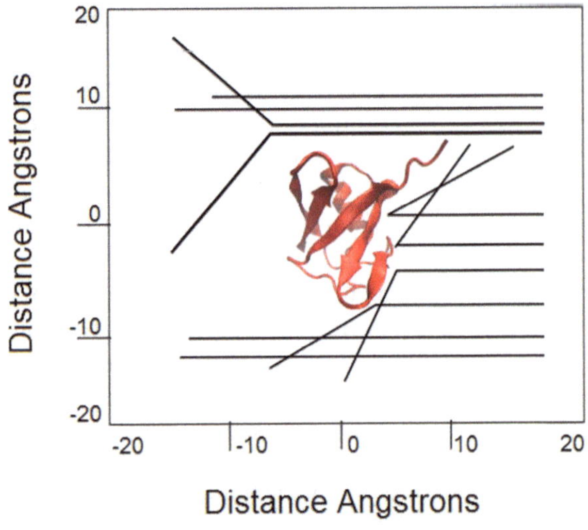

Fig. 2 The accurate estimation of CCS requires the calculation of all possible collision angles between a buffer gas and a target molecule, as illustrated here. (Figure adapted from ref. 22)

where μ is the reduced mass, g is the magnitude of the relative collision velocity, and b the impact parameter for all possible ion orientations defined by the three Euler angles θ, ϕ, and γ. Equation 2 can be calculated by numerical integration of the trajectories of buffer gas particles scattered off by the molecular ion propagated in an intermolecular interaction potential V [24]:

$$V(\theta,\phi,\gamma,b,r) = 4\sum_{i}^{n}\varepsilon_i\left[\left(\frac{\sigma_i}{r_i}\right)^{12} - \left(\frac{\sigma_i}{r_i}\right)^{6}\right] - \frac{\alpha}{2} \times \left[\left(\sum_i^n \frac{q_i x_i}{r_i^3}\right)^2 + \left(\sum_i^n \frac{q_i y_i}{r_i^3}\right)^2 + \left(\sum_i^n \frac{q_i z_i}{r_i^3}\right)^2\right] \quad (3)$$

The first term is a sum of Lennard-Jones 6–12 potential between the buffer gas and individual atom i of the molecular ion. The Lennard-Jones parameters, ε (depth) and σ (distance), are provided in the literature [16, 17]. The second term is the charge-induced dipole interaction with α being the gas polarizability, and x_i, y_i, z_i, and r_i, the relative position of each individual atom i of partial charge q_i. To properly compute the interactions between ions and molecular nitrogen (N_2) gas, these additional terms are required in Eq. 3: an ion–quadrupole interaction and the orientation of the linear N_2 molecule during the collision process [24]. This approach to the CCS calculation is called the Trajectory Method (TM) and provides an accurate prediction for the cross section of a given candidate geometry, although it is computationally expensive for sequential codes like Mobcal [17, 20, 21].

3 High Performance Collision Cross Section (HPCCS)

The CCS calculations for proteins and protein complexes under the TM approximation become computationally expensive given that at least 100,000 gas-target individual collision events are needed to accurately compute the CCS. Working with several candidate geometries for interpreting the experimental CCS, thousands of trajectories must be calculated. Thus, techniques to accelerate the computations, using the parallel technology available in current processors, become key to achieve accurate results in a timely fashion.

In the last decades, several features were added to the processors, like vector units and multicores, which enable programs to run tasks in parallel [25]. The HPCCS software is an efficient parallel code, which takes advantage of the current computing technology to allow users calculate CCS for large molecules, thus providing an excellent tool to extract molecular structural information. HPCCS is a parallelized, robust, clean, and user-friendly package written in C/C++ for CCS calculations under the TM approximation. Its main features are:

- The code was modularized, splitting the functions into files to facilitate support and debugging.
- Constant and global variables were declared on separate files for better support.
- All variables were declared as specific data types. In practice, this allocates only the necessary amount of memory for each variable, thus reducing the memory footprint.
- The arrays were allocated dynamically to use only the necessary amount of memory.
- The loop fusion technique was applied for faster execution.
- The Mersenne Twister [26] random number generator was used.
- HPPCS input is written with the PQR format, a user-friendly PDB-like coordinate file with atomic charges and radii parameters.

3.1 Compiling, Configuring, and Running HPCCS

HPCCS can be downloaded free of charge at https://github.com/cepid-cces/hpccs into a local directory.

To **compile**, the user should modify the `HPCCS` source directory in the `Makefile` (line 3):

```
INC_DIR = $HOME/HPCCS
```

On source folder, enter `make` to compile (*see* **Note 1**).

Works with `Intel` and `GCC` compilers. It is highly recommended to use `GCC 7.2` for better performance due to memory alignments.

The user should then **configure** HPCCS by editing the configuration file description, where all parameters must be split by tabs in one single line (`config/config.in`):

1. Number of Conformations → Specify how many conformations the input file contains (generally 1).
2. Number of complete cycles for average mobility calculation in `MOBIL2`. Default value is 10 if a number < 1 is supplied for this field or is not set.
3. Number of points in velocity integration in `MOBIL2`. Default value is 20 if this field uses a number < 1 or is not set.
4. Number of points for Monte Carlo integrations of impact parameter and orientation integrals in `MOBIL2`. Default value is 500 if this field use a number < 1 or is not set.

5. Number of target rotations for average potential calculation. If `ipr=0`, an unrotated configuration is used. Otherwise, `ipr` random rotations are employed. Default value is 1000 if this variable is not set.
6. Temperature must be specified; default value is 300 K if it is not set.
7. Specify buffer gas: set 1 for Helium (He), 2 for Nitrogen (N_2).
8. The file config.in should look as Table 1.

With this configuration, 100,000 simulations will be carried out (10 * 20 * 500). If a different total number of collision events is desired or needed, we suggested modifying only the fourth config variable, i.e., the number of Monte Carlo integration points (500).

The user should also **prepare the input file** (filename.pqr) for the target structure in the PQR format by modifying the desired PDB file. PQR is a modified PDB format, in which atomic partial charge (Q) and radius (R) fields follow the X, Y, Z coordinate fields in ATOM and HETATM records (*see* **Notes 2** and **3**).

To generate PQR format from a PDB structure, the user should use **PDB2PQR**, a `Python` software package that provides a utility for converting protein files in PDB format to PQR format [27] (*see* http://pdb2pqr.sourceforge.net/). Several force fields can be used to assign the atomic partial charges. In our most recent applications [22] and in Table 2, we have employed charges from AMBER03 [28].

The **protonation states** of the amino acid residues in a protein target may play an important role in the protein conformation and may affect the outcome of CCS calculations. Therefore, the user should assign charge states to ionizable residues. Physically sensible protonation states' assignments can be obtained using the H^{++}

Table 1
The file `config.in`

1	*10*	*20*	*500*	*1000*	*298.0*	*2*

Table 2
PQR file description. The following format must be adopted

	Atom number	Atom type	Amino acid type	Residue number	X, Y, Z coordinates			Partial charge	Radius
ATOM	1	N	MET	1	29.95	27.43	19.18	0.1592	1.824
ATOM	2	C	MET	1	28.88	28.41	19.4	0.0221	1.908

server [29, 30]. H++ outputs the completed structure in several common formats (PDB, PQR, AMBER inpcrd/prmtop) and provides a set of tools for analysis of electrostatic-related molecular properties. For more information about H++ server, *see* http://biophysics.cs.vt.edu/.

To **run a CCS calculation**, enter `./hpccs` filnename.pqr.

3.2 HPCCS Main Features and Outcomes

Approximately 95% of the CPU time in CCS TM calculations is dedicated to computing the intermolecular interaction potential function. This observation and the fact that the trajectories are independent for the computation of CCS inspired the development of the HPCCS parallel code. In HPCCS, a triple loop, which runs the collision simulations and calculates the scattering angles using the Monte Carlo method [31], was collapsed into a single loop using loop fusion [32]. Variables and arrays were privatized, so trajectory variables have their own memory positions and are not shared with other trajectory variables, turning the parallelization possible and easy to apply.

The OpenMP library [33] was used for parallelization: a powerful parallelization library that readily enables sequential code annotation by means of program pragmas [25]. The Vtune profiler tool [34] was used to find the most time-consuming parts of the code, pointing to the possible candidates for parallelization. To further improve HPCCS performance, arrays were memory aligned, enabling vectorization of the code [32, 35] based on the SIMD execution model [25]. This approach turned out to be an effective method for speeding up potential function calculations since all threads were allocated to run the collisions in parallel.

To validate the high performance of the HPCCS software, a test set of isolated proteins was used. All the calculations were executed on a cluster node consisting of Dual Intel Xeon E5 processor with eight cores each and AVX2, running at 2.4 GHz. The turbo boost was disabled to assure deterministic processor frequency, without any change during the experiments [36]. The Intel Parallel Studio 2016 compiler [34] was used, which has OpenMP support. Vtune was used to profile HPCCS along the entire program development, until the most time-consuming parts of the code were identified. As found in many applications, Hyperthread [37] was used to double the number of available threads, but speedups did not scale well in this case. Using only physical cores revealed better scalability and efficiency.

The description of long-range interactions, inherently present in Eq. 3, allows for more accurate collision cross sections for any system. Since the interactions must be calculated over all atoms, the computational time for each collision experiment increases linearly with the number of atoms. Therefore, the software parallelization capability is an important feature of HPCCS, which enables handling of large biomolecules within an attractive

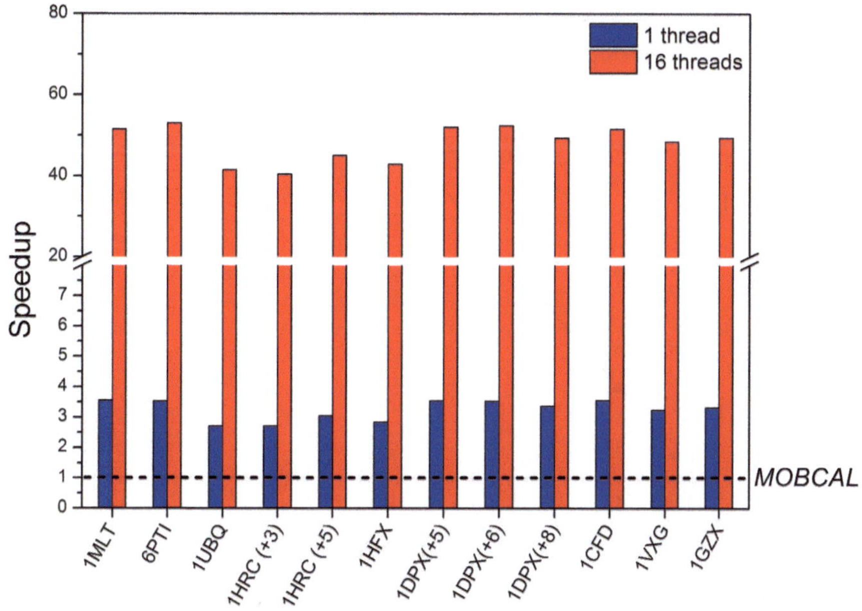

Fig. 3 Parallelization speedup for several proteins, considering 1 and 16 threads. Mobcal was used as a reference (dashed line) for speedup calculation, for both 1 and 16 threads execution of HPCCS. (Reproduced from ref. 22)

computational cost. Figure 3 shows the parallelization speedup for single proteins.

According to the results shown in Fig. 3, there is no significant difference in speedup for this set of proteins. Considering only optimizations applied to the sequential code, running with a single thread, the average speedup was 3.2, which confirms the efficiency of HPCCS software even when executing on a single thread. When 16 threads were used, the average speedup was 48.1, showing a considerable acceleration for CCS calculation.

The scalability efficiency of our system (16 physical cores), with respect to the number of threads, is shown in Fig. 4. Linear speedup was observed, while threads ran only on physical cores (i.e., 16 threads). Beyond 16 threads, speedup saturation was detected, meaning that Hyperthread technology had no significant impact on the acceleration of CCS calculations.

Figure 5 shows the increase in the total execution time with the protein size. For the largest protein considered here, containing 4392 atoms, the total execution time was approximately 6 and 20 h using HPCCS (1 thread) and Mobcal, respectively. The above execution times show that before HPCCS, it was unpractical to extensively use TM to perform CCS calculation. Therefore, HPCCS allows the user to accurately obtain structural information in a feasible time for large biomacromolecules.

Until now, CCS computations for larger biomolecular systems such as protein complexes were prohibitively costly at the TM level.

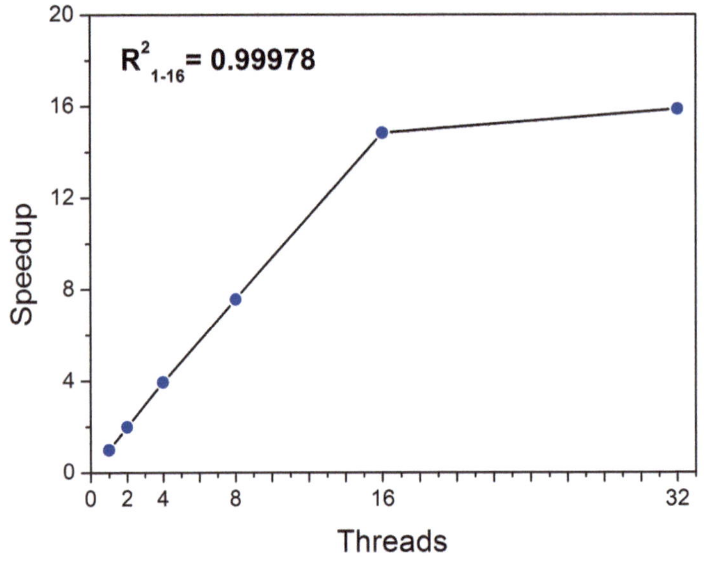

Fig. 4 Parallelization speedup versus threads number. *R*-squared considering only physical threads, 1–16, shows an excellent scalability of speedup for physical cores

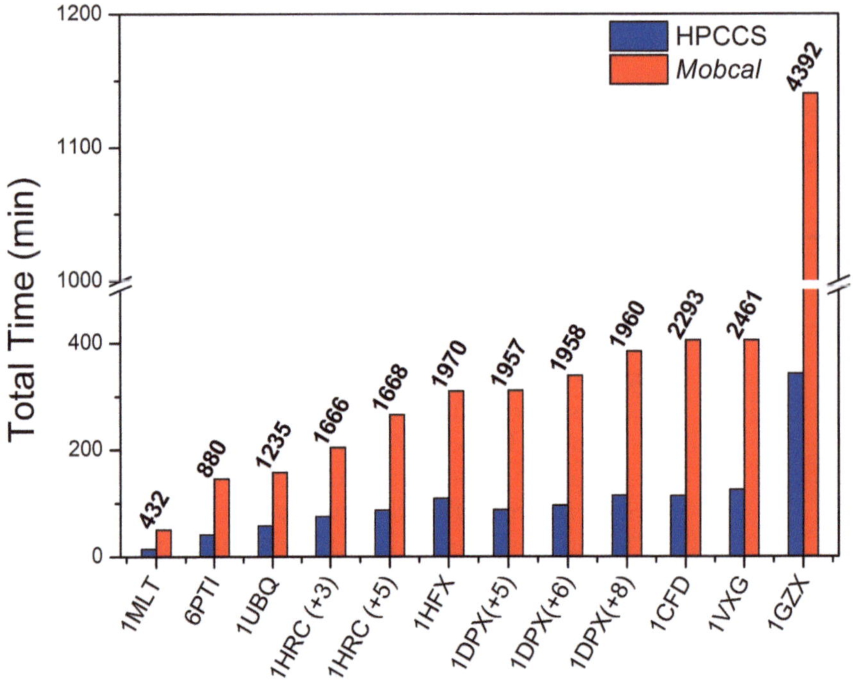

Fig. 5 Total execution time, in minutes, for selected proteins, considering the HPCCS (one thread) and Mobcal (serial) programs. The number of atoms for each protein is indicated. (Reproduced from ref. 22)

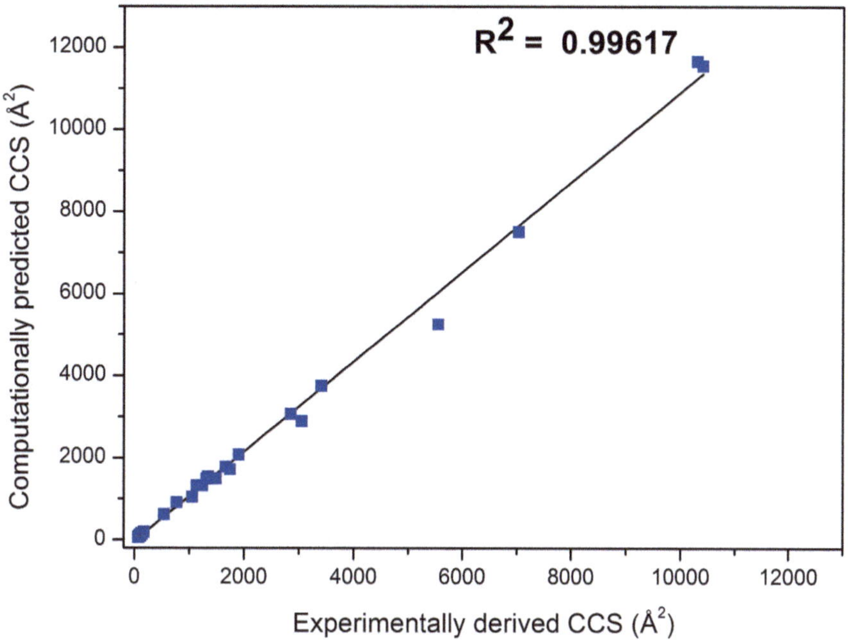

Fig. 6 Experimental versus computed CCS data using HPCCS, in Å^2, showing excellent agreement from small organic molecules to large protein complexes. (Adapted from ref. 22)

Reasonable results at low computational costs are obtained at the PA level. However, in addition to helium, gases such as N_2 and CO_2 are now being used as buffers in IM-MS experiments, resulting in different values for the collision cross sections. Since the PA method cannot differentiate buffer gases, HPCCS becomes an attractive option, capable of handling different buffer gases at the TM level with excellent performance.

Taken together, the results presented here indicates that HPCCS is capable of reproducing very well the experimental CCS values for a variety of molecular types, ranging from small organic molecules to protein complexes. Figure 6 depicts the experimental versus computed CCS data for all structures evaluated in [22], showing an excellent agreement between them. The high accuracy of our calculations and the relatively low computational costs indicate that HPCCS can be a valuable tool to aid the interpretation of experimental CCS data in the study of biomolecular structures.

IM-MS techniques have become highly valued as a tool for (bio)chemical analysis and can be profitably used both for analyte separation and structural investigation of a wide range of sample types. Despite its enormous potential, IM-MS data interpretation is often challenging in a variety of different scenarios, for instance, when multiple molecular conformers influence the resulting collision cross section, when the surface roughness of the target molecule affects CCS, when the specific interactions between the analyte and gas particles play a role (i.e., dependence on the buffer gas), or when the target is a large macromolecular system like protein complexes. In these cases, reliable and fast CCS estimates are needed in order to help in the interpretation of IM-MS data. HPCCS is the first parallel software that efficiently and accurately computes collision cross section for any set of atomic coordinates and, hence, fulfills this need.

4 Notes

1. Do not move the `hpccs` file to another directory because it needs the configuration and other files in the source directory.
2. Before constructing the PQR input file for HPCCS calculations, the PDB structures must be previously relaxed by means of energy minimization, available in most Molecular Dynamics simulation packages. Energy minimizations may be achieved via steepest descent using roughly 50,000 steps using a cutoff value of at least 10 Å.
3. For obtaining the geometry for small molecules, the user can use a variety of quantum chemical methods, ranging from semiempirical, to ab initio, to Density Functional Theory (DFT). The latter provides more accurate geometries. In our recent work [22], we used DFT at M06-2X/6-31+G(2d,p) level [38] for small organic molecules. Mulliken charges derived from these calculations were used to assign the individual atomic charges. Results were refined by single point calculations at the M06/6-311++G(2d,p) level using Gaussian09 [39].

Acknowledgments

This work was supported by a grant from the Sao Paulo State Research Foundation—Fapesp (Grant #2013/08293-7). Partial support from the Brazilian Research Council—CNPq is also acknowledged.

References

1. Mulder AA, Filatov M (2010) NMR chemical shift data and *ab initio* shielding calculations: emerging tools for protein structure determination. Chem Soc Rev 395:578–590
2. Egli M (2016) Diffraction techniques in structural biology. Curr Protoc Nucleic Acid Chem 65:7.13.1–7.13.41
3. Peters JP, Kowal EA, Pallan PS, Egli M, Maher LJ 3rd (2018) Comparative analysis of inosine-substituted duplex DNA by circular dichroism and X-ray crystallography. J Biomol Struct Dyn 36:2753–2772
4. Christensen AS (2010) Inferential protein structure determination and refinement using fast, electronic structure based backbone amide chemical shift predictions. Dissertation, University of Copenhagen; arXiv:1508.00808v1 [physics.chem-ph]

5. Abola E, Kuhn P, Earnest T, Stevens RC (2000) Automation of X-ray crystallography. Nat Struct Biol 7:973–977
6. Schneidman-Duhovny D, Rossi A, Avila-Sakar A et al (2012) A method for integrative structure determination of protein-protein complexes. Bioinformatics 28:3282–3289
7. Fenn JB (2003) Electrospray wings for molecular elephants (Nobel lecture). Angew Chem Int Ed 42:3871–3894
8. Wyttenbach T, Bowers MT (2011) Structural stability from solution to the gas phase: native solution structure of ubiquitin survives analysis in a solvent-free ion mobility-mass spectrometry environment. J Phys Chem B 115:12266–12275
9. Mikhailov VA, Mize TH, Benesch JLP, Robinson CV (2014) Mass-selective soft-landing of protein assemblies with controlled landing energies. Anal Chem 86:8321–8328
10. Liu L, Bagal D, Kitova EN, Schnier PD, Klassen JS (2009) Hydrophobic protein–ligand interactions preserved in the gas phase. J Am Chem Soc 131:15980–15981
11. Bush MF, Hall Z, Giles K, Hoyes J, Robinson CV, Ruotolo BT (2010) Collision cross sections of proteins and their complexes: a calibration framework and database for gas-phase structural biology. Anal Chem 82:9557–9565
12. Benesch JLP, Ruotolo BT, Simmons DA, Robinson CV (2007) Protein complexes in the gas phase: technology for structural genomics and proteomics. Chem Rev 107:3544–3567
13. Valentine SJ, Counterman AE, Clemmer DE (1997) Conformer-dependent proton-transfer reactions of ubiquitin ions. J Am Soc Mass Spectrom 8:954–961
14. Shelimov KB, Clemmer DE, Hudgins RR, Jarrold MF (1997) Protein structure in vacuo: gas-phase conformations of BPTI and cytochrome c. J Am Chem Soc 119:2240–2248
15. Shelimov KB, Jarrold MF (1997) Conformations, unfolding, and refolding of apomyoglobin in vacuum: an activation barrier for gas-phase protein folding. J Am Chem Soc 119:2987–2994
16. Shvartsburgc AA, Jarrold MF (1996) An exact hard-spheres scattering model for the mobilities of polyatomic ions. Chem Phys Lett 261:86–91
17. Mesleh MF, Hunter JM, Shvartsburg AA, Schatz GC, Jarrold MF (1996) Structural information from ion mobility measurements: effects of the long range potential. J Phys Chem 100:16082–16086; Erratum (1997) J Phys Chem A101:968
18. Wyttenbach T, von Helden G, Batka JJ, Carlat D, Bowers MT (1997) Effect of the long-range potential on ion mobility measurements. J Am Soc Mass Spectrom 8:275–282
19. Shvartsburg AA, Smith RD (2008) Optimum waveforms for differential ion mobility spectrometry (FAIMS). J Am Soc Mass Spectrom 19:1286–1295
20. Lanucara F, Holman SW, Gray CJ, Eyers CE (2014) The power of ion mobility-mass spectrometry for structural characterization and the study of conformational dynamics. Nat Chem 6:281–294
21. Bleiholder C, Wyttenbach T, Bowers MT (2011) A novel projection approximation algorithm for the fast and accurate computation of molecular collision cross sections (I) Method. Int J Mass Spectrom 308:1–10
22. Zanotto L, Heerdt G, Souza PCT, Araujo G, Skaf MS (2018) High performance collision cross section calculation – HPCCS. J Comput Chem 39:1675–1681
23. Shvartsburg AA, Schatz GC, Jarrold MF (1998) Mobilities of carbon cluster ions: critical importance of the molecular attractive potential. J Chem Phys 108:2416–2423
24. Campuzano I, Bush MF, Robinson CV, Beaumont C, Richardson K, Kim H, Kim HI (2012) Structural characterization of drug-like compounds by ion mobility mass spectrometry: comparison of theoretical and experimentally derived nitrogen collision cross sections. Anal Chem 84:1026–1033
25. Pacheco P (2010) An introduction to parallel programming. Morgan Kaufmann, MA
26. Matsumoto M, Nishimura T (1998) Mersenne twister: a 623-dimensionally equidistributed uniform pseudo-random number generator. ACM Trans Model Comput Simul 8:3–30
27. Dolinsky TJ, Nielsen JE, McCammon JA, Baker NA (2004) PDB2PQR: an automated pipeline for the setup, execution, and analysis of Poisson-Boltzmann electrostatics calculations. Nucleic Acids Res 32:W665–W667
28. Wang J, Wolf RM, Caldwell JW, Kollman PA, Case DA (2004) Development and testing of a general amber force field. J Comput Chem 25:1157–1174
29. Anandakrishnan R, Aguilar B, Onufriev AV (2012) H++ 3.0: automating pK prediction and the preparation of biomolecular structures for atomistic molecular modeling and simulations. Nucleic Acids Res 40:W537–W541
30. Gordon JC, Myers JB, Folta T, Shoja V, Heath LS, Onufriev A (2005) H++: a server for estimating pKas and adding missing hydrogens to

macromolecules. Nucleic Acids Res 33: W368–W371

31. Rubinstein RY, Kroese DP (2007) Simulation and the Monte Carlo method. Wiley series in probability and statistics. John Wiley & Sons, Inc., Hoboken
32. Wolfe M (1996) High performance compilers for parallel computing. Addison-Wesley Longman Publishing Co., Inc., Boston
33. Dagum L, Menon R (1998) OpenMP: an industry-standard API for shared-memory programming. IEEE Comput Sci Eng 5:46–55
34. Marowka A (2011) On performance analysis of a multithreaded application parallelized by different programming models using Intel VTune. In: 11th international conference on parallel computing technologies PaCT: parallel computing technologies, pp 317–331
35. Jeffers J, Reinders J (2013) Intel Xeon phi coprocessor high performance programming. Morgan Kaufmann, Waltham
36. Krieger E, Vriend G (2015) New ways to boost molecular dynamics simulations. J Comput Chem 36:996–1007
37. Marr DT, Binns F, Hill DL, Hinton G, Koufaty DA, Miller JA, Upton M (2002) Hyper-threading technology architecture and micro-architecture. Intel Technol J 6(1):4–15
38. Zhao Y, Truhlar DG (2008) The M06 suite of density functionals for main group thermochemistry, thermochemical kinetics, noncovalent interactions, excited states, and transition elements: two new functionals and systematic testing of four M06-class functionals and 12 other functionals. Theor Chem Accounts 120:215–241
39. Frisch MJ, Trucks GW, Schlegel HB, Scuseria GE et al (2009) Gaussian 09, Revision D.01. Gaussian, Inc., Wallingford, CT

INDEX

Giuseppe Paglia and Giuseppe Astarita (eds.), *Ion Mobility-Mass Spectrometry: Methods and Protocols*, Methods in Molecular Biology, vol. 2084, https://doi.org/10.1007/978-1-0716-0030-6,

M

N

O

P

S

T

U